DISCOVERING LIFE, MANUFACTURING LIFE

HOW THE EXPERIMENTAL METHOD SHAPED LIFE SCIENCES

Grenoble Sciences

Grenoble Sciences pursues a triple aim:
- to publish works responding to a clearly defined project, with no curriculum or vogue constraints,
- to guarantee the selected titles' scientific and pedagogical qualities,
- to propose books at an affordable price to the widest scope of readers.

Each project is selected with the help of anonymous referees, followed by an average one-year interaction between the authors and a Readership Committee, whose members' names figure in the front pages of the book. Grenoble Sciences then signs a co-publishing agreement with the most adequate publisher.

Contact: Tel.: (33) 4 76 51 46 95 – Fax: (33) 4 76 51 45 79
E-mail: Grenoble.Sciences@ujf-grenoble.fr
Information: *http://grenoble-sciences.ujf-grenoble.fr*

Scientific Director of Grenoble Sciences: Jean BORNAREL,
Professor at the Joseph Fourier University, Grenoble, France

Grenoble Sciences is supported by the French Ministry of Education and Research and the "Région Rhône-Alpes".

Discovering Life, Manufacturing Life is an improved version of the French book published by Grenoble Sciences in partnership with EDP Sciences. The Reading Committee of the French version included the following members:

- **Michel BORNENS**, CNRS Research Director, Curie Institute, Paris
- **Athel CORNISH-BOWDEN**, CNRS Research Director, Marseille
- **Roland DOUCE**, Professor at the Joseph Fourier University, Grenoble
- **Françoise FRIDLANSKY**, CNRS Researcher, Gif-sur-Yvette
- **Jean-Louis MARTINAND**, Professor at the ENS, Cachan
- **Jean-Claude MOUNOLOU**, Professor at the Paris-Sud University, Orsay
- **Valdur SAKS**, Professor at the Joseph Fourier University, Grenoble
- **Michel SATRE**, CNRS Research Director, CEA, Grenoble
- **Michel SOUTIF**, Honorary Professor at the Joseph Fourier University, Grenoble
- **Jean VICAT**, Professor at the Joseph Fourier University, Grenoble
- **Jeannine YON-KAHN**, CNRS Research Director, Orsay

Typesetted by **Centre technique Grenoble Sciences**

Front cover illustration: composed by **Alice GIRAUD** *with extracts from:*
© CNRS Photothèque/R. Melki: yeast prion's atomic structure. © CNRS Photothèque/ T. Launey, J.P. Gueritaud: Organotypic culture of rat's facial motor neuron. © CNRS Photothèque/Y. Bailly: rat's cortex neuron marked by the green fluorescent protein. © N. Franceschini: Head of the blowfly, *Calliphora*, as seen from the front, showing the two compound eyes. © N. Franceschini, J.M. Pichon and C. Blanes: Robot-Fly (Autonomous seeing vehicle)

DISCOVERING LIFE, MANUFACTURING LIFE
HOW THE EXPERIMENTAL METHOD SHAPED LIFE SCIENCES

Pierre V. VIGNAIS and Paulette M. VIGNAIS

Translation by Martha WILLISON

2010

Pierre V. Vignais†
Honorary Professor of Biochemistry
School of Medicine
Joseph Fourier University, Grenoble
France
†deceased sept. 2006

Paulette M. Vignais
Emeritus Director of Research CNRS
8 rue Nicolas Boileau
38700 La Tronche
France

"Science Expérimentale et connaissance du vivant, la méthode et les concepts, Pierre Vignais avec la collaboration de Paulette Vignais, Collection Grenoble Sciences - EDP Sciences, 2006"

ISBN 978-90-481-3766-4   e-ISBN 978-90-481-3767-1
DOI 10.1007/978-90-481-3767-1
Springer Dordrecht Heidelberg London New York

Library of Congress Control Number: 2010930423

© Springer Science+Business Media B.V. 2010
No part of this work may be reproduced, stored in a retrieval system, or transmitted in any form or by any means, electronic, mechanical, photocopying, microfilming, recording or otherwise, without written permission from the Publisher, with the exception of any material supplied specifically for the purpose of being entered and executed on a computer system, for exclusive use by the purchaser of the work.

Cover design: Alice Giraud

Printed on acid-free paper

Springer is part of Springer Science+Business Media (www.springer.com)

FOREWORD

Francis BACON, in his *Novum Organum*, Robert BOYLE, in his *Skeptical Chemist* and René DESCARTES, in his *Discourse on Method*; all of these men were witnesses to the scientific revolution, which, in the 17th century, began to awaken the western world from a long sleep. In each of these works, the author emphasizes the role of the experimental method in exploring the laws of Nature, that is to say, the way in which an experiment is designed, implemented according to tried and tested techniques, and used as a basis for drawing conclusions that are based only on results, with their margins of error, taking into account contemporary traditions and prejudices. Two centuries later, Claude BERNARD, in his *Introduction to the Study of Experimental Medicine*, made a passionate plea for the application of the experimental method when studying the functions of living beings. Twenty-first century Biology, which has been fertilized by highly sophisticated techniques inherited from Physics and Chemistry, blessed with a constantly increasing expertise in the manipulation of the genome, initiated into the mysteries of information technology, and enriched with the ever-growing fund of basic knowledge, at times appears to have forgotten its roots. In an epoch in which bioengineering is seen as an all-conquering "Open Sesame" that opens wide the doors to the temple of living beings, it may be salutary for the student or young researcher to look back upon the past and reflect, with respect, upon that which, in former times, inspired thinkers were able to achieve with the modest means at their disposal, means that were sometimes derisory when compared with contemporary technology. It may do them good to recognize that the harmonious edifice of current knowledge was built upon a solid foundation of fundamental discoveries made by famous thinkers of the past. They can only benefit from recalling the fact that audacious innovations based on experimental evidence, have, in their time, come up against the vindictiveness of an ill-prepared, ill-informed public that clings to its endless routine. One has only to remember that the introduction of asepsis and antiseptics into surgical practices in the 19th century, which was based on acquired proof of the existence of infectious germs, did not always proceed smoothly.

This book relates the circumstances that have, over time, led thinkers and philosophers to ask questions about living Nature, to dare to explore using experimental devices and to lay out beacons that have prevented them from falling into any traps that would be prejudicial to scientific progress. By basing the discussion on concrete examples that refer to easily-understood experiments, my intention has been to provide the reader with food for thought, not only with respect to the experimental method itself, its history and its role in the objective analysis of the mechanisms of living beings, but also with respect to the organization of science as it exists today, and how society regards it.

For those in charge of Administrations that program Science via ambitious projects, and for politicians who decide on budgets based on how high a profile in the media the societal impact of a particular line of research might have, the history of experimental science provides proof that it is often via less-trodden, and, at first view, unattractive pathways, and not via wide avenues, full of attractions, or blind submission to an apparently worthy conformity, that a researcher's curiosity and obstinacy lead to major discoveries; that carefully planned programs put together in pursuit of utilitarian objectives sometimes lead to bitter failure, resulting mostly in serious disillusionment, and, finally, that scientific creativity arises from the activity of the human mind, which is unable to handle restrictive meddling without becoming sterile. The history of experimental science is also there to highlight the fact that a scientific breakthrough is, generally speaking, the fruit of patient elaboration, which cannot readily be accomplished over the short term, and that, in the end, the training of a scientific researcher must include a long apprenticeship spent with competent masters able to transmit rigorous work methods, stimulate curiosity and inculcate a taste for taking risks in proceeding towards the unknown.

ACKNOWLEDGEMENTS

Any piece of writing requires evaluation by recognized experts, and this book is no exception. I wish to express my thanks to those of my colleagues who were willing to act as judges and provide me with their suggestions and their criticism: Jeannine YON-KHAN, Françoise FRIDLANSKY, Michel BORNENS, Athel CORNISH-BOWDEN, Roland DOUCE, Jean-Louis MARTINAND, Jean-Claude MOUNOLOU, Valdur SAKS, Michel SATRE, Michel SOUTIF and Jean VICAT. I would also like to thank Jean BORNAREL, scientific director of Grenoble Sciences, for his help, suggestions and support in the painstaking preparation of the manuscript. During this long process, Nicole SAUVAL acted as a particularly efficacious kingpin. Sylvie BORDAGE and her colleague Julie RIDARD showed their mastery in the final putting together of the French edition and Alice GIRAUD provided a touch of originality in the presentation of the cover of that edition. It is also impossible to forget all those who, at a time when the manuscript was nothing more than a rough draft, were kind enough to read it and share their impressions with me: Isabelle GEAHEL, for her often direct and always judicious comments, Jean GASQUET for his concern with grammatical forms, Gérard KLEIN for discussions that were both enlightening and full of sense with respect to science in today's world, Eric VADOT for his perspicacious and detailed remarks concerning culture in the ancient world, Alexandra FUCHS for her pertinent and professional opinions concerning the biology of the future, and Flavio DELLA SETA for his participation in the iconography of Chapter IV. This work is the fruit of a continuous collaboration with Paulette VIGNAIS, and owes its very existence to this constant, interactive collaboration. I would also like to thank Martha WILLISON for translating this work into English.

CONTENTS

Introduction 1

Chapter I
The roots of experimental science from Ancient Greece to the Renaissance 7

1. Scientific rationality in Ancient Greece 7
   1.1. From THALES to SOCRATES towards a new conception of Nature 8
   1.2. Emergence of the principles of logic in philosophy
        and of rationality in the sciences 16
   1.3. The Alexandrine period of greek science 23
   1.4. The post-hellenistic period - Hiatus or transition with the Middle Ages? ... 28
2. The philosophical and technological heritage of the Middle Ages 31
   2.1. The politico-economic context of the Middle Ages 31
   2.2. Philosophical-theological controversies in the 12th and 13th centuries,
        a critical look back over the Ancient World 34
   2.3. Alchemy and the technological revolution of the Middle Ages 39
3. Conclusion - At the dawn of the scientific method 45

Chapter II
The birth of the experimental method in the 17th and 18th centuries 49

1. The discovery of the circulation of the blood by W. HARVEY 50
2. How were the movements of the heart and the blood
   explained before HARVEY? 58
3. The first, faltering steps of experimental science applied to living beings 63
   3.1. From human anatomy to comparative anatomy systematics 63
   3.2. From microscopic tissular anatomy
        to the morphological description of animalcules 68
   3.3. The birth of physiology: seeking the experimental method 73
   3.4. First conceptual controversies concerning the use
        of the experimental method in living beings 82
        3.4.1. Refuting the theory of spontaneous generation 82
        3.4.2. The enigma of the regeneration of the hydra 84
        3.4.3. Refuting the idea of the pneuma as the agent of muscle contraction
               Birth of the idea of the reflex 86
4. The experimental method and its impact on the physical sciences
   in the 17th century 88
   4.1. A new theory of the cosmos 89

4.2. A new theory of movement 91
4.3. Proof of the existence of the vacuum... 93
4.4. Towards other revelations concerning the inanimate world.......... 97
5. Opening up chemistry to quantitative experimentation in the 18th century..... 100
   5.1. The birth of pneumatic chemistry
      and the study of the exchange of gases in living beings................ 102
   5.2. Gaseous exchanges in living beings... 103
   5.3. Measurement of animal heat and the birth of bioenergetics 111
6. Experimental science as seen by the philosophers
   of the 17th and 18th centuries.. 112
   6.1. Francis BACON and induction in scientific reasoning 113
   6.2. Robert BOYLE and requirements of experimental practice............ 116
   6.3. René DESCARTES and the cardinal principles of scientific research....... 118
   6.4. Contradictory currents in the philosophy of the sciences
      in the 18th century 120
7. Is there an explanatory logic for the birth of the experimental method?........... 124
   7.1. Sociopolitical crises.. 124
   7.2. A knowledgeable society.. 127
   7.3. A scientific quorum... 129
   7.4. Instrumentation, an integral part of the experimental method.... 131
   7.5. The enigma of the discovery of the experimental method
      and of its development in the West 132
8. Conclusion - The mariage of techniques and ideas... 134

Chapter III
The impact of determinism in the life sciences of the 19th and 20th centuries .. 139

1. The recognition of physiology as an experimental science in the 19th century 144
2. Determinism, the philosophical foundation stone
   of experimental physiology 146
   2.1. Claude BERNARD's determinist bible 148
   2.2. The many conceptual approaches of experimental determinism
      in the study of living beings.. 149
      2.2.1. The experiment to see what happens... 150
      2.2.2. The decisive experiment.. 153
      2.2.3. Serendipity and the unexpected discovery 155
      2.2.4. Advantages and traps involved in reasoning by analogy 157
      2.2.5. The part played by luck in the experimental method............ 161
3. The impact of technology on the life sciences in the 19th century 164
   3.1. Rationalization of operational physiology... 164
   3.2. The emergence of instrumental engineering
      adapted to physiological experimentation 169
   3.3. Application of analytical chemistry to physiological exploration 179

4. New disciplines in the life sciences in the 19th century
 and their methodological support 180
5. The idea of quantification in the life sciences... 186
6. A new experimental order for the life sciences in the 20th century 188
    6.1. A reasoned choice of model organisms... 189
    6.2. A breakthrough in techniques
         for exploring the functions of living beings 200
         6.2.1. Imagery of the infinitely small 200
         6.2.2. Enumerating and isolating macromolecular structures 209
         6.2.3. Isotopic labeling.. 215
         6.2.4. The instrument and the method
                The analysis of reality *via* the instrument 216
7. Opening up biological experimentation to reductionism 219
    7.1. The firsts steps in experimental reductionism: from the organ to the cell.. 219
    7.2. A cellular glycolysis:
         a prototype for a reductionist approach to exploring the metabolism........ 221
    7.3. Deconstruction and reconstruction of macromolecular complexes............ 222
    7.4. The birth of virtual biology - Modeling cell dynamics 225
8. The experimental method
   faced with contemporary trends in philosophy and in social life 229
    8.1. Confrontation between vitalists and mechanists. the emergence of
         organicism... 229
    8.2. *Novum Organum* revisited and contested.. 233
    8.3. Re-examining the process of the experimental procedure................. 235
9. Conclusion - Determinism and the expansion of the experimental method
   From the organ to the molecule 238

Chapter IV
Challenges for experimentation on living beings
at the dawn of the 21st century 241

1. The accession of biotechnology
   Towards a new paradigm for the experimental method............................ 242
    1.1. The genome explored... 243
         1.1.1. From molecular biology to genetic engineering..................... 243
         1.1.2. DNA becomes a molecular tool... 249
         1.1.3. DNA chips and protein chips - From genomics to proteomics........... 253
         1.1.4. From genomics to metagenomics 258
    1.2. The manipulated genome... 259
         1.2.1. DNA used as a construction material 259
         1.2.2. RNA interference: a new frontier
                in the manipulation of the expression of the genome........... 262
         1.2.3. The experimental transgression of the genetic code 264
2. Towards a mastery of the functions of living beings for utilitarian purposes.. 265

2.1. Manipulations of plant DNA
The challenge of genetically modified plants 266
2.2. Manipulations of human DNA and hopes for gene therapy 270
2.3. Stem cells and cloning 272
  2.3.1. The hope of stem cells 272
  2.3.2. The specter of cloning 277
  2.3.3. The bias of parthenogenesis in cloning 280
2.4. The "humanization" of animal cells for purposes of xenotransplantation 281
3. The progress of medicine face to face with the experimental method 283
  3.1. From empirical medicine to experimental medicine 284
  3.2. Contemporary advances in biotechnology
  The example of medical imaging 290
  3.3. From experimental medicine to predictive medicine 293
  3.4. The drug library of the future 295
4. Towards a global understanding of the functions of living beings 298
  4.1. Experimental demonstration of protein interactions 298
  4.2. Mathematical modeling of the complexity of living beings 304
  4.3. Biorobots and hybrid robots 312
5. The design and meaning of words in the experimental process 320
6. The experimental method, understanding of living beings and society 324
  6.1. Human cloning censured by codes of bioethics 326
  6.2. The patentability of living beings 329
  6.3. Animal experimentation versus the fight for animal rights 331
7. The place of the scientific researcher in the changing role of biotechnology 332
  7.1. Fundamental research faced with the metamorphosis
  in the experimental method 333
    7.1.1. A new strategy in the organization of research 334
    7.1.2. A new way of circulating knowledge 336
    7.1.3. A new horizon for cross-disciplinarity 338
  7.2. The experimental method taught and discussed 340
8. Conclusion Looking at the present in the light of the past 343

Chapter V
Epilogue 347

Bibliography 353

Index of personal names 371

Index of subjects 381

List of illustrations 395

Glossary 401

INTRODUCTION

"During the different periods of its evolution, the human mind has progressed successively through sentiment, reason and experience. First of all sentiment, imposing itself alone over reason, created the truths of faith, or theology. Reason having been mastered, it gave birth to scholasticism. Finally, experience, that is to say, the study of natural phenomena, taught Man that the truths of the outside world cannot be found, at first sight, in either sentiment or reason. These are merely indispensable guides, but to find truths, it is necessary to come down to the objective realities of things, with their phenomenal forms. In this way, according to the natural progression of things, the experimental method appears, summarizing everything and basing itself successively on the three legs of this immutable tripos, sentiment, reason and experience."

Claude BERNARD
Introduction to the Study of Experimental Medicine - 1865

To experiment is to interrogate Nature by means of various devices, under conditions that allow us to control parameters and obtain reproducible results. It involves gathering together the results obtained and judging whether or not they validate the hypothesis that preceded the experimentation. This experimentation is always associated with the idea that gave rise to it. It takes place within a well defined methodological framework, with well established rules, and it makes use of appropriate instruments. This approach to the understanding of Nature has been christened the **experimental method**. When applied to the study of living beings, the aim is to decode the mechanisms that govern the functions of organs and of their cells, that is, to decode the laws of Nature.

The use of the experimental method in the Life Sciences was born in Europe in the 17th century. The discovery of the circulation of blood in the body, made by William HARVEY, testified to its birth. At approximately the same time, the scientific method began to compel recognition in the physical sciences, with the demonstration of uniformly accelerated movement by GALILEO, and of the existence of the vacuum by Blaise PASCAL and Robert BOYLE. That the experimental method should be born so late in history does not fail to surprise people. In *Medea's Cauldron*, the epistemologist Mirko GRMEK tells us of his astonishment when, at the beginning of his medical studies, he learned that the circulation of the blood was only discovered in 1628, "I thought," he writes "that if I had been born in the ancient world, I would have had little difficulty discovering that blood circulates in the blood vessels of Man and the higher animals. This was both incorrect and pretentious; I would have had to have been born in the past with the education of a modern child."

Although the experimental method, as applied to the understanding of living beings, only appeared in the 17th century, it is nevertheless true that experimentation on animals took place well before this period. However, for the most part, such experimentation was sporadic, and had no major consequences. Certain experiments, either inadequately carried out or incorrectly interpreted, led, through fallacious arguments, to erroneous doctrines that survived for centuries. It is also true that the experimental method borrowed much from the logical approach of the philosophers of ancient Greece, who were able to conceive hypotheses, build reasoned arguments or refute arguments. The origin of the Universe, the movements of the stars in the sky, the principles that govern animal and plant life; all of these have been subjects for reflection on the part of thinking human beings since ancient times. The first explanations that were given were based on the supernatural, involving the intervention of cohorts of benign or wicked divinities. At the beginning of the 6th century BC, the advent of the Greek philosophers started a transition in ways of thinking, and new concepts appeared, leading to the formulation of new questions. A mode of thought that was based mainly on the idea of divine intervention in the ordering of the universe and the destiny of Man was replaced by a rational philosophy founded on a mechanistic approach to observable phenomena, without, however, departing from the theogonic tradition.

The Greek philosophers had the advantage of a vocabulary, syntax and writing system that could be used to describe their observations of the world that surrounded them and was accessible to their senses, and to express their ideas on the world of metaphysics. While people of ancient civilizations such as the Sumerians used special signs on clay tablets to represent objects, drawing an ear of wheat, for example, to represent wheat, or a head with horns to represent a cow, the Greeks used an alphabet where letters were translated as conventional signs that had equivalent sounds. Taken together, the existence of alphabetic writing, with its phonetic equivalents, a rigorous syntax and considerable fertility in the way in which words were derived, made it easier not only for well-read people of the period to communicate, but also for abstract **concepts** and subtly-nuanced thought to emerge. Speculations made by the Greek philosophers concerning phenomena in Nature, the structure of the Universe, and the idea of Life and of vital functions would be taken up again in the Middle Ages: ARISTOTLE and THEOPHRASTES, who had been observers of fauna and flora and had initiated the first logically-based classifications of beings in Nature, were to be rediscovered fifteen centuries later.

Medieval alchemists were certainly skilled technicians, but they were often guided by fanciful ideas. For many of them, experimentation was not based on any logical theory, but was mainly aimed at fulfilling extravagant dreams, such as the transmutation of metals into gold or the creation of a universal panacea for medical use. For others, alchemy had only a symbolic value. While medieval science in the West progressed with hopeless slowness, on the other side of the world, in China, amazing technical advances were being made, with such inventions as paper, printing, gunpowder and the compass, inventions that Europe would only acquire much later. Nevertheless, although, in a sudden intellectual and cultural leap forward during the 17th century, the West would invent the experimental method to help achieve an understanding of phenomena in Nature, Chinese science remained in the background, from this point of view.

In the West, the first universities appeared at the end of the 12th century. They spread rapidly throughout Europe. In 1500, there were approximately fifty of them. The teaching that was done initially was based on the writings of the Bible. Commentaries were also given of the works of the Greek philosophers, among whom ARISTOTLE held a privileged position. From the 13th century onwards, two original and distinguished thinkers, Robert GROSSETESTE and Roger BACON, both of them members of the Franciscan order, were questioning the value of teaching that was accepted without criticism, and endeavoring to bring recognition to the value of experimental fact as opposed to any preconceived idea. Their anti-establishment attitude was the prelude to a current of new ideas that were riddled with doubt and inquiry.

In the middle of the 16th century, a first scientific revolution was coming to the fore in the West, marked by the revival of astronomy, with a date of reference, 1543, when Nicolaus COPERNICUS published *De Revolutionibus Orbium Cælestium*: the Earth was no longer the center of the World, it revolved around the sun. It is a remarkable coincidence that in 1543, Andreas VESALIUS published his famous treatise on human anatomy, *De humani corporis fabrica*. This revolutionary movement, supported by cosmology and anthropology, grew throughout the 17th century, spilling over into the physical and life sciences. In short, until the Renaissance, science had remained essentially empirical. It corresponded to the Greek term ἐμπειρία which can be translated as either **empiricism** or **experience**, but an experience that is based on the observation of repeated phenomena, the causes of which produce the same effects.

From the 16th century onwards, the term **experience** took on a wider significance, in which phenomena that are observed in living beings are provoked by the experimenter. Conscious of the directive power of the experimental method, the Science of Nature, acquiring the status of an **experimental science**, moved away from medieval scholastic teaching and liberated itself from the woolly-minded, secretive practices of an alchemy that mixed magic and astrology. The novel and audacious experimental approach to mechanics made it possible to discover the laws of movement, and this marked a total break with Aristotlian physics. Using analogy, certain phenomena that are features of living beings, such as the circulation of blood in Man, and nerve reflexes, were explained on the basis of **mechanistic** concepts. Enthusiasm for this mechanistic philosophy rapidly became excessive. In counterpoint, alchemy, which was still active in the 17th century, fed an **animist** philosophy of Nature, in opposition to the mechanistic philosophy that was beginning to take over. Later on, the term animism gave way to the term **vitalism**, a fairly similar concept in which the soul is replaced by a vital principle, which can supposedly explain all physiological phenomena.

Somewhat later than in physics and the life sciences, at the end of the 18th century, chemistry was to undergo a decisive revolution, with the refutation of the phlogistic theory. The changeover from the qualitative to the quantitative in the study of animals or plants took place once the progress made in chemistry had made it possible to show, in terms of balanced equations, the modifications imposed on molecular species by the interplay of metabolisms. The 19th century can be seen as the flourishing, triumphal epoch of *in vivo* experimental physiology. The gains made by the study of the life sciences from the use of techniques inherited from physics and chemistry demonstrates the usefulness of expertise coming from different horizons.

With the almost exponential rise in the number of discoveries made during the 19th century, scientific knowledge began to be compartmentalized. By changing their label to **biology**, the Life Sciences laid claim to a new status. Chemistry began to distinguish itself from physics. Mathematics reigned over the whole ensemble. The eclecticism of the scholarly world, which was a characteristic of the 17th century, gave way to greater depth of detail, in well-defined sectors. Considerable technical progress was made, with the invention and perfecting of new instruments and new methods of exploration. At the same time as this technical progress was being made, new, often unexpected, ideas were emerging. Within the space of a few decades in the second half of the 19th century, the laws governing the transmission of hereditary characteristics were discovered, the theory of the evolution of living beings, based on natural selection, was put forward, and cell theory explained how the cell is the structural and functional unit of all tissues. From a Biology in which the two main axes were **anatomy** and **physiology**, sectors that up until then had barely been recognized, such as **genetics**, **embryology** and **biochemistry**, or those that hadn't even existed, such as **microbiology** and **immunology**, emerged and imposed themselves as major disciplines.

At the beginning of the 20th century, the exploration of living beings moved from *in vivo* to *in vitro*, with the use of perfused organs, cell homogenates and endocellular organelles. Major metabolic pathways were identified and the enzymatic reactions that take place in cell compartments were specified. The second half of the 20th century saw the birth and then the meteoric rise of **molecular biology**, emerging from a fusion between biochemistry and genetics. The experimental became molecular rather than cellular. Aided by techniques arising from the engineering sciences, **structural biology** examines the anatomy of living beings on an atomic level. The three-dimensional structure of thousands of macromolecules, particularly proteins, is determined and listed, giving rise to comparisons and classification into families and to the drawing up of phylogenetic trees that highlight the process of evolution. The decoding of the human genome sequence, an objective that was considered to be rather fanciful by certain people when it was first put forward during the final decades of the 20th century, has become a reality. Strengthened by an increasingly effective set of tools, genetic engineering, which involves "tinkering" with genes, is coming to be seen as an indispensable technology for many of those doing biological experiments.

The explosive increase in the amount of data about the genes of numerous animal and plant species, and about the proteins arising from the expression of these genes, has led to new areas of biology being given distinct identities, **genomics** and **proteomics**. There are multiple applications in areas that directly affect life in society. Beneficiaries include medicine, with new possibilities in the diagnosis of and prognostics for diseases; pharmacology, with the creation of new medicines; agriculture and stock-rearing, with the application of new know-how in genetics; and even certain areas of industry. Thus, at the beginning of the 21st century, the paradigm of the experimental method has found itself displaced. From a procedure in which a cell or an organ was required to give a response to variation of a single parameter, we have moved on to a procedure that involves a globalized, multi-parametrical approach, accompanied by modeling that aims to simulate cell dynamics. This new trend is based on the recent growth of **biotechnology**. Because of its immense potential and of possible repercussions in the public domain, sometimes affecting individuals in their private lives, biotechnology gives rise to heated debates in which the scientific world finds itself confronted with society at large. With biotechnology, the engineering sciences enter in force into the biological domain. The inventiveness of this form of engineering involves the exploration of issues arising from the study of the mechanisms of living beings, and of the appropriateness of techniques and instruments dominated by miniaturization, information technology and robotics. Both **discovery** and **invention** require ways of thinking and specific talents that must complement and support one another.

If we look at the stunning and sometimes chaotic progress made by scientific thought over the last five centuries, we might begin to ask ourselves what were the triggering factors that made it possible, at a particular moment in time in the western world, to invent and spread a logical and objective approach to the exploration of various enigmas in Nature. This new approach was the **experimental method**. What were the events that led up to its birth? Was there any influence from the social, political or religious context? Was the energy of a few individuals sufficient to initiate such a revolution? Or should we search for its origins in the long-ago past of philosophical thought? All of these questions lead us to reflect upon the way that science is "done" in modern times. As far as the life sciences are concerned, these questions are all the more meaningful in that the effects with respect to Mankind continue to be debated. As science advances, as its useful role is recognized, and as scientific knowledge becomes widespread throughout society, philosophy, which used to interest itself in the life sciences, now focuses on problems of ethics, particularly with respect to the reproductive functions that contemporary biology claims to be able to manipulate and modify. Bounced around between aggressive media attention and the need for openness, and subject to pressure to achieve that which is perceived to be useful, modern-day biological experimentation finds itself at a crossroads, solicited on the one hand by increasingly effective and enterprising engineering techniques, and confronted on the other by ethical and socioeconomic considerations. Faced with such questions, it is necessary to look back to and reflect upon the past.

Chapter I

THE ROOTS OF EXPERIMENTAL SCIENCE FROM ANCIENT GREECE TO THE RENAISSANCE

"In the genesis of a scientific doctrine, there is no absolute beginning. No matter how far back we go in the line of thinking that has prepared, suggested and announced this doctrine, we always arrive at opinions that have themselves been prepared, suggested and announced: if we cease to follow up this chain of ideas that have proceeded one from another, this does not mean that we have finally reached the initial link, but rather that the chain has plunged and disappeared into an unfathomable past."

Pierre DUHEM
The System of the World - 1913-1959

The discovery of the **scientific method** in the West, in the 17th century, doubtless owes a great deal to the sociocultural and politico-religious context of the time (Chapter II). Nevertheless, it would be unjust to forget the part in this inheritance that was played by the scientific and philosophical thought of the Ancient Greeks, which, after centuries of obscurity, was vigorously revived in the latter part of the Middle Ages. In the same way, it is not conceivable to study the history of the development of modern chemistry and its application to living beings without evaluating the technical know-how inherited from the medieval alchemists. Despite the illusory character of their goals – transforming metals into gold, or finding a universal remedy for all ills – alchemists had the merit of inventing apparatus, perfecting methods, and, having done this, discovering phenomena that were to become subjects for reflection and investigation.

1. SCIENTIFIC RATIONALITY IN ANCIENT GREECE

"If scientific culture is to be truly fruitful, it is necessary to have a living breath, an inventive genius, an instinct resembling that of the artist or poet. This is what the Greeks had, and what modern times have rediscovered."

Antoine Augustin COURNOT
Thoughts on the Progress of Ideas and Events in Modern Times - 1872

"In their search for what was permanent in the changing world of observation, the Greeks hit upon the brilliant idea of a generalized use of scientific theory, of assuming a permanent, uniform, abstract order from which the changing order of observations could be deduced. With this idea, of which geometry became the paradigm giving it the most precise expression, Greek science must be seen as the origin of all that has followed..."

Alistair Cameron CROMBIE
Augustine to Galileo: The History of Science AD 400-1650 - 1952

Arising in the 6th century BC from sources spread out over Asia Minor, southern Italy and Sicily, Greek science, which at that time was called philosophy, developed over ten centuries. After becoming installed for a certain time in Athens, it migrated to Egypt, and prospered in Alexandria. These successive migrations correspond roughly to three different periods of thought, known as pre-Socratic, Classical and Alexandrine, respectively. Throughout this long period, Greek philosophers continually questioned Nature. For them, observation was the rule and experimentation remained episodic. Nevertheless, given their way of asking judicious questions and laying down markers for thought based on the logic of their reasoning and the rigor of their dialectics, they deserve to be given credit for being the precursors of the scientific method that emerged in the West much later, at the end of the Renaissance.

1.1. FROM THALES TO SOCRATES
TOWARDS A NEW CONCEPTION OF NATURE

Our knowledge of the pre-Socratic period of Greek science (600 to 480 BC) is mainly based on doxographic documents gathered together centuries later. Pre-Socratic philosophy sought to explain the Universe in rational terms that were liberated from ancestral and supernatural myths. It took over from the philosophy of the Babylonians, Phoenicians and Egyptians. It should be recognized, however, that these peoples were the precursors of knowledge. **Astronomy**, which studies the movement of planets, was the basis for Babylonian science, **Arithmetic** was the basis for Phoenician science and **Geometry** was the basis for Egyptian science. The Phoenicians knew how to make glass, and like the Egyptians they knew how to prepare certain coloring materials and practiced the art of dyeing. They had mastered some of the processes involved in metal-working, an industry that was still in its infancy. The Greek mind, on the other hand, leaned more towards theoretical speculation than experimentation.

The profusion of ingenious hypotheses formulated by pre-Socratic philosophers, and their eclecticism in terms of scientific interest, demonstrate the spiritual vitality of this period. The first Greek pre-Socratic school arose in Miletus, a port in Ionia (western region of present-day Turkey) (Figure I.1). In Miletus, where merchandise from the East passed through on its way to the West, different civilizations rubbed shoulders, enriching one another in a political climate of great liberty. Here, in teachings and writings, attempts to answer questions that had preoccupied Man for centuries began to materialize: the movements of the stars, the cycle of the seasons and changes in Nature, birth, death, the meaning of life... While Babylonian, Phoenician and Egyptian science had been utilitarian, being interested, for economic reasons, in measuring time through determination of the moon's cycles, or in making survey calculations in order to delimit the surface areas of agricultural terrains, Greek science sought to explain the meaning of life, and of the activities that characterize it, through the observation of phenomena in Nature.

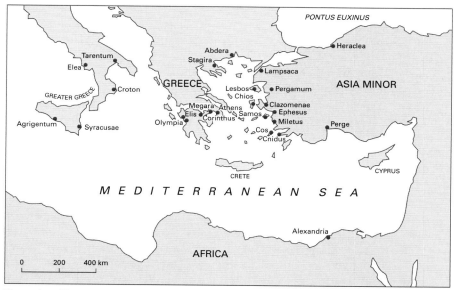

Figure I.1 - The world of Greek cities, with its trading posts in Asia Minor, Sicily and Italy

THALES (625 - 550 BC), ANAXIMANDER (610 - 545 BC), a disciple of THALES, and ANAXIMENES (580 - 530 BC), a pupil of ANAXIMANDER, taught in Miletus. These **Milesian philosophers** stated that "nothing comes from nothing" and that things in Nature arise from a **single fundamental principle** that can be transformed, a primordial element, which was **water** for THALES, the **boundless** for ANAXIMANDER and **air** for ANAXIMENES. Thus, despite the apparent diversity of materials making up the world, an idea emerged according to which materials are built from an elementary constituent. THALES stated that the Earth is carried by water, based on the fact that part of the Earth is submerged and that water is the origin of a cycle in which, pumped up in the form of vapor by the sun, it falls as rain. In fact, the concept of water as formulated by THALES included any liquid material, water representing a special case in that it can adopt a solid, liquid or gaseous state. ANAXIMANDER's concept of the boundless represented an idealized indefinable, holding the generating potentialities for air, water and fire. The idea of the curvature of the Earth is attributed to ANAXIMANDER, as is the first composition of a geographic map to be used by navigators. For ANAXIMENES, the air in the clouds gave rise to water that could solidify into ice or evaporate once brought near to fire, thus achieving a cycle of natural transformations. This problem of change, which took up THALES' idea once more, was discussed at a later date by HERACLITUS (550 - 480 BC), an Ionian philosopher from Ephesus. HERACLITUS chose **fire** as the primordial element because of its mobile and fleeting character.

If the things of the world are subject to change, is the information that we draw from observing them illusory? Already, the question of experimental doubt is raised. The Milesian philosophers also interested themselves in the origins of the human species. Thus, ANAXIMANDER put forward the idea that Man arose from inferior animal species. Distinguishing their ideas from those of myths in which thought is subordinate to fabulous tales that show Man as being only a passive witness to cosmic phenomena, and from those of a theogony that shows the caprices of the gods influencing the course of events in an arbitrary manner, the Milesians dared to state that "the Universe can be understood and rationalized on the basis of fundamental principles" and that attempts to understand it required an **objective detachment** on the part of the philosopher. These completely new ideas began to sprout forth all at once from a small group of men. They became the basis of a philosophy that, for centuries, would enlighten the Ancient World.

Curiously, at the same time as the birth of the brilliant civilization of Ancient Greece, on the other side of the Earth, in China, during the ZHOU dynasty, a theory was being taught that postulated the existence of five elements considered to be forces; wood, fire, earth, water and metal. It was thought that these elements might interact in a circular manner, wood generating fire, fire generating earth, earth generating metal, metal generating water, and water generating wood. In this conception of the Universe, according to the teachings of CONFUCIUS (6th-5th centuries BC), Man is likened to a microcosm made up of the same five elements that make up the macrocosm.

At the end of the 6th century BC, Ionia was taken over by the Persians, and this led to the exile of the Greek philosophers, who moved to the cities of southern Italy and Sicily. This was the case for PYTHAGORUS (570 - 480 BC) whose name remains surrounded by legend. At the age of 40, PYTHAGORUS founded his school at Croton, in southern Italy. With these Pythagorians came the emergence of a mechanistic mode of thought, similar to that of the Milesians, the originality of which was that it was based on calculation and geometry. The famous theorem states that "in a right triangle, the area of the square on the hypotenuse is equal to that of the sum of the squares on the other two sides," reminding us that PYTHAGORUS was a fervent believer in mathematics. **Numerology** achieved an almost magical cult-like status arising from the manipulation of numbers. Thus the number 1 being set down as the primordial matrix, we observe that the sum of the first four numbers is equal to 10 and that the sum of their cubes is equal to 100. A perfect number is defined as an integer that is the sum of its proper positive divisors. Thus 6 is a perfect number because it can be divided by each of the numbers of the sum, 1, 2, and 3. The Pythagorians invented proof by contradiction. The visual evidence found later in EUCLID's *Elements* (3rd century BC) could be added to the requirements of reasoning. Numbers were "**geometrized**", being represented by points aligned in squares, rectangles, triangles or more complex figures. **Everything became measurable**. By highlighting a relationship between the frequency of sounds made by vibrating cords and the length of these cords, the Pythagorians managed to digitize music, and, by extension, phenomena in Nature: numbers, considered as elements of things, create a harmony in Nature that is close to divine perfection. Thus the idea that planet Earth is a sphere arose from the postulate that the sphere is a perfect geometrical figure. The Pythagorians are also credited with the idea of earth, known as a dense body, being the fourth primordial element in Nature, along with the air, water and fire of the Milesians. The three elements of the philosophers of Ancient Greece, earth, water and air, can be seen as the counterparts of our current conception of the three states of matter, solid, liquid, and gas. The Pythagorian PHILOLAOS of Croton (5th century BC) gave us the idea of a cosmos with a center occupied not by the Earth, but by a central fire, around which the celestial bodies, including the Earth and the Moon, rotated, and, by extension, the idea of a corruptible, sub-lunar world to which the Earth belonged and of an incorruptible world beyond the Moon. This concept reduced the Earth to the status of a simple planet in the solar system.

The Milesians had tried to explain Nature based on single principles. The Pythagorians had thought that they could recognize a structure founded upon mathematics, in Nature. As for the **Eleatics**, they required even more rationality in the understanding of living beings. Using HERACLITUS' idea that conclusions drawn from observation are made faulty by the state of perpetual change that exists for things in Nature, the Eleatic school laid down a true doctrine of scientific **doubt**. XENOPHANES (570 – 480 BC), founder of the Eleatic school in Elea, southern Italy, was opposed to Greek theogony. As a basis, he used the Milesian doctrine of the primordial element, attributing significance to this that was no longer tangible, as with water and air, but abstract, designating it as the **One**. In his *Poem on Nature*, the poet philosopher PARMENIDES (515 - 450 BC), one of the most famous Eleatics, developed an idea that was later taken up by PLATO (428 - 348 BC), that is, that perception is only an illusion; only reason is credible. ZENO of Elea (490 - 420 BC), a disciple of PARMENIDES, founded the art of **dialectics**, the undercurrents of which are the principle of contradiction and reasoning by the absurd. ZENO is known for the famous paradox of ACHILLES and the tortoise: even though it is very slow, the tortoise, having departed before ACHILLES, can never be caught up by the speedy ACHILLES. The paradox is built upon the postulate that, if the distance to be covered is divided into equal intervals, each time ACHILLES crosses a given interval, the tortoise will have left it the instant before.

In contrast to PARMENIDES, EMPEDOCLES of Agrigentum (490 - 438 BC) rehabilitated the role of the senses and observation. There are not one, but four, primordial elements, **water, air, fire** and **earth**, considered to be the reflection of certain states of matter, and the roots of things in Nature that are accessible to our senses. While incorporating solid bodies into the notion of earth, gaseous bodies into the notion of air and liquid bodies into the notion of water, EMPEDOCLES considered that beings and things are made up of a mixture or a combination of the four elements in proportions that determine their specificity. Thus wood contains some of the earth element, as it is heavy and solid, some of the water element, which it exudes when it is heated, and some of the air and fire elements, because it gives off smoke and produces flames when it burns. This theory of the four elements was accepted and taught until the Renaissance, and even beyond. It was popularized by geometric symbols represented on public buildings. These symbols are still visible on the remains of some monuments (Figure I.2).

Figure I.2 - Symbols for the four elements, air, fire, earth and water, represented on columns of the Benedictine cloister at Monreale, in Sicily
(photograph by Henri PAYANT)

EMPEDOCLES imagined that at the birth of the Universe, the four elements were mixed, and that plants and animals arose from these elements being selectively attracted. He supported the idea that animal and plant life presented analogies with respect to their morphologies and the functions of their organs and tissues. Thus birds' feathers and fishes' scales were considered to be analogous, as were the leaves of trees and animal fur. Despite its rational tendencies, EMPEDOCLES' philosophy still continued to allow the Supernatural a share in the laws of Nature.

ANAXAGORUS of Clazomenae (500 - 428 BC), a disciple of ANAXIMENES, adopted a **pluralist theory** that led him to postulate the existence of a vast crowd of bodies, a mixture of similars and contrasts, that had formed from the beginning of time, and had been present there, witnesses to the chaos of the origins and to an arrangement by ordered reasoning. In this system, an understanding of things depends on the **evaluation of contrasts**: thus cold is only felt as a contrast to hot. ANAXAGORUS postulated that all was born of all, but within a certain order: air is born from fire, water from air, earth from water, stone from earth and fire from stone, and thus a new cycle can be begun. Based on the observation that a goatskin filled with air cannot be compressed, he arrived at the conclusion that air is a body. PROTAGORAS of Abdera (490 - 410 BC) and GEORGIAS of Leontium (487 - 430 BC), who were contemporaries of ANAXAGORUS and EMPEDOCLES, were sophists whose doctrine foreshadowed that of SOCRATES, in which Man is introspective. For them, the conception of the things that surround us was based on sensory perceptions, all of which had a value and a meaning.

With LEUCIPPUS of Abdera (5th century BC), who was a disciple of ZENO, and DEMOCRITUS (460 - 370 BC), who was LEUCIPPUS' student, we see the emergence of a radically **mechanistic** view of phenomena in Nature. In contrast to the immobile world implicit in the postulates of their predecessors, they put forward the idea of a world of movement: inanimate objects and living beings are the result of an **assembly of atoms** considered to be invisible, indivisible corpuscles that arise from a single substance, which differ only in their size and shape and which move and catch onto each other in a transitory manner. Between atoms and groups of atoms, the **something** ($\delta\acute{\epsilon}\nu$), there is the void, the **non-something** ($\eta\delta\acute{\epsilon}\nu$). The idea of vacuism, adopted by EPICURUS (341 - 270 BC), was disputed by ARISTOTLE (384 - 322 BC) who advocated the continuity of matter. This argument between **vacuists** and **plenists** would last more than twenty centuries. DEMOCRITUS had postulated that the shapes of atoms, angular or round, for example, or convex or concave, would give their arrangement a **specific function** that would take account of the diversity of known bodies. Thus, angular atoms were assigned an acid character and rounded ones a sweet character.

Although the **atomic theory** of LEUCIPPUS and DEMOCRITUS was very different in its conception from that put forward much later by DALTON (1766 - 1844), it nevertheless shows the logic of a reasoning that aims to discover how the Universe is structured. Given their lack of knowledge concerning chemical species and molecular interactions, the Greek atomists showed a remarkably creative imagination, and one can only admire their prophetic train of thought. Later on, EPICURUS would reiterate the postulate that the world, including living beings, is made up of atoms, and that the soul that dies with the body is also made up of atoms. In *De Natura rerum*, LUCRETIUS (97 - 55 BC), Latin poet and materialist epicurean, held that everything in Nature can be explained on the basis of natural laws drawn from the interaction between atoms. The controversy came to the fore again in the 17th century with René DESCARTES (1596 - 1650) and Pierre GASSENDI (1592 - 1655). Although he expressed himself with the reserve necessary to the moral prudery of the period in which he lived, GASSENDI was nonetheless a believer in atomism. In an interesting essay on Greek science, *Nature and the Greeks* (1954), the Austrian physicist Erwin SCHRÖDINGER (1867 - 1961) suggested that DEMOCRITUS' atomic theory was abandoned at the beginning of the Christian Era due to the temptation of extending it to cover the soul. In fact, if we consider that the movement of atoms responds to strictly material physical laws, no room is left for free will.

At its beginnings, the **understanding of living beings** was closely linked to the practice of medicine, which included the observation and classification of the symptoms of diseases, of their evolution, and of therapies using plant extracts. Anatomical examination of Ancient Egyptian mummies has shown the existence of surgical operations, often tricky ones, during this period of history, for example amputations, trepanation, and the reduction of fractures. Nevertheless, magic and recourse to divine beings were an integral part of medical arts. Although the period of the Milesian philosophers was one of **Temple Cures**, in which a patient took part in incantatory ceremonies, it was also marked by the rise of the **rational**, which began to overtake the **magical**. Thus, in a treatise *On the Nature of Man* which is part of the *Hippocratic Corpus*, it is possible to read that the sacred illness (i.e., epilepsy) is no more sacred than any other illness. To support this thesis, the author highlights the fact that when an autopsy is carried out on a goat that had suffered from epileptic nervous problems, the brain is found to have a fetid odor, and he concludes that the cause of the sacred illness is an alteration of the brain. Nevertheless, Temple Cures continued to operate until the 4th and 5th centuries AD. At this late date, we find the cult of the Greek god of medicine, ASKLEPIOS (who became AESCULAPIUS for the Romans), mixed in with the cult of the early Christian saints.

Among the most prestigious of the doctors of pre-Socratic Greece were ALCMEON of Croton (6th century BC), EMPEDOCLES and HIPPOCRATES. ALCMEON is credited with having carried out the first dissection of the human body. Far from the scientific method applied to living beings, the observation of dissected cadavers showed arteries that were empty of blood, leading to the conclusion that the arteries carried air. This error would be perpetuated for centuries. The veins, which still contained blood, were considered to be the sole elements of the circulatory system. Despite these erroneous conclusions, some interesting anatomical discoveries were made.

ALCMEON showed a tube that allowed communication between the tympanic cavity and the nasopharynx, a tube that would be rediscovered in the 16th century by the Italian anatomist EUSTACHIUS (Chapter II-2). ALCMEON recognized that thought took place in the brain, when, up until that point, thought was supposedly located in the heart. He was also responsible for the idea that health is a state of perfect harmony between all of the substances that make up the body and that illness results from a breakdown of this harmony. Curing illness involved restoring the equilibrium. This gave rise to the ideas of humoral pathology developed by HIPPOCRATES (460 - 377 BC).

The *Hippocratic Corpus*, the work of HIPPOCRATES and his many disciples, includes around sixty treatises. *Epidemics* is remarkable for the precision of the clinical analyses reported day by day. They comprise the first example of systematic observations carried out at a patient's bedside, with mention of symptoms that, taken together, make it possible to define an illness, monitor how it evolves, describe a typical crisis in this evolution and formulate a prognosis. In the treatise *On the Nature of Man*, details are given of the **doctrine of four humors**, blood, phlegm, yellow bile and black bile. This doctrine would survive through the Middle Ages and even beyond. It postulated that there is a balance between the four humors in a healthy person. Any imbalance caused by an excess or lack of one of these principles leads to illness. The Hippocratic doctrine also called upon the **theory of the four elements**, air, earth, water and fire. Fire is considered to be enflamed air from which is derived the *pneuma*, an essential principle of life that is found in the heart and circulates in the vessels. A second principle of life is **heat**; blood from the liver brings the necessary heat to the heart and keeps it constant. These notions, though outdated, are nevertheless a first attempt to explain the functioning of the organism at a point in history in which nothing was known of physiology. In the treatise *On the nature of the child*, a typically experimental exploration is offered, to follow the development of embryos in fertilized chicken eggs. Starting with twenty or so fertilized eggs, one is opened each day. In this way, the embryo's appearance and growth can be observed and followed up.

In his NOBEL lecture (1928), the pastorian Charles NICOLLE (1866 - 1936) reminded his audience that the Greek historian THUCYDIDES (460 - around 395 BC), famous author of the *History of the Peloponnesian War*, gave an astonishingly accurate description of the major epidemic (possibly typhus) that hit Athens around 430 BC. His description ended with the remark that those who survived were resistant to renewed infection, which foreshadowed the concept of acquired immunity. At the same period as that described above, remarkable developments were being made in medical practice among the Maya, the Aztecs and the Inca and, above all, in China, with therapies making use of a rich pharmacopoeia based on medicinal plants. In China, particular importance was given to the study of the pulse, diagnosis, prognosis and therapy depending on an analysis of it. Thus, in the Ancient World, in different regions that were far apart from one another, medicine played a predominant role among Man's preoccupations. Curiously, the same is true for astronomy, as if the study of the stars and that of one's own body proceeded, for the thinking human being, from the same exacting, harrowing thirst for an understanding the mysteries of the self, faced with the Universe.

1.2. EMERGENCE OF THE PRINCIPLES OF LOGIC IN PHILOSOPHY AND OF RATIONALITY IN THE SCIENCES

Following the scientific empiricism of pre-Socratic Greece, there was a period that was distinguishable for its rapid intellectual and cultural development, and that brought forth new concepts concerning ways of thinking and methods of reasoning. This period, known as the Socratic period, may be regarded as a necessary springboard, which, albeit at a distance of several centuries, led to the emergence of the scientific method in the West. In Athens, SOCRATES (470 - 399 BC), awakener of ideas, known for his maieutic arts, inaugurated the so-called **classical period** of Greek philosophy (480 to 320 BC). Our knowledge of his ideas comes mainly from the writings of PLATO and XENOPHON (430 - 355 BC) which have come down to us, despite many trials and tribulations. Both PLATO, who was a disciple of SOCRATES and founder of the Academy, and ARISTOTLE, who was himself a disciple of PLATO and founder of the Lyceum, developed and discussed the doctrines of the pre-Socratic philosophers, with particular emphasis on the principle of **causality**. Their theories would have an effect on the philosophers of medieval scholasticism, and, well after these, on the discoverers of the scientific sethod in the West, in the 17th century. As PYTHAGORUS had done with numerology, PLATO used **geometry** to "mathematize" the Universe. Originally, for the Egyptians, geometry (γέα, earth and μέτρον, measurement) was a technique for surveying agricultural terrain, arising from the need to mark out areas of land after each flooding of the Nile. How important PLATO considered geometry to be is shown by his having "Let no one who is ignorant of Geometry enter here" inscribed over the door of his Academy.

Geometrical science proceeds by **deduction** and deductive reasoning is an *a priori* **approach** to knowledge. The gulf would widen with the **inductive method**, which proceeds from the singular to the general, and the *a posteriori* **approach** used in the experimental sciences. In *The Timaeus*, PLATO explains that the structures of things and of beings result from the **assembly of polyhedra** of increasing complexity – tetrahedron, hexahedron, octahedron, dodecahedron, and icosahedron – based on very simple elementary figures, **equilateral triangles**. The four elements are associated with four of these figures, fire with the tetrahedron, earth with the hexahedron, air with the octahedron and water with the icosahedron. Herein lies a barely believable premonition. At the end of the 20th century, high-resolution electron microscopy and X-ray diffraction showed that the protein coats of certain viruses have an icosahedric structure, made up of an assembly of twenty equilateral triangles. Following the example of geometrical figures that may be combined to make more complex ones, or, in contrast, simplified, PLATO supposed that one element could be transformed into another: "By concretion, water becomes earth and stones. When it becomes more fluid, it is transformed into vapor and air. Burning air becomes fire. Squeezed and condensed air becomes clouds and mist. Compacted, clouds and mist flow as water and from this water earth and stones arise once again." These ideas already contain the seeds of medieval alchemy.

By PLATO's time, it had been recognized that next to astral bodies that appeared to be fixed, that is, the stars, there were astral bodies that were visibly mobile. These were named **planets** (from πλανήτη, vagabond), or errant stars. EUDOXUS of Cnidus (406 - 355 BC), a contemporary of PLATO, is known for his hypothesis concerning an immobile Earth and celestial bodies carried on homocentric spheres rotating within each other. He achieved a good degree of accuracy in determining the synodic and zodiacal revolution durations of Venus, Mercury, Mars, Jupiter and Saturn. PYTHEAS (4th century BC), an explorer and astronomer, is said to have invented the **gnomon**. Thanks to the gnomon, a ten-meter high obelisk, the shadow of which in the horizontal plane was noted every day of the year, it became possible to determine the periods corresponding to solstices and equinoxes.

In order to explain Man's relations with the outside world and the sensations experienced, DEMOCRITUS had supposed that there were phantoms given off by sensitive objects. DEMOCRITUS' phantoms led PLATO, in *The Republic*, to formulate his famous theory of **image-ideas**, which is exemplified in the **allegory of the cave**. Some men are captives in a cave, the entrance to which, open to the light, extends the length of the facade. These men have been chained up since they were children, with their faces turned towards the wall opposite the cave opening. Outside, there is a fire that has been lit in the distance, which shines. Travelers with their baggage pass in front of the cave. Their shadows are cast upon the wall of the cave. The chained men look at these shadows, and, prisoners of their senses, think that they are real things. However, these shadows are simply the reflection of reality. True reality escapes our senses, and can only be understood by the mind. This is what the scientific method would confirm much later, showing how scientific discoveries may be true, despite appearances. Thus, the Sun that rises in the morning and goes down in the evening gives the visual impression of traveling around the Earth, and it would require the great boldness of COPERNICUS (1473 - 1543) to claim otherwise.

The allegory of the cave led to the concept of **ideas**. Visible and audible things are the reflection of Ideas, which are external to our senses. In explaining material phenomena that are accessible to our senses, Ideas are above real things. PLATO declared that at birth, the body receives a soul that has already contemplated Ideas and which, in part, contains them, these Ideas reappearing spontaneously during a person's lifetime. They are sometimes called **intuitions**. In a critical analysis of the notion of knowledge, *The Theaetetus*, a young boy, THEAETETUS, guided by SOCRATES' questions, discovers, without preconceived ideas, the relationship that exists between the side of the square and its diagonal. For PLATO this is the proof of a recollection of knowledge acquired during previous existences. PLATO's dialogues come under three headings: logic, physics and ethics, the term physics encompassing all research and speculation about the Universe, for both the animate and inanimate worlds. Over time, physics was to be enhanced by discoveries and inventions that would help to trigger its split with the Life Sciences.

In contrast to PLATO, ARISTOTLE advocated observation and **pragmatism**. He considered the mathematical speculations of the Pythagorians and the idealistic analyses of PLATO on the structure of beings and the theory of Ideas to be dialectical futilities. For PLATO, the notion of Man arose from a **universal concept** of humanity, independent of the regard given to particular men and to their differences. ARISTOTLE, as a realistic pragmatist, considered men along with their peculiarities, while recognizing that there were common characteristics making it possible to relate them to a universal entity, humanity. Briefly, he distinguished between **primary substances**, corresponding to concrete objects and **secondary substances**, with their abstract and general characteristics. These two ideas of the world would clash in the Middle Ages with the classical "quarrel of the universals" (Chapter I-2.2) and were extended by THOMAS AQUINAS (1225 - 1274) to cover the ideas of **existence** and **essence**.

ARISTOTLE made **logic** the required instrument of scientific procedures. The treatise *The Organon*, brings together texts relating to logic, with a fundamental contribution, the **syllogism**. The syllogism was defined by ARISTOTLE as a series of three propositions in which, certain things having been stated, something other than these things necessarily results as a conclusion. The predicate of the conclusion is the major term, while its subject is the minor term. The major and minor terms are linked to the middle term. Thus we have the classical syllogism, "all men are mortal, SOCRATES is a man, therefore SOCRATES is mortal," in which the **middle term** is man, the **major term** is mortal and the **minor term** is SOCRATES. The principle of **causality** was also discussed in *The Organon*. According to ARISTOTLE, there are four essential causes for the birth of something: the **material** cause (the material of which the thing is made, for example, marble for a statue), the **formal** cause (the form that the sculptor gives to the marble), the **efficient** cause (the sculptor) and the **final** cause (the representation of a man by the statue). The form has a particular importance for the living being; the being is only contained as a potential in the material, and it is the form that allows the potentiality to be actualized in the real. The **final cause**, or, in other terms, the principle of **finality**, was a leitmotiv of ARISTOTLE's philosophy. Modern science has substituted the principle of **functionality** for the principle of causality, giving attention to the link, often hidden, between two, frequently distant, phenomena. In *Analytical Seconds*, ARISTOTLE discusses induction and deduction: "we learn either by **induction**, or by **deduction**. Deduction is included in the universal truths, induction is included in specific truths. Nevertheless," he adds, "it is impossible to acquire the contemplation of the universal truths, if not by induction."

ARISTOTLE took up and developed EMPEDOCLES' theory of four elements. He added to it, making use of old ideas, associating qualities in pairs; hot, cold, dry and wet. Thus, fire is hot and dry, air is hot and wet, water is cold and wet, earth is cold and dry. Furthermore, he imagined the possibility of one element being transformed into another by modification of one of its qualities. This postulate was taken up again in the Middle Ages, by alchemists who strove to transmute base metals into gold. To EMPEDOCLES' four elements, ARISTOTLE added **ether**, an incorruptible principle located in an inaccessible region of the sky. While terrestrial bodies were subject to the modifications studied by physical science, celestial objects were supposedly incorruptible. In the **world of the sky**, the planets show a perfect rotational movement that can be explained by divine control. On the other hand, in the corruptible **sublunary world**, bodies follow a rectilinear movement, either downwards for the elements earth and water, or upwards for the elements air and fire. Why does a stone that is released from a certain height fall vertically? For ARISTOTLE, as this stone is not in its **Natural Place**, it has "the potential" to occupy the natural place that it is deprived of. Having been released, the stone moves towards its natural place, the center of the Earth, by a movement that is qualified as **natural**, the source of which is in the stone itself, pushing the stone towards a place where it can rest, downward movement being the natural movement of "heavy" objects. If smoke escapes upwards, that is the place to which its lightness calls it, its natural place, upward movement being the movement of light bodies. Thus each body, depending on whether it is heavy or light, when free from all constraint, moves towards its Natural Place, i.e., upwards or downwards. Because of this, its potential is transformed into an act. **Sphericity, circular movement, attraction towards a Natural Place; these were the three fundamental principles of Aristotlian physics.**

ARISTOTLE was particularly interested in the **movement** resulting from an **initial impulse**. How can we explain the fact that a stone projected forward continues its movement for a certain length of time after being given a push? The cause given for this is the existence of a motive principle, air associated with the object in movement. ARISTOTLE assumed that air closes behind the object in movement and propels it forward. In short, movement could be considered as being a phenomenon of replacement. This is far from the principle of inertia that would only be formulated in the 17th century and would be taught thereafter as unquestionable fact.

When dealing with the animate world, that of beings that are born, move, multiply and die, ARISTOTLE saw the **soul** as the source of all movement and all change. As in the inanimate world, any modification that, in the animate world, led to the transformation of **"potential into act"** reflects an **intention directed towards an aim**, i.e., **a fundamental teleonomic principle**. In his *On the Soul*, ARISTOTLE alludes to a "vital principle", the **soul**, which animates the body. The body comprises only a potential, and the soul gives it the power to exist, that is, live life with all its attributes. ARISTOTLE made a distinction between three sorts of soul, the vegetative soul, the animalistic soul and the rational soul. Functions such as nutrition, growth and reproduction were attributed to the **vegetative soul**, which was the least evolved, and common to animals and plants. The **animalistic soul**, possessed by animals, gave the latter the power to perceive and to move: located in the heart, it was supposed to generate the warm breath, or *pneuma*, that vitalized the body. As for the **rational soul**, also located in the heart, it was specific to Man, whose actions are conditioned by reasoning. Refusing platonic soul-body dualism and the possibility of the soul "traveling" outside the body, ARISTOTLE stated that the soul is inherent to the life of an organism. While the body comprises that material of the being, the soul is its form, like the "pilot in his ship". Taking the eye as an example, ARISTOTLE considered that if the eye was a whole animal, then sight would be the soul. The eye is the material of sight, and if it disappeared, the eye would no longer be an eye, or "if not, by homonymy, an eye of stone". What is true for one part of the living body is true for the whole; the soul is inseparable from the body.

In the Ancient World, the texture of Space was a subject of speculation. ARISTOTLE stated that Space was full, within the limits of a Universe that was finite. The idea of the vacuum was thought to be impossible and was therefore rejected, based on a logic that considered that things and Space are a single entity. Everything was **continuous**. The **plenist** theory was very popular during the Middle Ages, spread by such terse sayings as "Nature abhors a vacuum."

ARISTOTLE had established himself as a **scientific theorist** in *The Organon*. He established himself as a **scientific practitioner** in his treatises on Nature. More than five hundred different species of animals, insects, fish, reptiles and mammals, are listed and classified in his *History of Animals*. Many of these were sent to him by the brilliant general who had been his pupil, ALEXANDER the Great (356 - 323 BC), from countries that the latter had conquered in the Middle East. The animals' characteristics and habits are described with considerable shrewdness and good sense. A differentiation is made between fishes and the cetaceans, whales, porpoises and dolphins, which are marine mammals that have lungs and nourish their young with milk. ARISTOTLE was the **first naturalist** who set out to classify animals. He distinguished two major classes, according to whether animals contained red blood (ἔναιμος) or did not (ἄναιμος). The first class contained Man and viviparous quadrupeds, i.e., mammals, including cetaceans, seals and bats, birds, and oviparous quadrupeds (lizards, turtles, batrachians and fishes (subdivided into cartilaginous and boney). Animals without blood were mollusks, crustacea, shellfish and insects. This classification was to last, almost unchanged, until the 18th century. The classification of representatives of the animal kingdom by the English naturalist John RAY (1627 - 1705) was largely inspired by ARISTOTLE's, particularly with respect to the classification of animals into two groups according to whether or not they contain blood. According to Georges CUVIER (1769 - 1832), the major divisions and subdivisions of the animal kingdom proposed by ARISTOTLE "were astonishing in their accuracy." Reasoning on the basis of the complexity of functions, ARISTOTLE explained how, from plants to animals, Nature raises itself by a continuous transition, which, from a philosophical point of view, implies the existence of a hierarchy. This idea was to be taken into account by Christian theology in the Middle Ages, with its conditional acceptance of Aristotlism.

ARISTOTLE's treatise *Parts of Animals* was the first attempt at a **comparative anatomy** to appear in zoology. Analogies are established between the structures and functions of different organs – for example, between fish gills and mammal lungs, or between fish fins and bird wings – and commented upon in terms of evolution. In *The Generation of Animals* and *The Movement of Animals*, ARISTOTLE looked at **reproduction, growth** and **locomotion**. Two principles emerge from all of these works:

1. there is an equilibrium between the different parts of the body, an idea that would be specified twenty centuries later by Etienne GEOFFROY SAINT-HILAIRE (1772 - 1844) under the name of the "principle of the equilibrium of the development of organs",
2. Nature does nothing in vain.

If we consider the eye as an example, its structure explains the faculty of seeing. In this way, ARISTOTLE takes up the theory of the Pythagorians who tried to explain vision according to a "principle" coming out of the eye and going to probe the form of the object that is looked at. For ARISTOTLE, this hypothesis is justified by the convex shape of the eye, which is in contrast to the concave shape of the ear that functions as a receptacle for sound. This **teleological conception** was to be disputed later by LUCRETIUS, who postulated that the sense organs, eye, tongue, and ear, appeared spontaneously. It was after they had appeared that they were used for sight, speech and hearing. LUCRETIUS' antifinalism led him to write, in *De Natura rerum*, that "the eyes were not made for sight, but we become aware of sight because we have eyes." Nevertheless, the aura of ARISTOTLE was such that his finalist theory was taken up by medieval scholastics, and taught until the Renaissance.

Although there are many highly accurate analyses in ARISTOTLE's work, there are also errors scattered throughout his writings. How could it be otherwise at a period in time when the understanding of living beings was still in its infancy, and where imagination made up for experimentation? Thus, the heart was considered to be a generator of heat, the blood it contains coming to the boil, overflowing and spreading through the vessels. The spontaneous generation of animals from silt was an admitted fact. These errors persisted through twenty centuries, before being subjected to the screening of the scientific method, and then refuted. Nevertheless, ARISTOTLE remains one of the most prestigious reference points in an Ancient Greek culture that was open to Nature. He asked the right questions. Even if the answers that were found were sometimes incorrect, there is merit in these questions having been asked. Later on, answers were given that were more judicious and more in conformity with the nature of things, while, at the same time, audacious ideas were ripening, ideas that were opposed to tradition and beliefs.

Aristotlian thought, as carried through into the Middle Ages and beyond, often appeared to be dogmatic, which led to disputes, or even refutations, making it necessary to re-think certain areas of the physical and life sciences that were taught as inviolable truths, and to carry out a salutary revision of deeply rooted errors. Why did ARISTOTLE's aura continue to shine for so many centuries? This was doubtless partly due to the fact that his thought was tuned into the idea that naive Man forms of Nature *via* his senses. ARISTOTLE's theory of **Natural Places**, towards which things seem to move spontaneously, explaining, for example, the way smoke rises towards the sky or a stone falls to the ground, are in agreement with a simplistic, widely-peddled form of reasoning, a simple reflection of perceived sensations.

Greek thinkers did not consider "manipulating" Nature in order to obtain information about it, for errors and illusions are latent in any manipulation. ARISTOTLE's physics is based on observation, and does not allow for experimentation. The Catholic Church of the Middle Ages, *via* the writings of THOMAS AQUINAS, needed no urging to support Aristotlian thought. ARISTOTLE's discussions on the difference between basis and form, between the material of things that speak to our senses and the mysterious principle that is the essence of this material and cannot be understood by our intelligence, were used to explain certain passages of the *Bible*. Aristotlism, one of the underpinnings of medieval scholastic teaching, would continue to base itself on observations of Nature and to draw up theories by induction, often based on a single observation, sometimes aided by an *a priori* concept.

At the death of ALEXANDER the Great, ARISTOTLE, no longer benefiting from protection, a prey to public condemnation and accused of impiety, had to exile himself. This exile marked the end of the classical period of Greek philosophy. Both PLATO and ARISTOTLE, as seen through their teachings and their writings, are still, even today, considered to be giants of philosophical and scientific thought, with its two poles, the former giving primacy to the idea, the latter giving it to the concrete. Their conceptual approach to phenomena in Nature, the projective scope of their ideas, and the soundness of their reasoning would, for a long time, affect the way in which the complexity of living beings was seen, and would provide the keys for understanding them.

The momentum that ARISTOTLE had given to the natural sciences was added to, in Athens, by his pupil THEOPHRASTES (372 - 287 BC), a botanist who took over the Lyceum. In two remarkable works, *History of Plants* and *Causes of Plants*, THEOPHRASTES described hundreds of plants, some of which came from tropical regions. He proposed a system of classification into trees, bushes, shrubs and grasses that would be taught throughout the Middle Ages. He was interested in the effect of a plant's milieu on its development, and, for the first time, gave a clear description of germination, i.e., the formation of the plant from seed, at the same time creating terms having a specific scientific meaning for a grouping of words; for example, περιχάρπιον (pericarp) designates the envelope that surrounds the fruit (χάρπιον). Because of all his contributions, THEOPHRASTES is considered to be the **father of botany**. The interest in plants shown by THEOPHRASTES and his successors was based on the search for properties that would help in the treatment of illnesses, which led to the early development of a pharmacopoeia that was rich is remedies of plant origin. Although he recognized the idea of finalism inherited from ARISTOTLE as the basis of his botanical system, THEOPHRASTES admitted that there were exceptions to the rule. Such a qualification was not without its philosophical influence and after-effects. At the same period, ZENO of Citium (335 - 264 BC) created the Stoic school in Athens. The master taught under an arcade (στοά), giving rise to the term stoicism, which was given to a doctrine that advocated reason, without any deistic preconceptions, as the basis for understanding.

1.3. THE ALEXANDRINE PERIOD OF GREEK SCIENCE

ALEXANDER the Great died in 323 BC, and his empire was divided up among his generals. The province of Egypt fell to one of them, PTOLOMY SÔTER (around 360 - 283 BC). Alexandria rapidly became an intellectual center, and would remain so for seven centuries. It was a shining light of the Ancient World, due to the presence of a museum, equipped with teaching and dissection rooms, an observatory, a rich botanical garden and a menagerie of rare animals, and a library that was said to contain nearly 700 000 volumes. There was an army of scribes, occupied with copying manuscripts. The PTOLOMEIC dynasty reigned for three centuries, ensuring political and financial support for a scientific elite. This was the first example of a large-scale link between **political power** and the **promotion of science**. What this period, known as the **Alexandrine Period**, illustrates, is the development of an **industrial technology** associated with abstract concepts. As Jacques BLAMONT (b. 1926) says in *The Digit and the Dream* (1993), "for the Greeks, whose greatest thinkers had put manual work and artisans at the bottom of the social scale, the result was an unexpected coming together of the use of instruments and the most ethereal of speculations, of figures and dreams. This is," adds BLAMONT, "the most novel, most astonishing, most mysterious aspect, that there could be such a close, fruitful coupling between the mechanical arts, the mastery of which was handed down by word of mouth, without the use of writing, and the science that was cultivated by bookworms who showed little taste for applications." The remarkable economic activity of Alexandria, which generated a financial boom, added a not inconsiderable weight to the political support given to this unprecedented intellectual and technological boom. This city had, in fact, become a very active center for commercial exchanges and one of the most flourishing markets in the East.

In parallel with this boom in physical science, and its applications to different areas of industry, life science had undergone a revival and was having the audacity to look at **Man**. The first school of human anatomy was created in Alexandria by HEROPHILUS (330 - 260 BC) and ERASISTRATUS (310 - 250 BC), and the teaching here was based on the practice of **dissecting** human cadavers. HEROPHILUS distinguished between arteries that beat, and veins. He measured the pulse rate, evaluating the time by means of a clepsydra, or water clock. He described the ventricles of the brain, and distinguished nerves from tendons. He individualized the initial part of the small intestine where the pancreatic duct and the bile duct come into. The term δωδεκαδάκτυλος (twelve times the width of a finger) was used to denote the topographical particularity of this region of small intestine. Latinization of this Greek term led to the word used today, duodenum. ERASISTRATUS became interested in the phenomenon of digestion, and recognized that the heat of the stomach is unable to cook food, which was what was believed at the time. He made the brain the operational center of the nervous system and established a relationship between the functioning of the brain and intelligence. He also tackled the problem of the role of the heart in the circulation of blood, which was still a mystery at this time, without managing to resolve it. Despite these advances, there were a few errors, partly due to the temptation to attribute a structure to different types of "**spirits**". It is from the time of ERASISTRATUS that we can date the belief that the nerves are hollow tubes in which animal spirits circulate, a belief that lasted until the time of DESCARTES. Sticking to the observation of cadavers, ERASISTRATUS postulated, as had ALCMEON, that the blood circulates only in the veins and not in the arteries, which were supposed to carry air, an error that would be corrected by GALEN. Thus, the pulmonary artery which takes "black" blood to the lungs was wrongly named the arterial vein, while the pulmonary vein which contains "red" blood was christened, again wrongly, the venous artery. The tracheal artery was named thus because it was considered to be a large air channel.

The teachings of HEROPHILUS and ERASISTRATUS, which had provided a broad outlook on the fundamental ideas of anatomy and physiology, were succeeded by an empirical school, for which the art of healing was of more importance than a scientific understanding of the human body, then by a "pneumatic" school, which considered that the most important element was the "vital breath", or **pneuma**, that was supposed to bring energy to all parts of the body, and finally by an eclectic school which found its sources in all the doctrines of the previous centuries. It would seem that the trend in medicine during this period was more theoretical than practical, and that the understanding of anatomy did not make much progress, the practice of dissecting cadavers having fallen into disuse.

In the 2nd century AD, Claudius GALENUS (better known in English as GALEN, 131 - 201) came to the fore, not only as a doctor (along with HIPPOCRATES, the greatest in Ancient Times), but also as an experimenter. Born in Pergamum in Asia Minor, he went to Alexandria, which had become the capital of Hellenistic culture, in order to study anatomy, and then he returned to Pergamum, where he practiced as a surgeon for gladiators. At 33 years of age, he set himself up in Rome. Having become physician to emperor MARCUS AURELIUS (121 - 180), he acquired an extraordinary reputation. GALEN carried out many dissections, and wrote a great deal. Most of his dissections were carried out on apes, and for a long time it was thought that the descriptions he made of these dissections referred to Man. GALEN also carried out vivisections on pigs and dogs, exploring functions that had remained inadequately understood up until that time. He described the innervation of the muscles of the larynx by nerves that were "as thin as hairs" and noted that if these nerves were ligatured or sectioned in an animal, the animal became aphonic. This explained why, if a lesion of these nerves occurred when carrying out the excision of a hypertrophic thyroid in a patient with a goiter, it often led to paralysis of the voice. GALEN showed that sectioning the spinal chord of a dog in different locations led to paralysis of different parts of the dog's body. Sectioning between the first and second vertebrae led immediately to the death of the animal, due to shutdown of the heart and respiratory movement. The fact that the ureters play a role in the elimination of urine from the bladder was proven by a simple ligaturing experiment. From these experiments, it was possible to conclude that any alteration in a function arises from a defect in the functioning of an organ, and that any lesion to an organ leads to an alteration in the function for which this organ is responsible. GALEN recognized that arteries contained blood, and not air, as ERASISTRATUS had written that they did. ERASISTRATUS' error resulted from dissections carried out on cadavers. GALEN was able to refute this mistake by showing that if an artery is exposed in a living animal, and if this artery is tied off above and below a point where an incision is subsequently made, the artery is found to be filled with blood. Like ERASISTRATUS, GALEN studied the problem of the circulation of the blood, without, however, managing to solve it. He wrongly postulated that the septum that separates the right and left ventricles of the heart is full of pores through which the blood circulates. This error would be taught as dogma for centuries. Not being able to come up with a simple explanation for the mechanism involved in the circulation of the blood, GALEN made use of verbal palliatives in order to give a semblance of logic to this mechanism, dipping into the writings of PLATO and ARISTOTLE and the Hippocratic corpus, and resorting to the abstract idea of spirits that the blood takes on in different locations within the organism: **natural spirits** in the liver, **vital spirits** in the heart, and **animal spirits** in the brain (Chapter II-2). The solution would be found fifteen centuries later by William HARVEY (1578 - 1657), who used a rigorous approach that signaled the birth of the scientific method in the life sciences (Chapter II-1). Nevertheless, Galenic doctrine, which was conveyed by Arab scholars, including the famous physician AVICENNA, was to act as an inviolable reference for the Medieval world.

ARISTOTLE had been an observer. GALEN was both **an observer** and **an experimenter**. The thinkers of this period were well aware that any experimentation on a living animal inevitably modifies certain parameters. However, no matter how judicious and meticulous their experimentation was, it left doubts concerning the validity of interpretations of it. A proof of this is the famous experiment concerning the origin of arterial pulsation. GALEN made a small longitudinal incision in an artery near to the skin, in a goat. He put a finger-length reed into it. The artery was ligatured at the place where the reed was put in. GALEN noted that the region below the ligature stopped beating and that the reed stopped moving. He concluded that there had been an interruption in the "pulsatile virtue" of the artery. ERASISTRATUS had carried out the same experiment. In contrast to GALEN, he had observed beating of the reed. He deduced that the pulse was carried by the contents of the artery. Such contradictions threw **doubt on experimentation**, which was still in an embryonic phase, all the more so because, with rare exceptions, Greek philosophers, naturalists and doctors, had been and remained reluctant to carry it out. In addition, tradition dictated that manual work be reserved for slaves. Also, techniques used in the mechanical arts were considered as being actions directed against the forces of Nature, as ruses or tricks for conquering Nature. The Greek term "μηχανάομαι" can be translated both as "make" and as "make up a ruse". Experimentation, which was considered to be a device, was, for a long time, an **object of suspicion**, and it is not surprising that, despite the power of Greek thought, and despite the logical support that underpinned it, this thought was only rarely put to use to carry out experiments.

Alexandria was home to many thinkers whose names have been preserved for posterity; the physician STRATO (335 - 265 BC), who continued on from ARISTOTLE, mathematicians EUCLID, known for his *Elements of Geometry*, which included thoughts concerning Arithmetic and the theory of numbers and APOLLONIUS of Perga (262 - 190 BC), author of a work on *Conic Sections*, as well as ARISTARCHUS of Samos (310 - 230 BC), a pupil of STRATO. The latter was the first to formulate the revolutionary hypothesis that the Sun and the stars are immobile, that the Earth revolves around the Sun and that it also rotates around its own axis. The Earth was reduced to the status of a planet and Man was no longer at the center of the Universe. This sacrilegious idea led to ARISTARCHUS being accused of impiety. ARISTARCHUS' heliocentric theory was to be rejected in favor of the geocentric theory of HIPPARCHUS of Nicaea (2nd century BC). In order to maintain a certain consistency, the latter theory required the use of models that were made all the more complicated by the fact that astronomers' observations were becoming more numerous and more accurate. HIPPARCHUS made most of these observations in Alexandria. He invented the astrolabe, an instrument that makes it possible to measure the position of the stars and their apparent height. In 146 BC, Greece came under Roman rule. The Roman period took over from the Hellenistic period. Nevertheless, for several more centuries, Alexandria remained the intellectual, scientific and cultural center of the Ancient World.

In the 2nd century AD, Claudius PTOLEMY (90 - 168) published a complete theory of the geocentric system, and produced an encyclopedia, *The Almagest* (η μέγιστη), which was a masterly synthesis of the astronomical knowledge of this period, and included, in particular, a star catalogue. In *The Almagest*, PTOLEMY, referring to the work of both APOLLONIUS of Perga and HIPPARCHUS, explained certain anomalies in the orbits of the planets, which, in theory, were circular, by the effect of a double movement. He assumed that a small circular movement, the epicycle, was superimposed on a large circular movement, the deferent, which took place around the Earth. *The Almagest* would remain a credo for the scholarly world for nearly fifteen centuries.

While astronomy was subject to the curiosity of many thinkers, geography remained something of a poor relation. One of the greatest geographers of Ancient Times was ERATOSTHENES (273 - 192 BC). He estimated the circumference of the Earth to within 140 km and he gave us the idea of the obliqueness of the ecliptic. Greek mathematical science reached its peak with DIOPHANTUS (3rd century), author of a presentation on algebra, PAPPUS (4th century), who was known for his works on mathematics and physics, in which is found the first definition of the center of gravity, and HYPATIA (370 - 415), a famous lady mathematician and philosopher.

The Alexandrine period was no stranger to the development of the mechanical arts, due to famous inventors such as ARCHIMEDES (287 - 212 BC), PHILO of Byzantium (3rd century BC) and HERO of Alexandria (1st century). HERO's automata foreshadowed the flute player and the "digesting" duck of VAUCANSON (1709 - 1782) (Chapter II-6.3). In his *Pneumatica*, PHILO described an instrument that he called a thermoscope, which was able to respond to heat or cold by a movement of liquid. The principle was based on the expansion of air when heated (Figure I.3).

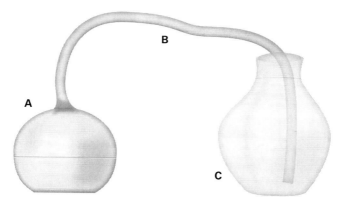

PHILO's thermoscope comprises an empty lead globe (**A**) to which is welded a curved tube (**B**) that descends into a vase filled with water (**C**). If globe **A** is placed in the sun, the air it contains is heated and expands, and the increase in volume leads to some of the air being given off through the water in vase **C**. If globe **A** is moved out of the sun's rays, water from vase **C** climbs tube **B** and descends into globe **A**. If globe **A** is placed in the sun again, water returns to the vase, and so on, for as many times as the experiment is repeated.

Figure I.3 - Diagram illustrating the principle of PHILO's thermoscope (3rd century BC)

Here we can see a glimmer of the scientific method. However, it was only in the 18th century that temperature would be measured with precision by alcohol or mercury thermometers (Chapter II-3.3). It is a generally accepted and noteworthy fact that the scientific method finally took over in the West, after the Renaissance, from the moment when people became convinced that the validity of an experiment depends on the **exactitude of the measurements** carried out and on the **evaluation of the margin of error**.

With the Roman occupation, which occurred around the turn of the first millennium AD, the intellectual dynamism that had livened the city of Alexandria began to fade. From the 3rd century onwards, sporadic riots broke out, linked to the religious disagreements that marked the beginnings of Christianity, and, later, Islam. During one of these riots, in 415, HYPATIA was killed. The library of Alexandria was pillaged and partly burned. The Golden Age of Alexandria came to an end in the 7th century. Thus came the disappearance of a brilliant civilization that had illuminated the Ancient World.

1.4. THE POST-HELLENISTIC PERIOD
HIATUS OR TRANSITION WITH THE MIDDLE AGES?

During the reign of emperor CONSTANTINE (270 - 337), Byzantium, which had become Constantinople, acquired the status of a capital city, and Christianity was decreed to be the State Religion. Constantinople attracted Greek men of letters and became the new seat of science. After the Roman Empire was split into two in 395 AD, Constantinople would remain the capital of the Eastern Empire for around ten centuries, until it was occupied by Ottoman troops in 1453. In 529, emperor JUSTINIAN (482 - 565) issued an unfortunate decree forbidding non-Christians to teach. Greek scholars, accused of impiety, had to expatriate themselves to Syria. At the same date, in Italy, the first monastic order, that of the Benedictines, was founded. Monasteries flourished in the West. They housed centers for learning in Latin, where theology centered on biblical tradition was the basis of the **philosophical culture**.

Battered by the barbarian invasions that occurred successively from the 4th century onwards, the Western Roman Empire, undermined by internecine squabbles and corruption, fell apart. Scientific thought entered a long sleep.

Historians have tried to understand why the scientific momentum that was carried forward by the Greek thinkers gradually slowed and then died. The birth of Christianity, which was more oriented towards the metaphysical, and placed more emphasis on the spiritual, and which, later, would impose its dogmatic orthodoxy on medieval teaching, cannot really be blamed, when considering this period in time. At most, it is possible to note the lack of interest of the first Fathers of the Church, from ORIGEN (185 - 254) to AURELIUS AUGUSTINUS (saint AUGUSTINE)

(354 - 430), in the theogonic theories of the Greek philosophers, whose often contradictory opinions were in opposition to the unitary character of the Christian dogma of Genesis, even though the theory of Ideas, the only true reality for PLATO, was to be assimilated, by the Church, with divine reality. A more credible hypothesis, which also provides food for thought for the science of today, is that which implicates the responsibility of political authorities. Rome was, in fact, more interested in the applications of discoveries than in the discoveries themselves. What was important for the Roman administration was to construct roads, bridges and aqueducts to improve communications, in order to ensure that the legions were supplied with arms needed to carry their wars beyond Rome's frontiers and to conquer the world, to issue laws that ensure that the city functioned correctly and to search at all times for an immediate profit. Why bother, for example, creating and financing schools of medicine and widening the field of knowledge of the domain of living beings, when the treatises of HIPPOCRATES and GALEN were already available? Of what possible use were philosophers who thought about abstract science, for a society that was fundamentally pragmatic?

At the beginning of the 7th century, the newly-born Islam was imposing itself as an all-conquering religion. In 642, Alexandria, which had been the leading light of Ancient Civilization at the beginning of the Christian Era, became subject to Arab domination. Arabic became the language spoken by scholars. It would be the main vehicle for transmitting the knowledge of Ancient Greece to the West of the Middle Ages. From the 8th to the 12th centuries, Arab science took over from Greek science. In the Middle East, it was located first in Damascus, and then spread out from Baghdad. It moved to Cordoba after the conquest of Spain by the Omayyads. This period produced some well-known scholars who left their mark, particularly in the areas of mathematics and medicine. IBN MÛSÂ AL-KHUWÂRIZMÎ (783 - 850), an Arabo-Persian mathematician, popularized the decimal writing system for numbers inherited from India, and introduced the formulation of equations with one or two unknowns that was later to be called **algebra**, a word derived from the Arabic *al-jabr*. He described a process for extracting square roots. In Europe, AL-KHUWÂRIZMÎ's Latinized name became algorism, and then **algorithm**, designating any system of calculation used to solve problems. The notions of sinus, cosinus and tangent, inherited from the Indian mathematician ARYABHATA (6th century) were added to Arab trigonometry. AL-BÎRÛNÎ (973 - 1048), another famous mathematician, is known for his work in geometry, carrying on the Greek tradition. In the Life Sciences, IBN ISHÂQ (809 - 877) translated the medical treatises of HIPPOCRATES and GALEN. AVICENNA (IBN SÎNÂ) (980 - 1037) whose *Canon of Medicine* was re-printed thirty-six times, was a reference in the Middle Ages, as was AVERROES (IBN RUSHD) (1126 - 1198), a philosopher, physician, translator, commentator and admirer of ARISTOTLE, from whom he extracted the rationalist and materialist aspects. The teaching of Averroism in France was not to occur without creating major conflicts with the ecclesiastical authorities.

IBN AL-NAFÎS (1210 - 1288) showed the flow of blood that went from the heart and to the lungs. This was the **pulmonary circulation** that would be rediscovered by the Spaniard Miguel SERVETO (1509 - 1553) three centuries later (Chapter II). Since Ancient Times, there had been questions about light, about the organ of vision, the eye, and about optical illusions. These queries had remained without answers or had been given erroneous explanations. For GALEN, who followed on from ARISTOTLE's ideas, the eye that looks at an object emits a vibration that spreads in a straight line and is reflected on the object. In Cairo, around 1000 AD, the physician IBN AL HAYTHAM (965 - 1039), a specialist in optics, dissected the eye and studied its structure. He understood correctly that the eye captures the light reflected by objects. Nevertheless, it was necessary to await the 13th century, with Robert GROSSETESTE and Roger BACON, of the Oxford school, before a truly experimental approach was applied to studying the propagation of light, thanks to the use of mirrors and perfectly polished lenses (Chapter I-2.2).

In the 12th century, medicine became the domain of Arabic-speaking Jewish physicians. The most well-known, MAIMONIDES (1135 - 1204) served the sultan SALADIN (1138 - 1193). His treatises, which presented a pharmacopoeia describing the therapeutic use of several hundred plants, was an inspiration for many doctors in the Middle Ages. From the 11th century onwards, the Arab influence penetrated such intellectual centers in southern Italy as the school of medicine in Salerno. Salerno had already acquired widespread fame. Founded in the 7th century by Benedictine monks, it stood out because of the absence of magic and astrology. Dissections of animals were carried out systematically there. During the 12th century, edicts appeared that forbade the monks from practicing medicine, which passed into the hands of the laity.

The memory of Greek civilization was preserved thanks to manuscripts that were saved from the disaster of Alexandria's library, and, for a large part, recovered from other sources, including the University of Edessa in Mesopotamia and the House of Wisdom in Baghdad. The patriarch of Constantinople, NESTORIUS (380 - 451), who had been exiled to Edessa for heresy, founded a University there, bringing together a large number of the works of the Greek philosophers that had been dispersed around the Mediterranean world, and having them translated into Syriac. Later, the caliph of Baghdad, AL-MAMÛN (9th century), once again started collecting available Greek texts, assembling them in the House of Wisdom and having them translated into Arabic. These manuscripts, which, fortunately, had been spared from destruction, comprised the lion's share of the works that made their way to Spain.

The 11th century saw the beginning of the reconquest of Spain by the Christians. In the 11th century, Toledo became a particularly active intellectual center that was livened by the presence of such major scholars as GERARD of Cremona (1114 - 1187) and ROBERT of Chester (12th century). It was in Toledo, but also in Salerno in Italy and Palermo in Sicily, places of contact between Arab culture and the Christian world, that part of the Hippocratic corpus and also works by PLATO, ARISTOTLE, PTOLEMY, EUCLID, APOLLONIUS, ARCHIMEDES, and GALEN were **translated from Arabic into Latin**. Thus, channeled by these translations, part of the heritage of the Greek thinkers, which had traveled *via* the Arab world, came to be transmitted to France and circulated through the Christian West. Unhappily, successive translations, sometimes with questionable interpretations made by the translators themselves, and the presence of numerous errors, affected the value of the texts and constituted in themselves factors for doubt. When Constantinople was taken by the Turks in 1453, Byzantine scholars were forced to leave and to take refuge in Italy. They brought with them the original manuscripts of ARISTOTLE, PLATO, PTOLEMY, and a few other Greek philosophers, who would be translated directly into Latin, without the risk of major errors.

2. THE PHILOSOPHICAL AND TECHNOLOGICAL HERITAGE OF THE MIDDLE AGES

The period of western history known as the Middle Ages covers approximately ten centuries, from the fall of Western Roman Empire in 476 to the fall of the Eastern Roman Empire in 1453, the date on which Constantinople was conquered by the Turks. The end of the Middle Ages also coincided with the discovery of the Americas by Christopher COLUMBUS (1451 - 1506) in 1492 and the period of the great maritime explorations. This ten century period is characterized by a mosaic of political, economic, cultural and scientific events that comprise not only its richness, but also its complexity and its ambiguity. Despite its chaotic appearance, this period, towards the end, showed a dynamic that was driven by a mixture of cultural curiosity and audacious ideas, which would take a concrete form in the Renaissance and herald the scientific revolution of the 17th century.

2.1. THE POLITICO-ECONOMIC CONTEXT OF THE MIDDLE AGES

During the first centuries of the Middle Ages in the West, invasions, wars and political troubles had given rise to a deep economic slump and the almost complete disappearance of an intellectual life. The reign of the Carolingians was marked by a cultural revival in the West, supported by a stated political will. CHARLEMAGNE (742 - 814) called upon the theologian ALCUIN (735 - 804), of Irish origin, and other scholars, to set up, in the monasteries, a scholastic reform that was based on an active pedagogy involving commentaries of texts and dialogues. Teaching was done in Latin. This would be called **scholastic teaching** (from the Latin *scola*, or school). ALCUIN founded the first monastic school in Tours, and this was considered to be the parent of French monastic schools. During the tormented period of the High Middle Ages, the monastic schools had the merit of preserving and transmitting a base of cultural knowledge.

The monk GERBERT OF AURILLAC (938 - 1003), author of a treatise on geometry, became pope in the year 1000 under the name of SYLVESTER II (999 - 1003), and gave his support to the intellectual structure that was in the process of trying to rebuild itself. He was responsible for introducing so-called Arab figures, from 1 to 9, into current practice. The figure 0 was still missing, but it would later be imported from India. Pierre ABÉLARD (1079 - 1142) was one of the enlightened thinkers of this period. For ABÉLARD, the idea of God was distinct from His presence in the works of His creation. In other words, Nature operates according to laws that are specific to it. In order to understand these laws, the English scholastic philosopher ADELARD OF BATH (1070 - 1150), one of the divulgers of Arab science, advocated the use of objective reason.

During the 12th and 13th centuries, medieval civilization was remodeled by a **technical revolution**, the importance of which has been compared to that of the industrial revolution in the 18th and 19th centuries. A warming of the climate favored the growth of agriculture. There is no doubt that the abundance and improved quality of the food played a major role in the population growth that characterized this period and in the development of urban crafts and trade. Agriculture benefited from technical advances: an improved plow replaced the swing plow and the harness was improved. The yoke was adapted for use with horses, which replaced oxen for heavy work. Water and wind energies were harnessed by means of paddle wheels and windmills. There were more than five thousand hydraulic mills in England in the 13th century. It was possible to measure time accurately from the beginning of the 14th century by means of clocks with weights equipped with an escape mechanism. At this time, the preparation of different acids, such as nitric, sulfuric and hydrochloric, as well as that of alcohol, was already well understood. Glass lenses were available to merchants selling spectacles that made it possible to correct the problems of long-sightedness, which increase with age. A not insignificant and unexpected result was the age at which people retired from active life, in well-to-do society. The textile industry was helped by improvements in weaving looms. In 1338, Florence had more than 200 textile workshops. Navigation made considerable progress thanks to improvements in rigging. The compass, with its magnetic needle on a pivot, had its origins in the Far East, and appeared in the West in the 13th century, making maritime exploration over large distances possible. The manufacture of paper, which had been in use in China since the 2nd century AD, was introduced into Europe in the 12th century. However, the most important revolution from the intellectual and cultural point of view remains the invention of **printing**. Here again, the Chinese were the precursors. BI SHENG (11th century) has been credited with the production of thousands of writing characters in terra cotta to be used for the printing of books. Made of clay, the characters were fixed to an iron plate using a resin, prior to printing. Towards the 12th and 13th centuries, the terra cotta characters where replaced with wooden ones, and then with characters made of tin. Around 1450, typographical printing made its appearance in Europe, moving with incredible rapidity from North to South in a couple of dozen years. In Mainz, Johannes GUTENBERG (1400 - 1468) developed a typographical process based on the use of reusable, movable metal characters, each character having an inverted letter in relief. He used a special ink that allowed printing on both sides of a page of paper. A screw press equipped with a large plate was used to apply the block with its characters to the sheet of paper. In 1455, 180 copies of the famous 42-line Bible were produced.

This accumulation of inventions that occurred over the few centuries marking the end of the Middle Ages, and which upset the established order, had its counterpart in the sustained effort made by the religious authorities of the period to promote a certain form of intellectual culture. In 1179, the third Council of Latran had decided to attach a school to each cathedral. In these **Episcopal schools**, teaching passed from the hands of the monks to those of the secular clergy, who, being in contact with the people, were in charge of educating them. The first groupings of educating clerics and free auditors were the precursors of the creation of the first Universities in the West, under the aegis of the Catholic Church. At the end of the 12th century, the Universities of Oxford and Bologna were created, and, in the following century, those of Paris, Montpellier, Padua, Naples, Salamanca and Toulouse. The Universities of Cambridge, Grenoble, Prague, Heidelberg and Cologne followed in the 14th century. These Universities, which were attached to cathedrals, developed within an open urban environment, according to a corporatist mode, bringing together masters and students, and benefiting from a statute and relative autonomy, with election of the rector, decisions made by assemblies of teachers and pupils, and exemption from taxes. Nevertheless, the Church maintained the final word in the management of the Universities. Thus, the autonomy of the University of Paris, which was conferred in 1231 by a papal bull, *Parens scientarum*, issued by pope GREGORY IX (1145 - 1241), was subject to the protection of the pope, and, because of this, to his supervision. It was common for students to move from one University to another, which was a major factor in the spreading of knowledge, and also a means of comparing teachings and discovering excellence. In Italy, the development of the first Universities coincided with that of the city-states, benefiting from an autonomous status due to **charters** that allowed them to administrate, increase their prosperity and finance the sciences and the arts as they wished. This was the case for Venice, which presided over the particularly successful flowering of the University of Padua.

In the 13th century, the Faculty of Arts was the center of gravity of the University of Paris. The teaching of the seven liberal arts that was given here was spread out over two cycles. The first cycle was the *trivium*, which was mainly literary, and included grammar, rhetoric and logic. The second cycle was the *quadrivium*, which was more scientific, and included arithmetic, geometry, music and astronomy. The *trivium* and the *quadrivium*, which were a kind of foundation course, prepared students for the specialized faculties of medicine, law and theology. Pre-Renaissance medieval teaching tried to reconcile the biblical tradition with the enormous amounts of information coming from the rediscovered philosophers of Ancient Greece, which opened minds to a knowledge not only of logic, but also of physics and cosmology. Nevertheless, the recurring problem of this mainly theoretical teaching remained the absence of consideration for manual work. The term "liberal arts" meant "the arts of free men" as opposed to the "**mechanical arts**". In response to more and more direct criticism, there were the first stirrings of reform. From the 12th century onwards, the administration of the Saint Victor school, founded in Paris by Hugues DE SAINT VICTOR (1096 - 1141), decided to add the teaching of the mechanical arts to the traditional teaching of liberal arts, covering, in particular, techniques of the forge and of construction, weaving, navigation, agriculture and hunting.

Far from being an obscure amalgam of outdated ideas and verbiage, medieval scholasticism, with its turbulence, its dialectics and its questioning, was a prelude to the creative thought of the Renaissance. In the schools of medicine, the teaching of anatomy, which had followed Galenic doctrine to the letter, saw the dawn of ideas of reform. Since the time of ERASISTRATUS and HEROPHILUS, there had been practically no dissections of human cadavers. The anatomy of Man was deduced from the observations made since GALEN on animals: monkeys and pigs. It was only at the beginning of the 14th century that the first dissections on human cadavers were authorized. The Great Council of Venice, which was the patron of the University of Padua, allowed them at a rate of one per year. The practice spread to Bologna, and then to France, first to Montpellier and, at the very beginning of the 15th century, to Paris. After this it became more general. Works on anatomy from this period show the master reading and commenting on the treatises of GALEN in Latin, while an aid uses a pointer to show the organs that appear under the scalpel of the attendant who is actually dissecting the cadaver. The teaching of sciences, which, since the Middle Ages had been carried out almost viumexclusively by ecclesiastics, was handed over to the laity during the Renaissance. It should be noted that from the 13th century onward, the famous school of medicine in Salerno, Italy, had already become completely secularized, and that by the Renaissance, scientific eclecticism had prevailed, and thus the Life Sciences interested not only doctors, zoologists and botanists, but also physicists, mathematicians and philosophers, a mixture that would prove to be a hotbed for the growth of ideas.

2.2. PHILOSOPHICAL-THEOLOGICAL CONTROVERSIES IN THE 12TH AND 13TH CENTURIES, A CRITICAL LOOK BACK OVER THE ANCIENT WORLD

"Almost all of our intellectual culture is Greek in origin. A deep understanding of these origins is an indispensable condition for our release from its overly-powerful influence. In such a case, an ignorance of the past is not only undesirable, it is quite simply an impossibility. One may not be informed about the doctrines and works of the great masters of Antiquity, of Plato and of Aristotle ; one may not even have heard their names; one is nevertheless dominated by their authority. Not only does their influence extend to us via their

successors, both ancient and modern, but the whole of our thought; the categories into which it falls, the forms of language that it uses (and which govern it), all of this is, for the most part, an artificial product, and, above all, the creation of the great thinkers of the past."

Theodor GOMPERZ
The Greek Thinkers - 1928 (translated from the French edition)

From the 13th century onwards, an animated debate took place within the Universities, known as the quarrel of the **Universals**. In medieval scholasticism, the term "universals" meant general ideas, often linked to theology and accepted as inviolable truths, both universal and "realistic", for example, the idea of divine creation. This trend combined with the Platonic theory of **Ideas**, i.e., of abstract concepts such as the idea of the animal, the idea of the tree. In contrast to this, nominalists argued that only individuals or things considered according to their particularities were concrete realities, and that by observing singular facts and using inductive reasoning, it should be possible to work back to general principles. ABÉLARD won renown in this quarrel with a somewhat half-hearted theory, named **conceptualism**, according to which universal concepts proceed from an abstraction generated by thought, which did not exclude a pathway from the particular to the general. A century later, THOMAS AQUINAS, "the angelic doctor", like ABÉLARD, adopted a moderate attitude, admitting that while the observation of the specific characteristics of beings and of things is the basis of our understanding of the world, these specific characteristics cannot be dissociated from the general characteristics and from the concepts that arise from them. Whether to opt for the inductive reasoning of nominalism or the deductive reasoning of the Universals obviously determined the way in which problems were to be met with and solved. Nominalism dominated European thought from the 16th century onwards. It would assist the flowering of the scientific method in the 17th century.

In the 18th century, when experimental science recognized the need for an organization of species, in order to explain the incredible diversity of the living world, a certain form of the quarrel of the universals broke out again. BUFFON (1707 - 1788), who held that "in Nature there are only individuals, and kinds and species are only products of the imagination," was opposed by systematicians who tried to group the innumerable species of animals and plants into hierarchies, thus providing comparative anatomy and physiology with the tools necessary to allow them to progress, and providing researchers with rational choices of experimental models.

In 1210, the Church, threatened to excommunicate those who spread the ideas of ARISTOTLE, which were considered to be too materialistic, using the pretext of the inappropriateness of these ideas with respect to the biblical tradition of Genesis. This measure only fanned the flames of the curiosity of scholars, and was repealed by GREGORY IX in 1231. It is in this context that the Franciscans Robert GROSSETESTE (1170 - 1253) and Roger BACON (1214 - 1294), and the Dominicans ALBERT the Great (around 1193 - 1280) and THOMAS AQUINAS were both witnesses to and actors in discussions about whether ARISTOTLE's theories concerning the laws of Nature were well-founded. Robert GROSSETESTE, who was master of studies at the University of Oxford, where he had Roger BACON as a pupil, tended towards neoplatonic ideas. Influenced by GROSSETESTE, BACON turned towards the mathematical and physical sciences and wrote several treatises on optics and acoustics. He gave a detailed description of the anatomy of the eye. He considered that an understanding of the phenomena of physics required the use of the **mathematical tool**, an idea that is close to that of contemporary modelers of the mechanisms of living beings. Foreshadowing OCKHAM, he postulated that Nature takes the shortest possible pathway in its actions, and obeys principles of economy. Being an enthusiast of the **scientific method**, and distancing himself from the scholastic tradition, he argued that the truths of Nature were made inaccessible by obstacles connected with teaching methods, i.e., incompetent teachers, old habits and a profound ignorance hidden beneath unproductive verbosity. At a time in which intellectual work was still considered to be in opposition to manual work that was seen as servile, Roger BACON stood out because of his insistence on the necessity for men of science to experiment with their own hands. He did his own polishing of the lenses and mirrors that he used in his optics experiments. The idea of the focal point in spherical mirrors is attributed to him. Roger BACON knew Pierre de MARICOURT, known as "Pilgrim" de MARICOURT (13th century), a military engineer who was very keen on metallurgy, and who became interested in the properties of magnetite, or lodestone, which has the particularity of attracting iron. In 1269, Pierre de MARICOURT published *Epistola de magnete*, in which he gave such precise descriptions of the experiments he had carried out on magnetite, that Roger BACON cited him as a model man of science and promoter of innovative ideas. With his confidence in scientific progress, Roger BACON prophesized that one day it would be possible to construct ships without oarsmen, flying machines and instruments capable of lifting enormous loads.

From 1267 - 1268, as requested by Pope CLEMENT IV (1200 - 1268), Roger BACON published three works, *Opus majus*, *Opus minus* and *Opus tertium*, in which he put forward an anti-establishment philosophy. Certain aspects of PTOLEMY's astronomy were disproved. Roger BACON advocated the need for experimentation in order to reach an understanding of the laws of Nature. In the Opus majus, he wrote, "there are three ways of knowing the truth: authority, which can only produce faith when it is justified in the eyes of reason, reasoning, for which the most certain conclusions leave something to be desired if they are not verified, and, finally, experience, which is sufficient unto itself." Further on, he also wrote, "reasoning completes the question, but it does not provide proof and does not remove doubt, and it does not allow the mind to find repose in the conscious possession of the truth, unless that truth is discovered by experience." Prosecuted for theological deviancy, and made suspect by his apology for the scientific method and his criticism of scholasticism, both of which made him a dangerous deviationist, he was condemned by the Paris theology faculty in 1277. He remained in prison until 1292.

At the end of the Middle Ages, Robert GROSSETESTE and Roger BACON were symbols of the rising critical spirit that had begun to undermine Aristotlian and Galenic traditions.

ALBERT the Great, whose real name was Albert von BOLLSTADT, studied theology, mathematics and medicine in Padua. In 1223, he entered the Dominican order. First he taught in the German states, and then he taught in Paris from 1245. He commented on the works of ARISTOTLE, particularly those concerning logic, ethics, politics and metaphysics, while distancing himself from certain Aristotlian theories such as that of the internal motor of movement. ALBERT the Great's aim was to provide a basis for theology in science. His work covers a huge area. It deals with the physics of the globe and astronomy, minerals, animals, botany... and it contains numerous instances in which old theories are disproved. The pertinence of his comments in his treatise on botany, *De vegetabilibus*, and his treatise on zoology, *De animalibus*, which was partly inspired by ARISTOTLE's *History of Animals*, led him to be recognized as the greatest **naturalist** of the Middle Ages. By restricting theology to the study of religion, he was a decisive figure in the split that was beginning to occur between the natural sciences and the supernatural.

THOMAS AQUINAS, who was a pupil of ALBERT the Great, tried to reconcile Christian orthodoxy and the science of Ancient Greece. In order to achieve an agreement between Aristotlian philosophy, which considered that the world was eternal, and the biblical tradition of a world created by God, THOMAS AQUINAS substituted a God who was both a creator and a mover for ARISTOTLE's God, who was the mover of the eternal world. In *De Ente et Essentia*, he developed the theory that essence, while latent in existence, appears at the same time as existence. With THOMAS AQUINAS, **science** and the **theology** become two completely **distinct entities**: science is fed by observation and experience, both of which proceed *via* reasoning, and theology brings a revealed truth that escapes human reason and must be accepted as an article of faith. THOMAS AQUINAS was the author not only of the famous *Summa Theologica*, but also of scientific works such as *The Nature of Minerals*, *Meteorology* and even *Commentaries on Aristotle's Physics*.

ALBERT the Great and THOMAS AQUINAS had the merit of bringing ARISTOTLE's philosophy into the Christian cosmology of the Middle Ages. The consequences were far from negligible. In fact, in Christian doctrine, men are free and equal before God, and manual work is no longer only for slaves, but is to be shared by all men. Armed with a logical basis, within a framework that nevertheless remained limited, and under the control of theological doctrine, the technical arts were able to develop new bases for scientific experimentation, thanks to new processes and new instruments.

In Paris, ARISTOTLE's works were taught and commented on in a critical way in both the Faculty of Arts and the Faculty of Theology, which gave rise to a rivalry that was a source of conflict. In 1277, the archbishop of Paris, Etienne TEMPIER (ca. 1210 - 1279), condemned 219 articles that were taught in the Faculty of Arts as being heretical. In particular, the so-called double or two-fold truth proposal was targeted. According to this proposal, certain assertions may be considered as being true if one follows the logic of reasoning, but they may be considered as being false if reference is made to the Scriptures, i.e., if it becomes a matter of faith. This condemnation had only a short-lived effect. ARISTOTLE's theories continued to be commented upon and discussed from both the religious and the scientific point of view. TEMPIER's condemnation had unexpected, far-reaching repercussions. By casting doubt on ARISTOTLE's theories, it helped to make a breach in scholastic teaching.

John DUNS SCOTUS (1274 - 1308), a theologian of Scottish origin, nicknamed "Doctor Subtilis", is recognized for his famed arguments in favor of the theory of Universals. He was one of the first to abandon the Aristotlian concept of the Universe, rejecting the distinction between celestial and terrestrial physics, and postulating that everything in the Universe is made of a common material. Jean BURIDAN (1295 - 1385) who taught physical sciences at the University of Paris between 1330 and 1340, and was twice its rector, criticized ARISTOTLE's theory about movement. In 1350, he came up with the new concept of *impetus*. By this term BURIDAN meant the force communicated to an object, a stone for example, in the form of an intrinsic motive force. The *impetus* depended on the quantity of matter contained in the stone and the speed it had acquired. With time, the *impetus* weakened, which led to the stone falling towards the ground. ARISTOTLE had stated that the air had a role in the movement of the stone. BURIDAN denied this role for air and postulated that the motive power imprinted on a moving object maintained its movement, once the moving object had been separated from its motive force. The *impetus* was likened to a sort of internal force that remained linked to the body, a concept that has an Aristotlian flavor. Later, physics would separate the notions of inertia and quantity of movement. Extrapolating the *impetus* theory to include heavenly bodies, BURIDAN considered that, at the time of the Creation, God had given them such an impulse that they would continue along their paths for eternity. In contrast to ARISTOTLE, who made a distinction between celestial and sublunary mechanics, BURIDAN wrote that there was no difference between celestial mechanics and the mechanics of objects present on the surface of the terrestrial sphere.

A contemporary of BURIDAN's, Nicolas ORESME (1320 - 1382), who was a master in theology, bishop of Lisieux and councilor to French king CHARLES V (CHARLES the Wise, 1338 - 1380), asked the same sort of questions as BURIDAN concerning the possibility of the Earth rotating about its axis, the rotation resulting from an initial *impetus*. When the objection was made that if the Earth rotated on its own axis, the point of impact of a stone falling from the top of a tower could not be vertically below its starting point, ORESME remarked that the stone followed the movement of the atmosphere, which rotated with the Earth. Another major controversial figure of this period was William of OCKHAM (1280 - 1349). After studying at Oxford, OCKHAM taught in Paris. An enthusiast of the DEMOCRITUS's theory of atomism, he was banned from the Franciscan order for heresy. His name is often cited with respect to the **principle of parsimony** (OCKHAM's razor), which would be taken up again by DESCARTES. This principle stipulates that the fewer assumptions are made in putting forward an argument, the more persuasive that argument is. OCKHAM was a **nominalist** whose philosophy was in opposition to that of DUNS SCOTUS. He felt that science should interest itself in things that are tangible. There was a differentiation between intuitive knowledge and abstract knowledge. While the latter leads to gratuitous assertions, intuition refers to objects or facts that are directly perceptible. For OCKHAM, as for his predecessors, science and theology are areas of thought that are fundamentally distinct from one another, and which should be kept apart when one is trying to decipher the laws of Nature. Experience based on **empiricism** must be one of the fundamental underpinnings of science. Taking his arguments even farther, OCKHAM maintained that the existence of God cannot be proven because the concept of "God" does not possess an empirical foundation. Belief in God is therefore an act of faith. Given this increasing flow of controversial ideas, it is not surprising that the theologico-philosophical approach to astronomy that was common during this period gave rise to reservations on the part of such renowned astronomers as the Austrian Georg PEURBACH (1423 - 1461) and his German disciple, known as REGIOMONTANUS (1436 - 1476).

At the end of the Middle Ages, when the bases upon which human knowledge is founded were the subject of open discussion, the momentum had been given, opening the way for the **critical and skeptical spirit of the Renaissance**. The laity took up their places in the Universities that had formerly been controlled by clerics. Italy rapidly became a center of attraction for intellectuals hungry for knowledge. To this was added the **liberal spirit** that reigned in such Italian Universities as Bologna, Padua and Pisa, making them hot beds of common-sense-based thought, from which experimental science would be born. **Science and philosophy had begun to be distinguished from one another**. The gulf between them would continue to widen in the 17th century, and Science, armed with its experimental tools, would become oriented towards the study of physicochemical mechanisms, while Philosophy turned its attention to the area of the mind and ideas.

2.3. ALCHEMY AND THE TECHNOLOGICAL REVOLUTION OF THE MIDDLE AGES

Medieval alchemy, equipped with material that was certainly rudimentary, but was nonetheless efficacious, played a decisive role in the birth of the scientific method, not only in chemistry, but also in the physical and life sciences. It helped to bring together the theoretical knowledge of the philosophers and the manual expertise of technicians such as those who extracted and purified metals. Starting from the idea that any substance can be transformed into another, alchemists were sure that it should be possible to ennoble base metals. This idea, which was already well anchored in the ideas of the philosophers of the Greek school in Alexandria, had its origin in the science of dyeing, involving a set of techniques used to gild, silver, and color objects. ZOSIMOS of Panapolis, who lived in the 4th century AD, carried out both distillation and a process that he called digestion (Figure I.4). The apparatus used for digestion comprised a clay vessel, to the neck of which a metal strip was attached. This strip was covered with substances to be experimented on. The whole assembly was covered with a bell jar. The apparatus was placed in an oven. Affected by the heat, the substances deposited on the metal strip melted. The liquid droplets collected in the bottom of the vessel were vaporized. The vapors reacted with the residues that were still present on the strip, leading to changes in color that were interpreted in terms of transmutation.

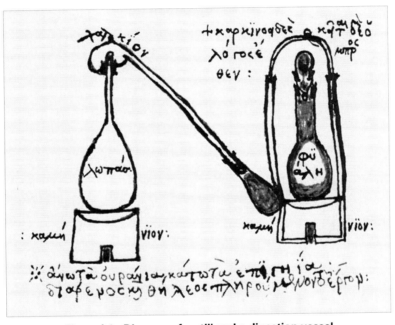

Figure I.4 - Diagram of a still and a digestion vessel used by Greek alchemists of the Alexandrian school
(16th century copy of an alchemy treatise written by ZOSIMOS in the 3rd century BC)

Arab science inherited techniques from the Alexandrian school. These became the pivotal point of an experimental policy that was given the name **alchemy**, which welcomed, with the same enthusiasm, ideas that had come from China because of more and more frequent commercial exchanges. Cloaked with a veil of confidentiality, alchemy, even at its beginnings, was distinguishable for its curious **mixture of mysticism and experimentation**. The first alchemists considered that metals were made up of two entities, sulfur and mercury, with the reservation that these entities were the mainly symbolic representation of certain properties of a particular metal. Thus, it was admitted that mercury was a stable entity that gave metals their brilliance and ductility and that, in contrast, sulfur was a yellow-colored principle that was easily decomposed. It was postulated that mercury and sulfur existed in various proportions, and with different degrees of purity, in metals. Finally, it was claimed that it was possible to increase or decrease the brilliance, volatility or even the color of a particular body, with the exception of gold, which was considered to be the most perfect metal.

Concurrently with metallurgical experimentation, alchemists became interested in the manufacture of electuaries, including the famous theriac, a panacea capable of curing all ills. Curiously, Chinese alchemy developed in an almost parallel manner to Greek and Arab alchemy, with the same chimerical aim of developing an elixir for a long life. Emphasis should be placed on the richness of the Chinese pharmacopoeia of this period and the attention that was given to a knowledge of medicinal plants in the teaching of medicine.

JÂBIR IBN HAYYÂN, who was known by the name of GEBER (721 - 815) is considered to be the father of **Arab alchemy**. PARACELSUS (1493 - 1541) called him the master of masters of the chemical sciences. In a work that was later translated into Latin, *Summa perfectionis*, GEBER distinguishes the two trends that existed in alchemy, the unrealistic hope of transmuting metals and the illusion of having found a panacea with the **red elixir** prepared from dissolved gold. In another treatise, *De Investigatione magisterii*, he writes that "claiming to extract a body from another that does not contain it is madness. However, as all metals are made up of mercury and sulfur, it is possible to add to them what is missing or remove from them that which is in excess. To achieve this, we use the appropriate means. Here are those that experience has taught us; calcination, sublimation, decantation, dissolution, distillation, coagulation and fixation." The calcination of a metal corresponded to the oxidation reactions of modern chemistry: lead, copper or iron, subjected to a flame in the presence of air, were transformed into lead, copper or iron oxides, which were indiscriminately named "lime" at this time. Among the reagents used were salts, alums, vitriols, glass, borax, vinegar and fire. GEBER added a third entity to mercury and sulfur, arsenic. The four basic elements; water, earth, air and fire, were not forgotten. The belief that lasted until the 18th century was that they are present in different proportions in mercury, sulfur and arsenic. For symbolic reasons, GEBER and the alchemists who followed him attributed more importance to that which was volatile, which made them consider sublimation as an operation of a higher order. From this period onwards, recipes were given for the preparation of *"liqueurs"* that were enriched in potash and in soda from terrestrial and marine plants, in sulfuric acid (oil of vitriol) by the heating of alum, in nitric acid by the treatment of saltpeter with oil of vitriol. An illustrious successor to GEBER was the chemist and physician RHAZES (AL-RÂZÎ) (865 - 923) to whom is attributed the discovery of alcohol by the distillation of wine.

The chemical know-how of the Arabs spread to western Europe both *via* Spain, which was conquered in the 8th century, and *via* Sicily, which was conquered the century after. Along with this know-how came some terms that are still used today, such as alkali and alcohol. Alchemists' dispensaries were created and developed throughout the Middle Ages. These had only modest tools, which, even in the 13th-14th centuries, were limited to bellows furnaces, oil baths, retorts and stills (Figure I.5).

Figure I.5 - An alchemist's dispensary in the Middle Ages
(from a plate held in the French National Library - BnF)

In the **western alchemical tradition**, we find natural philosophers and physicians, such as Roger BACON and ALBERT the Great. In Montpellier, Arnauld DE VILLENEUVE (1235 - 1311), who was a doctor and regent of the school of medicine, perfected the procedures of distillation. He gave detailed descriptions of the operations required to obtain a certain number of products, including alcohol, which was already known, and was called *"eau-de-vie"*, or water of life. He provided a process for preparing turpentine. Another alchemist of the beginning of the 14th century, Ramon LULL (1232 - 1316), who was of Catalonian origin, emphasized the idea of **quintessence**, a subtle principle that was present in many products, to which he gave a specific quality. Thus, brandy was considered as the quintessence in which resided the virtue of wine. Here we have the idea of **active principles** being present in natural substances, principles that modern chemistry would endeavor to purify, and which would prove highly useful in therapeutics. However, in the Middle Ages, therapies were mainly empirical, and often useless. Epidemics ravaged the world and medicine remained powerless. In the middle of the 14th century, the Plague, starting from Asia, invaded Europe. It killed twenty-five million people. Five centuries later, the organism that spread the Plague was discovered, and it became possible to fight this unmasked enemy.

At the beginning of the 15th century, Basil VALENTIN (15th century) perfected preparations of antimony, which he used for therapeutic purposes. This was the beginning of **iatrochemistry**, or chemistry applied to medicine, one of the most famous representatives of which was PARACELSUS. In Basle, PARACELSUS occupied one of the first chairs of chemistry to be founded. He said that the real aim of chemistry was not to make gold, but to cure disease. Irrespective of the antimony salts recommended by VALENTIN, PARACELSUS recommended the internal use of mercury salts in the treatment of syphilis. Mercury salts were used until the 20th century.

In 1542, Jean FERNEL (1497 - 1558) published *De naturali parte medicinae*, a work on physiology that came to be considered as authoritative. Medicine of the period, which was mainly human and moral in its outlook, nevertheless benefited from the progress involved in the prophylactic measures, based on reasoned observation (opening of quarantine stations, disinfecting of suspect objects), that were carried out with the aim of stopping the spread of infection. Surgical practices were called into question. Medieval surgery had had notable success with both innovations and rediscovered practices (suturing of wounds, suturing of tendons, cataract operations). This was a prelude to an era of surgical rebirth, with Ambroise PARÉ (1509 - 1590) and the great anatomists of the Italian school (Chapter II-2). PARÉ started out at the Hôtel-Dieu in Paris. Employed as a surgeon in the Piedmont military campaign of 1537, he was soon confronted with the treatment of wounds arising from firearms such as pistols and harquebuses, which had just begun to appear in the fields of battle. Because of the generally-accepted idea that gunpowder was poisonous, it was common practice to cauterize wounds by applying boiling oil. One day, having run out of oil, PARÉ treated the wounded with cold dressings. Expecting the worst, he was surprised to see that the wounded who had been treated this way healed better and with less pain than those who had received the classic boiling oil treatment. Following this experience, he published *Method of Treating Wounds made by Harquebuses and other Fire Arms*. This work created a lot of noise, and even scandal, as it was written in French and, moreover, went against the teachings of the masters of the period. PARÉ's cause was not helped by a frankness that manages to surprise us even today: "it is no great thing," he said, "to leaf through books, to gabble and natter in a chair of surgery, if the hand neither toils nor is put to use by good sense."

A contemporary of PARACELSUS, the German AGRICOLA, whose real name was Georg BAUER, (1494 - 1555) wrote the first work on metallurgy that we know of, *De re metallica*. He observed, as did Jean REY (1583 - 1645) a few dozen years later, that lead increases in weight when it is exposed to air. However, he wrongly deduced from this that metals are alive and grow as plants do. During the same period, Bernard PALISSY (1510 - 1590), who is considered to be the founder of **technical chemistry**, discovered a process for manufacturing enamels. His many treatises show an original and inventive mind, close to experimental fact, and concerned with details.

PARACELSUS, AGRICOLA and PALISSY were succeeded by scholars who made the transition to modern chemistry: Jean REY, who pointed out the increase in the weight of metals due to oxidation, Jan Baptist VAN HELMONT (1577 - 1644), who discovered the "gas sylvestre" (carbon dioxide) and the "gas of salt" (hydrochloric acid), Johann GLAUBER (1604 - 1668) who gave us the preparation of sodium sulfate (GLAUBER's salt), Robert BOYLE (1626 - 1691) and Edme MARIOTTE (1620 - 1684) who showed, independently, the relationship that exists between the volume of a gas and its pressure, Johann KUNCKEL (1638 - 1703) who isolated phosphorus from urine, Joachim BECKER (1635 - 1682), author of a new theory concerning three types of earth, vitrifiable, inflammable and mercurial, Wilhelm HOMBERG (1652 - 1715) who discovered boric acid (or HOMBERG's salt), Georg Ernst STAHL (1660 - 1734), who promoted the theory of affinities in chemistry and also the unfortunate **phlogistics theory** that arose from BECKER's idea of inflammable earth. Over the centuries, the alchemists' instrumentation was perfected. **New apparatus** was created and old apparatus was improved: furnaces equipped with bellows, oil baths, crucibles, cupels, retorts, and glass or ceramic receptacles.

The alchemical tradition would last until the end of the 17th century. It had so much influence that geniuses like NEWTON were enthusiasts. It had helped to fill out the catalogue of bodies whose properties are accessible to the senses. Even quite late in time, it was thought that these properties were **identifying realities**. Bodies that presented certain characteristics in common were identified as belonging to a same entity, even if all the evidence showed that they were of a different nature and different extraction. Thus, for BECKER, the fact that quartz and diamond are transparent, as vitrifiable earth was, by definition, was a proof that quartz and diamond were part of the same vitrifiable earth. Given the limited facilities available to science, it was difficult for him to evaluate the complexity of the material and to envisage the idea that a supposedly simple body could hold several others, of different types.

Despite obvious technical advances, medieval alchemy did not succeed in imposing a credible experimental method. One of the reasons for this shortcoming was the difficulty involved in designing reliable measurement instruments and in specifying that which, in experimentation, needed to be measured. Another reason was the problems involved in escaping the secret and mystical character of any experimentation. During the 17th century, the alchemical approach to an understanding of things in Nature, which was blemished by animistic beliefs and cabalistic practices, gave way to a mechanistic approach, associated with an objective outlook.

3. CONCLUSION - AT THE DAWN OF THE SCIENTIFIC METHOD

More than twenty centuries were to go by, peppered with trials and tribulations, between the first awakening of a consciousness of the physical reality of phenomena in Nature by Greek thinkers, and the moment when, suddenly springing into life, the bases of the scientific method were laid down and delimited by physicists, naturalists, doctors and philosophers in the 17th century. The flowering of the scientific method in the 17th century in the West was the result of **several legacies**: a legacy of scientific thought and the rules of logic from the philosophers of Ancient Greece, a legacy of concepts ranging from mathematics to the natural sciences that in time crossed Eurasia, a legacy of technical and instrumental innovations provided by medieval artisans and alchemists, and finally, the legacy of a current of argumentation against medieval scholasticism that took hold from the beginning of the 13th century and was crystallized during the Renaissance.

From the 6th century BC onwards, Ancient Greek philosophers were putting forward the idea that the **Universe is structured** and that its origin was some primordial element that could be likened to water, air, earth or fire, depending on the different currents of thought. It was acknowledged that these elements could be transformed into one another, and that the phenomena of Nature were dependent on these transformations. Inherent to this notion, which cut itself off from any recourse to creation myths, was the theory of the **conservation of matter**: nothing comes from nothing. Following this, **Pythagorian philosophy**, impregnated with the science of numbers, helped to mathematize the Universe. The **atomist philosophy** proposed to structure matter into indivisible particles, or atoms, separated by emptiness. ARISTOTLE used **empiricism** to oppose PLATO's theories concerning the **geometrization** of things in Nature. As an attentive observer of the form and habits of animals, and of the anatomy and functions of their organs, ARISTOTLE is considered to have been the first naturalist. He also gave us the **fundamental rules of logic**. A rationalized idea of medicine, known as a Hippocratic concept, represented another way of understanding the laws of the living world. In contrast to the medical practices of the previous centuries, **Hippocratic medicine**, based on the observation of a patient's symptoms and of their evolution, wished to be stripped of any notion of having recourse to magic.

Knowing nothing of the laws of physics and chemistry, the Greek philosophers of the 6th and 5th centuries BC dared to formulate a set of propositions whose apparent logic, excluding any supernatural causality, appeared to them to explain phenomena in Nature. Nevertheless, they did not move on from speculative theory and the world of ideas to experimental verification. This is partly explained by a prejudice against manual practices, which were considered to be servile, and partly by the fear of artifacts linked to experimentation that inevitably alter the order imposed by Nature. PLATO and ARISTOTLE were two of Ancient Greek philosophers and scholars who, for centuries, left an indelible stamp on western thought,

PLATO searching for the reality behind appearances, and ARISTOTLE extolling the virtues of empiricism and observation.

The period of the Alexandrian school stood out as the Golden Age of culture and science in the Ancient World. In Alexandria, in a manner that was without precedent in history, a state became involved in a voluntarist and enlightened policy with respect to support for talents arising from different backgrounds, bringing them together in the same place and providing them with the appropriate conditions for them to work effectively. From the 3rd century BC until the first centuries AD, Alexandria produced geographers, astronomers, surveyors, doctors, and anatomists. At the beginning of the Christian era, GALEN, despite often major errors, gave anatomy and physiology such a momentum that Galenic doctrine was able to traverse the whole of the Middle Ages, almost without any major obstacles. **Here we have one of the first examples of a situation in which abstract concepts and technical culture combined and created the conditions for the emergence of an experimental science.** Wars, invasions, political and confessional vicissitudes, and a gradual loss of interest in pure science on the part of the successive powers, helped to smother an otherwise promising scientific momentum.

The Arabs inherited from Greek culture. They spread it. They translated a large number of Greek works and added judicious comments to them. From the 12th century onwards, these texts, translated from Arabic to Latin, would penetrate into France and spread throughout Western Europe. The philosophical and scientific propositions contained in these texts, particularly Aristotlian thought, helped to build **Medieval Scholasticism**. The Middle Ages is the period in the history of the West during which we see a slow ripening through the centuries of a type of thought that had apparently been asleep, without any intellectual arguments being freely expressed. The knowledge of this period, which was initially centered on theology, was noticeably modified by the contributions of the philosophers and scholars of Ancient Greece. In the nascent Universities, it was structured around an understanding of the liberal arts, including logic, arithmetic and geometry, encouraging minds to equip themselves with methods for reasoning and thought suitable for increasing their consciousness of the realities of the inanimate and living worlds. However, medieval teaching did not establish a sharp distinction between science, which is the subject of objective knowledge, and philosophy, which incorporates psychology, morality and metaphysics. The synonymy of science and philosophy began to become blurred with the Renaissance, but did, nevertheless, survive until the 17th century.

From the 13th century onwards, when circles of scholars came together as forerunners of future Academies, theories that had been peddled throughout the centuries without verification were subjected to the screening of criticism without indulgence. Intellectual life blossomed under this intense activity. This was the beginning of a changing of the guard, marked by audacious questioning and notable technical advances. **Alchemists** took part in this scientific resurgence, developing ingenious procedures and manufacturing original apparatus. Nevertheless, the esotericism of their language and secretiveness with which they surrounded their experiments were obstacles to the credibility of their practices.

The Renaissance, a turbulent period of transition, still held in thrall by superstition, but developing the will to rise above this and raise the curtain of obscurantism, and remarkable for the freshness of thought that did not hesitate to criticize the superannuated dogmas of medieval scholasticism, **was distinguishable by a constructive imagination that made everything seem possible to Man**. The art of being able to think correctly when off the beaten track and on a pathway without any known destination, the genius involved in designing instruments that made it possible to go beyond observation by the senses, passionate, generous and visionary patrons, more and more numerous and assertive contacts between scholars of different persuasions; all of these factors together ensured that the Renaissance was the necessary step towards the final emergence of the scientific method and its application to the exploration of phenomena in Nature. In this crucible, containing a mixture of often-contradictory currents of ideas based on metaphysical questions concerning **why**, spurred on by a curiosity that clearly asks questions concerning the **how**, experimental science finally bloomed and expanded rapidly in the West.

Chapter II

THE BIRTH OF THE EXPERIMENTAL METHOD IN THE 17TH AND 18TH CENTURIES

"The irreplaceable merit of the 17th century was not that it saw more than its predecessors, more or less correctly, but that it looked at the world with new eyes, with principles that would remain established. This is where it can and should be called the initiator of modern science."

René TATON
Modern Society - 1995

An inventory of the discoveries that accumulated during the 17th century, and of the ways in which they were made, shows that there was a sudden transition with respect to the preceding centuries. This transition was a break with the past leading to consequences that were so decisive for contemporary science that we talk in terms of a **scientific revolution**. What were the motive forces? These certainly included a particular form of civilization that was specific to the West during this period, and which associated liberalism and a thirst for knowledge, audacious questioning of established dogma and a rejuvenated vision of the world, emancipation from prejudices and the enterprise spirit, rigor in reasoning and the construction of apparatus able to instrument emerging ideas, and also, no doubt, the rapid development of craft and commercial enterprises that contributed to the enrichment of society and the economic growth of States. This mix of ingredients led to the explosive dawn of scientific renewal. If such a revolution had not taken place in the preceding centuries, this was no doubt because the mix had not yet reached a critical threshold, and because the scientific world was not prepared for it. Nevertheless, this renewal was partly dependent on a past that goes back to Ancient Greece, and certainly cannot ignore the Middle Ages (Chapter I). Even if the bases of understanding inherited from this past were tainted with errors, sometimes major ones, ideas were formulated that answered pertinent questions about the observable world. The major breakthrough of the 17th century was the recognition of the fact that Nature is subject to **laws** and that these laws respond to **mathematical rigor**.

In the 17th century, **experimental science** included both the **inanimate world** and the **world of living beings**. The models of these two worlds were certainly different, but naturalists and physicists were preoccupied by the same requirements, i.e., the **accuracy of measurements** and the **quantification of experimental data**. The full meaning of the **discovery of the circulation of the blood** by HARVEY in 1628 can be understood when we consider the parallel advances made in experimental physics, with, in particular, the **discovery of accelerated movement** by GALILEO (1564 - 1642) and the **demonstration of the existence of the vacuum** by BOYLE and PASCAL (1623 - 1662). Lagging behind physics and the life sciences, chemistry started its own transformation at the end of the 17th century. It became modern chemistry after refuting the theory of phlogistics and adopting a rational nomenclature for the elements, at the end of the 18th century. It provided reliable methods for analysis and dosing. The life sciences benefited from these. In order to take into account the historicity of the facts, this chapter groups together the knowledge acquired in the different domains of experimental science, physics, chemistry and biology, using the experimental method, during the 17th and 18th centuries, and shows them in parallel.

1. THE DISCOVERY OF THE CIRCULATION OF THE BLOOD BY W. HARVEY

"We can see how the discovery of the circulation of the blood is first, and perhaps foremost, the substitution of one concept, designed to "render coherent" specific observations made on the organism at various points and at different times, with another concept, that of irrigation, which is imported directly into biology from the domain of human technology. The reality of the biological concept of the circulation presupposes the abandon of the commodity of the technological concept of irrigation."

Georges CANGUILHEM
Understanding Life - 1965

W. HARVEY
(1578 - 1657)

William HARVEY was born in 1578 in Folkestone, England, to a family of shopkeepers. The oldest of nine children, he was the only one to undertake a medical career. After a few years spent at the University of Cambridge, HARVEY went to Padua in 1597 in order to learn medicine. At this time, Padua was one of the most prestigious centers of western science, not only in medicine, but also in physics and astronomy. New ideas, even revolutionary ones, were welcomed and discussed without prejudice. In Padua's school of medicine, anatomy lessons were no longer simple scholastic commentaries on the treatises of GALEN and ARISTOTLE. They were accompanied by the dissection of cadavers. HARVEY followed this teaching. FABRICIUS d'Acquapendente (1533 - 1619) – originally Hieronymus FABRICIUS ab Acquapendente – had just discovered the venous valves.

What was the function of these valves? Did they hold the secret of the movement of blood in the body? The pathway that William HARVEY decided to follow was now laid out. Later on, when carrying out the dissection of human cadavers and the vivisection of warm-blooded and cold-blooded animals, he worked to understand the role of the contractions of the heart in the movement of blood around the body, and the way in which the venous valves intervened in this movement. His famous tourniquet experiment, carried out on the arms of volunteers, determined the respective places of the arteries and veins in the blood network, which led to the concept of a circular movement of blood in the body. In order to understand the **immense importance, for physiology, of the discovery of the circulation of blood**, it is necessary to remember that at this period the common belief was that blood was formed directly from the products of digestion, in the liver, and that it flowed slowly through the vessels from the heart in order to irrigate the extremities, without returning (Chapter II-2).

Having graduated as a doctor of medicine, HARVEY returned to England in 1604 in order to practice. In 1609, he was appointed head physician at St Bartholomew's hospital in London, and in 1615 he took over the chair of anatomy at the Royal College of Physicians in London. His teaching was based on the dissection of cadavers, as had been common practice in Padua. HARVEY was also interested in pathological anatomy, a science that tried to link the well-characterized symptoms of a particular disease with an alteration in a tissue or an organ, highlighted by a *post mortem* dissection. His growing fame brought him to the attention of royalty. In 1618, king JAMES I (1566 - 1625) chose HARVEY to be his personal physician. It was during this period that HARVEY wrote, in Latin, about the results of his experiments and his own ideas concerning the heart and the movement of blood in the body. The work was printed in 1628 in Frankfurt-on-the-Main, with the title *Exercitatio anatomica de motu cordis et sanguinis in animalibus* (Figure II.1). When JAMES I died, his son CHARLES I (1600 - 1649) came to power. HARVEY continued in his role as the king's physician. He stayed with CHARLES I until the latter was condemned to death by Oliver CROMWELL (1599 - 1658). His loyalty to the royal family made him very unpopular with the public. His lodgings were destroyed and his manuscripts burned.

De motu cordis et sanguinis analyzes the movements of the heart and arteries, and then provides proof for the existence of a circular motion of the blood, which leaves the heart and returns to it. Straight away, a commonly held error of the period is mentioned "at the moment when the heart hits the chest, an impact that is felt from the outside, the ventricles distend and the heart fills with blood [...]. In fact," wrote HARVEY, "the impact of the heart corresponds to its contraction (systole) and its emptiness. That which had been thought to be the diastole is the systole. The heart is actually active not during the diastole, but during the systole". Resorting to the vivisection of sheep and dogs, he noted that: "When an artery is cut, no matter where it is, the blood is seen to spurt out, sometimes further and sometimes nearer to the wound. The strongest jet corresponds to the diastole of the arteries at the exact moment when the heart hits the thoracic wall, in such a way that at the moment when the contraction and the heart systole occur, the blood is driven into the arteries. These facts," he concluded, "demonstrate that, contrary to received opinion, **the diastole of the arteries corresponds to the systole of the heart.**"

HARVEY's work on the circulation of the blood represents one of the first reasoned forays of the experimental method into the world of living beings.

Figure II.1 - Frontispiece of HARVEY's work on the circulation of the blood
(from Guilielmi HARVEI - *Exercitatio Anatomica de Motu Cordis et sanguinis in Animalibus*, Francofurti, 1628)

HARVEY went so far as to give a lesson in **comparative anatomy**, showing that fish and frogs have only one ventricle and, instead of an atrium, they have only a pouch located at the base of the heart, filled with a large quantity of blood. It was obvious that the blood arrives *via* the veins and leaves *via* the arteries. HARVEY observed that "all animals have a heart, not only, as pointed out by ARISTOTLE, large warm-blooded animals and those that have blood, but also smaller ones, who do not have any blood, such as crustaceans, slugs, crayfish, water fleas, and many others. Even in wasps and flies," wrote HARVEY, "using a magnifying glass, I have seen a heart beat at the extremity of their bodies." He was aware of the existence, in the human fetus, of the foramen ovale, which allows the two auricles to communicate, and of the ductus arteriosus that joins the pulmonary artery to the aorta. At the moment of birth, the two points of communication between the venous blood and the arterial blood disappear, due to obstruction. The result is that venous blood is carried to the lungs by the "arterial" vein (pulmonary artery). From the lungs, it goes into the venous artery (pulmonary vein) in order to go into the left auricle. Pulmonary circulation is thus assured from the time of birth.

Here were the proofs of the circulation of blood in Man and mammals. First of all, the high speed movement of the blood in the vessels necessitated recycling. Based on the rhythm of the cardiac contractions and an estimate of the minimum amount of blood expelled at each contraction, HARVEY calculated that **in half an hour, the volume of blood flowing through the heart is much greater than the quantity of blood contained in the body**. His reasoning continued thus: "In one day, the heart, with its systoles, transmits more blood to the arteries than food could provide: the blood therefore does not come from the products of digestion." Another argument was as follows: "We open the jugular artery in a sheep or a dog, taking care with the jugular vein. Immediately, the blood spurts out, and, in a short time, all of the arteries and veins of the body are emptied. Why then, when we open a cadaver, is so much blood found in the veins and so little in the arteries? Why is there a large amount in the right ventricle and hardly any in the left ventricle? The simple explanation," wrote HARVEY, "is that the blood of the veins cannot pass in transit through the arteries without going through the lungs and the heart. However, after the death of an animal, the lungs having ceased to move, the blood can no longer move from the pulmonary artery to the pulmonary vein and from there into the left ventricle." This was followed by the elegant and crucial **tourniquet experiment**. Two types of **ligatures** were carried out, a tight ligature and a loose ligature. "The ligatures are tight," remarked HARVEY, "when we no longer feel the arteries below the ligature beat [...]. The ligatures are slack when the arteries are still beating weakly below the ligature, which is what happens during blood-letting." Loose pressure makes veins located below the tourniquet stand out, but not those that are located above it. In addition, the pressure of a finger on one of the veins below the tourniquet interrupts the blood flow upstream from the point of pressure (Figure II.2).

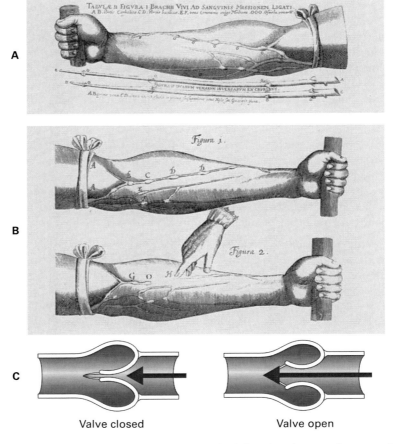

A - In a figure from FABRICIUS'S book, moderately tightening the tourniquet on the arm makes dilatations appear in the veins. These dilatations correspond to the locations of the valves (ostioles) as shown by the "turned back" veins (from FABRICIUS - *De Venarum Ostiolis*, 1603).

B - Tourniquet experiment carried out by HARVEY. The pressure of the arm by means of a lightly tightened tourniquet leads to the swelling of valves B, C, D downstream from the tourniquet (on the forearm side) (Figure 1). If a finger is pressed onto the dilated vein at H, this leads to the vein subsiding upstream, while the turgescence is maintained downstream (on the hand side) (Figure 2) (from Guilielmi HARVEI - *Exercitatio Anatomica de Motu Cordis et sanguinis in Animalibus*, Francofurti, 1628).

C - Diagram showing the operation of the venous valves. The valves are opened by the pressure of the blood returning to the heart, but close to prevent the venous blood descending to the extremities of the body. This explains the one-directional flow of venous blood from the extremities towards the heart.

Figure II.2 - Swelling of the valves of the veins due to pressure of the arm

How should these observations be interpreted? The blood arrives *via* the arteries into the veins, and not in the opposite direction. "It should not be said," added HARVEY, "that we are doing violence to Nature by carrying out this experiment, [...] because, in removing the finger as quickly as we can, we will see the blood return rapidly from the lower parts, filling the vein."

How does the blood return to the heart *via* veins from the extremities of the body, and how do the veins carry out their function? Why doesn't the blood flow back from the heart towards the veins? The explanation lies in the anatomical structure of the **venous valves** (Figure II.2). FABRICIUS D'ACQUAPENDENTE had discovered these valves, but was unable to find out what their real use was (Chapter II-2.2). A few observations and the application of common sense provided HARVEY with the correct answer: "The valves," he wrote, "are so arranged that they always prevent the venous blood from the heart returning either upwards to the head or downwards to the feet. In contrast, they leave the way wide open for blood traveling from small veins to larger ones." The venous valves have a **one-way action** that forces the blood to flow in one direction, from the extremities towards the heart, and that prevents the blood from flowing in the reverse direction, from the heart to the extremities of the body. If the blood expelled by the heart cannot be sent *via* the veins towards the limbs, then it must necessarily flow *via* the arteries; from which it is possible to conclude that there is a **circular movement of blood**, driven by the beats of the heart in the arteries, traveling from the arteries to the veins *via* anastomoses, and then returning to the heart *via* the veins. If the veins swell at the extremity of a limb to which a lightly tightened tourniquet has been applied, this is because, by this action, the venous blood flow has been blocked, while the arterial flow has not been modified.

Supported by different arguments, experimental proof was thus given, which led to the conclusion that the movement of the blood in the body is organized around the beating of the heart "in the simplest, most logical and most harmonious fashion." The blood cannot help circulating because the heart pumps considerable quantities of it. It can only pass from the arteries to the veins because the veins swell with blood when a lightly tightened tourniquet is placed on an arm. To conclude, **the blood completes a large circuit, starting from the left ventricle of the heart, traveling *via* the arterial system, and returning to the right auricle *via* the venous system**. The one-way mechanism of the valves in the veins prohibits the return of the blood from the heart *via* the venous system. From the right auricle, the blood passes into the right ventricle, from whence it is sent to the lungs *via* the pulmonary artery. From there, it travels along the pulmonary vein to the left auricle, and then to the left ventricle. A **double blood flow circuit** is thus accomplished (Figure II.3). All that remained was to determine how the blood passed from the arterial system to the venous system. HARVEY suggested the existence of anastomoses between the arterial system and the venous system, located at the organs and the extremities of the limbs, which would be demonstrated by the Italian physician and anatomist Marcello MALPIGHI (1628 - 1694) about thirty years later.

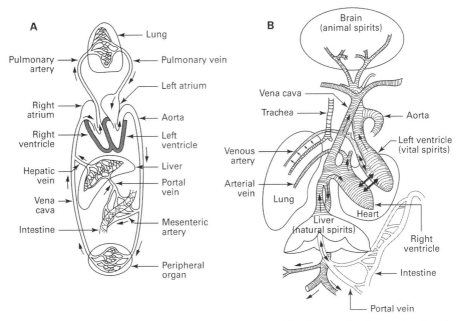

A - According to HARVEY, the circulation of the blood is the result of two circuits, the pulmonary circuit (minor circulation), by which the blood is oxygenated, and the general circuit (major circulation) that supplies the different parts of the body with oxygenated blood. Note the movement of the blood in the auricles and ventricles of the heart.

B - Schematic interpretation of GALEN's theory. The vein ramifications leaving the intestine bring the products of digestion to the liver. The liver transforms them into blood. GALEN's theory postulated that part of the blood formed in the liver travels through the venous system to irrigate the body, while another part is directed towards the right ventricle of the heart. The right ventricle supposedly communicates with the left ventricle by means of channels (large arrows), allowing a continuous exchange of blood between the two ventricles. The diagram shows that according to GALEN's theory, the air breathed in by the lungs *via* the trachea passes into the venous artery and is sent *via* this artery to the left ventricle of the heart. The blood that is there is pneumatized. The aortic artery leaves the left ventricle, carrying the pneumatized blood throughout the body. Briefly, according to GALEN, there are two parallel currents for blood irrigation, a venous one, from the liver, and an arterial one, from the heart. The idea of a blood circuit is absent. Note the production of three types of spirit, located in the brain (animal spirits), the heart (vital spirits) and the liver (natural spirits).

Figure II.3 - Circulation of the blood, according to HARVEY's theory (A) and according to GALEN's older theory (B)

In 1661, MALPIGHI used a microscope to visualize fine vessels in the lung of a frog. These vessels, **known as capillaries, connected the arteries and the veins.** This demonstrated the physical reality of the hypothetical anastomoses suggested by HARVEY. In 1674, Dutchman Anton VAN LEEUWENHOEK (1632 - 1723) confirmed the existence of the blood flow within the capillaries, using the gills of the tadpole as material for observation. The idea of capillaries spread rapidly to include mammals. **The capillaries were therefore the vessels that completed the arterio-venous circuit.** Here was the irrefutable proof of the existence of the circulation of the blood.

De motu cordis et sanguinis was considered to be a frontal attack on the dogma that had been taught since the time of GALEN, involving the slow and irreversible irrigation of the organs with blood. Criticism was not slow in coming. The first sallies were sent forth by the English physician James PRIMEROSE (1592 - 1659) who was joined by the Italian doctor Emilio PARISANO (1567 - 1643). The most virulent attacks, however, came from Gui (Guy) PATIN (1601 - 1672), dean of the Paris Faculty of Medicine and the Parisian anatomist Jean RIOLAN (1580 - 1657), who had been physician to HENRI IV of France, and then LOUIS XIII. RIOLAN assumed that the blood moved on its own, and that it prevented the heart from drying out! As for the heart, it was supposed to provide the blood with heat and "spirit". In response to the objections of his detractors, HARVEY repeated the decisive proofs for the circulation of the blood in *Exercitationes academicae de motu cordis et sanguinis*, which was published in Cambridge in 1649. Based on an indomitable Galenism, the Paris Faculty of Medicine's opposition to HARVEY's theory became so sharp that, in 1672, LOUIS XIV (1638 - 1715) decided to create a parallel teaching of anatomy at the *"Jardin du Roi"* which, at the Revolution, would become the Natural History Museum. The teaching of the major blood circulation was handed over to a young surgeon, Pierre DIONIS (1643 - 1718).

René DESCARTES took HARVEY's side against his detractors. For DESCARTES, the circuit made by the blood in the body was in agreement with his **mechanistic** theory of the operation of the organs in Man and animals, and fit in well with the operation of a machine. Nevertheless, in the 17th century, the idea still survived that the heart is the true source of heat that maintains the temperature of the body. DESCARTES had become an apostle of this idea. In his *Treatise on Man*, which appeared after his death (1662), he wrote: "the flesh of the heart contains one of these lightless fires that makes it so hot and so ardent that as blood enters one of the two chambers or cavities within itself, it swells up and dilates." A comparison was made with the vaporization of water or milk poured drop by drop into a very hot vessel. DESCARTES concluded that "the fire that is in the heart of the machine that I describe to you has no other purpose there than to dilate, heat and subtilize the blood." In other terms, the heart was considered to be a **kettle**, and the expulsion of the blood contained in it became the result of the heat given off by this kettle. From this came the erroneous conclusion that the cardiac diastole (and not the systole) was responsible for the movement of blood in the arteries. It was only at the end of the 17th century that the function of the heart as a **pump**, whose actions drive the blood into the arteries, was finally accepted.

Judicious observations, simple but well-interpreted experimentation and bold thought that excluded presuppositions led HARVEY to discover the circulation of the blood. He did not stop there. Taking an interest in embryology, he published *Exercitationes generatione Animalium* in London in 1651, where, for the first time, the magic words were written: *"Ex ovo omnia"*. The theory of spontaneous generation, which was widely believed at the time, was openly argued against in this work. Here, HARVEY described the comparative anatomy of the reproductive organs in numerous vertebrates and invertebrates, and discussed theories of embryonic development from the fertilized egg, openly coming down on the side of **epigenesis** rather than **preformation**.

2. HOW WERE THE MOVEMENTS OF THE HEART AND THE BLOOD EXPLAINED BEFORE HARVEY?

"GALEN had given medicine an orientation that would have allowed it to make considerable progress, if, on the one hand, he had not himself been imprisoned within a teleological doctrine, and if, on the other hand, the decadence of the spirit of research at that time – a strange collective phenomenon involving a decrease in critical faculties that arose from the politico-social situation – had not made of his system a "noli me tangere" that robbed it of the greatest of virtues, that of being a teacher and a guide along the pathway of experience."

Arturo CASTIGLIONI
History of Medicine - 1927 (translated from the French edition of 1931)

In order to understand why the discovery of the circulation of the blood was a revolution in the experimental approach to living things, it is necessary to give a brief overview of the theories and experiments that had followed one another since ancient times and which became crystallized, with GALEN's successors, into a dogmatic form, studded with major errors, and spread in an intangible manner by medieval scholasticism (Chapter I-2). Since the period of Ancient Greece, the heart had been considered as a source of life. In the embryo, ARISTOTLE recognized a "point" that was animated by contractile movements, and which grew and structured itself to become the heart. For HIPPOCRATES, the heart was a source of heat. Based on the results of questionable experimentation, ARISTOTLE subscribed to the false idea, inherited from Hippocratic doctrine, that there was a mixture of air and blood in the heart. Blowing air into the trachea of a cadaver, he saw air go into the heart, an error that was no doubt due to a deterioration of the lungs. In agreement with the theory of HIPPOCRATES, ARISTOTLE thought that the heart was a heat-generating organ that was able to transform foods and make them similar to blood. GALEN perpetuated these errors.

GALEN had wrongly asserted that the wall between the right and left ventricles was perforated and that the two ventricles could intercommunicate. Was this error the result of observations made on the hearts of animals such as tortoises or lizards, in which the circulation of the blood is significantly different from that in mammals, these observations being extrapolated to include mammals and Man? Whatever the reason, this idea of **interventricular pores** continued to be put forward, without verification, for centuries. This **almost insurmountable obstacle** blocked the progress of any hypothesis that might allude to a blood circuit within the body.

One of the fundamental postulates of GALEN's theory with respect to the movement of blood in the body was a belief in the existence of a principle, the ***pneuma***, a notion inherited from ARISTOTLE (Chapter I-1.2). For GALEN, the *pneuma* represented the essence of life. The *pneuma* included three types of spirit; **animal spirits**, with their seat in the brain, **vital spirits**, which resided in the heart and **natural spirits**, which were found in the liver, the latter being considered as a major center of exchange. This postulate gave rise to the notion of different stages, which can be summarized as follows (see Figure II.3). During the first stage, the digestion products from the stomach and intestine traveled along the so-called portal vein to the liver, where they were transformed into **blood** that was **still imperfect**, loaded with **natural spirits**. From the liver, the blood was taken to the right-hand side of the heart by the vena cava. From the right-hand side of the heart, part of the blood returned to the liver to be renewed, and another part went to the lung *via* the "arterial" vein (pulmonary artery) to feed this organ. A final part of the blood oozed through the pores of the interventricular wall into the left ventricle, which was considered to be the seat of the innate heat. It was here that the blood, "heated" by the heart, was mixed with the air coming from the lungs *via* the venous artery (pulmonary vein). The blood, cooled in this way, and loaded with **vital spirits** from the heart, overflowed, because of the heat inherent to this organ, into the blood vessels and thus spread slowly through the body. In the brain, the blood acquired **animal spirits**. These spirits circulated through the nerves, which were likened to hollowed tubes, and they were responsible for the control that the nerves exert over the muscles. In agreement with the postulate that there were vital spirits and natural spirits, GALEN stated that there were two blood networks, a **venous network** centered on the liver and an **arterial network** centered on the heart, and that the two types of blood, the heavy, thick venous blood and the "slender and subtle" arterial blood, flowed in **parallel directions** to **slowly irrigate and feed the different parts of the body**. Thus GALEN missed the major and minor circulation.

Looking back from a present in which the tiniest molecular nooks and crannies of the functions of living beings are analyzed, it cannot be over-emphasized that a mere four centuries ago, the circulation of the blood was still taught in a completely erroneous way, based on Galenic theory. The decoding of the mechanism of the circulation of the blood by HARVEY is, without a doubt, the event that symbolizes the **recognition of the science of living beings as an experimental science**. We should not, however, be unjust and fail to recognize that which GALEN gave to science. He refuted ERASISTRATUS' theory (Chapter I-1.3) and proved that the arteries are filled with blood that is "pneumatised" and "subtle" and different from venous blood. He recognized the existence of the sigmoid valves at the heart orifices. He differentiated the systole (contraction) from the diastole (dilatation) by removing the sternum in animals so as to expose the heart. He showed that the movements of the heart are independent of cerebral activity and of respiration. The heart and brain are connected by means of the carotid arteries, the jugular vein and nerves. In the animal, it is only necessary to section the nerves or bind the blood vessels to observe that the heart continues to beat. It is possible for anyone to verify that breathing rapidly several times or holding one's breath does not modify the cardiac rhythm. With GALEN, we see a glimmering of the use of an animal experimentation that aimed to connect anatomical details with overall physiology. Instead of demonizing Galenic theory, as certain people are content to do, considering him as having put a brake on progress, would it not be more reasonable to blame a certain form of intransigent obscurantism, deprived of any critical spirit, which, during the Middle Ages, led to an insistence on following, to the letter, that which had been written by GALEN centuries before, and turning it into an inviolable reference?

In 1553, an opuscule entitled *Christianismi Restitutio*, dealing with theology, appeared. In this, curiously, a few pages were devoted to the movement of the blood in the human organism, and, surprisingly, to the description of a circulation of the blood *via* the lungs. The author, a theologian of Spanish origin, was Miguel SERVET (SERVETO). How did SERVET reach this point? The Holy Scripture, wrote SERVET, says that the soul is in the blood and that the soul is the blood itself. As this is so, in order to understand how the soul is formed, it is necessary to understand how the blood is formed. What is more, in order to understand how the blood is formed, it is necessary to understand how it moves. Thus, starting with an inquiry concerning the soul and the blood that is supposed to animate it, SERVET was led to the postulate that there is a **pulmonary circulation**. Inevitably, he had to come to terms with GALEN's theory of spirits. How did the vital spirits form? They result, said SERVET, from the mixing, in the lungs, of air with blood that the right ventricle of the heart expels *via* the arterial vein (pulmonary artery). The "pneumatized" blood, he added, is taken to the left half of the heart by the venous artery (pulmonary vein). SERVET asked himself why this was so, and came up with the **novel idea that the interventricular wall was not perforated**. This was not the first time that the existence of GALEN's interventricular pores had been questioned. An Arab doctor, IBN AL-NAFIS, had already observed that there was no interventricular communication, and he had suggested that the blood was purified by its passage through the lungs (Chapter I-1.4). Nevertheless, tradition credits SERVET with the first description of the **"minor circulation"**, i.e., the **pulmonary circuit**, by which the blood leaves the heart to return to it, irrigating the lungs on the way. From the theological point of view, *Christianismi Restitutio* cast doubts on the doctrine of the Trinity, and SERVET was declared to be a heretic and subversive. In 1553, SERVET, wishing to discuss the grievances against him with CALVIN (1509 - 1564), went to Switzerland. He was arrested and enjoined to renounce his beliefs. He refused, and was condemned to be burned at the stake. In Geneva, on the 27th of October 1553, he was burned alive, with his works. Six years after the death of SERVET, Realdo COLOMBO (1516 - 1559) was to re-describe the pulmonary circulation and, once again, refute the theory that there are pores in the interventricular wall. A first breach had been made in the Galenic theory of the blood flow. Nevertheless, it would be nearly a century before HARVEY formulated the concept of the "**major circulation of the blood**". During this time, there was an accumulation of observations that were not immediately explained, but which would prove useful. These came out of the **Italian anatomy school**.

In the 16th century, northern Italy, with the Universities of Padua, Pisa and Bologna, was coming to be considered as one of the most active intellectual hotbeds in Europe, with anatomists whose names have been passed down to posterity, such as Bartolomeo EUSTACHI (EUSTACHIUS) (1500 - 1574), Realdo COLOMBO, Andreas VAN WESEL (VESALIUS) (1514 - 1564), Andrea CESALPINO (CAESALPINUS) (1519 - 1603), Gabrielle FALLOPIO (FALLOPIUS) (1523 - 1562), and FABRICIUS D'ACQUAPENDENTE (who taught HARVEY). The **dissection of human cadavers**, which was now authorized, revealed that many of the descriptions and anatomical drawings published by GALEN and attributed to Man were in fact deliberate extrapolations made from mammals, monkeys and pigs in particular.

VESALIUS
(1514 - 1564)

In Padua, VESALIUS, a Belgian physician, was one of the first to cast doubts on the writings of GALEN on the basis of dissections carried out on human cadavers. After having undergone classical studies in Louvain, VESALIUS was admitted to the Faculty of Medicine of Paris at 19 years of age. He spent three years there before returning to Louvain where he presented his thesis in medicine. In 1537, he left Louvain for Padua. On the 5th of December 1537, he was accepted as a doctor in the Padua school of medicine. The following day, he was appointed to the chair of anatomy and of surgery, where he remained until 1542. In 1543, in Basle, he published his major work, *De humani corporis fabrica libri septem*. This work, written in Latin, was divided into seven books, with some two hundred figures and several dozen anatomical plates. At barely thirty years of age, VESALIUS was contesting the authority of GALEN, an authority that had been recognized and added to by AVICENNA and AVERROES, illustrious doctors and anatomists whose treatises had been followed to the letter in Medieval schools of medicine. VESALIUS could not confirm the existence of the interventricular pores reported by GALEN. Having been cautious in the first edition of *De humani corporis fabrica* (1543), in which he was willing to admit that perhaps, after all, there might be invisible interventricular pores, he stated in the second edition (1555) that the interventricular wall of the heart is as thick, compact and dense as any other part of the heart and that interventricular pores do not exist. However, the belief in Galenic doctrine continued to hold sway, to such an extent that VESALIUS was called a libeler by his peers, and had to leave Italy. His career as an anatomist and teacher was destroyed. His subsequent wanderings led him to the court of the Holy Roman Emperor, CHARLES V (1500 - 1558) whose personal physician he became, also becoming the personal physician of CHARLES' son, PHILIP II of Spain (1527 - 1598). VESALIUS had a considerable influence on the medicine of his time. While recognizing the value of GALEN's work overall, he denounced its weaknesses, and, because of this, he helped to destroy the blind confidence that had been given to GALEN's work up until then.

In 1571, Andrea CESALPINO, who was teaching in Pisa, gave a detailed description of the passage of the blood from the right half to the left half of the heart *via* the lungs, and called this the pulmonary circulation (without citing COLOMBO and SERVET). He drew attention to the fact that when a moderately tightened tourniquet is put on an arm, a swelling of the veins is always observed below, and never above, the ligature, which, he added, was well-known to those who practiced bleeding as a treatment, and always carried it out below the ligature. These tourniquet experiments would be repeated and explored in detail by HARVEY. CESALPINO had an encyclopedic knowledge. He wrote a treatise on plants, the classification of which was based on their fruiting organs, flowers, fruits and seeds. Accused of atheism because of his daring theories, he was only saved by the protection of pope CLEMENT VIII (1536 - 1605).

In 1562, FABRICIUS D'ACQUAPENDENTE succeeded CESALPINO in the chair of anatomy at the Padua school of medicine. He is credited with having noted, in the veins, the existence of **valves**, a sort of one-way flap in the direction of the heart, which he called **ostioles**. The **arteries** do not have these (Chapter II-1). In his *De venarum ostioli* (1574), he postulated that the valves are there to "prevent the blood from accumulating in the feet and hands, which would have the worrying consequence of the upper parts of the members suffering from a lack of food." According to him, "the valves act as sluice gates like those that control the distribution of water in mills;" not completely forbidding the blood flow in one direction, they simply prevent all the blood from passing rapidly into the extremities of the body. Thanks to this hypothesis, FABRICIUS saved the Galenic theory of a slow irrigation of the extremities of the body by the blood, but did not reach the truth. In *De motu cordis et sanguinis* HARVEY, FABRICIUS' pupil, states that it was in listening to his master speak about valves, in Padua, that he suddenly had the intuition that the blood flow from the extremities to the heart *via* the veins corresponds to part of the flow of the major circulation. "I began to ask myself," he writes "whether there might not be a movement of the blood that was circular in some ways. This is what I later found to be true." This stroke of genius on HARVEY's part was the end result of a series of observations that cast doubt on the Galenic theory of the slow irrigation of the body by the blood, without invalidating it. This is often how it is with great discoveries: facts accumulate, but it is still necessary to spot the link that will suddenly make everything come together and provide a harmonious explanation.

3. THE FIRST, FALTERING STEPS OF EXPERIMENTAL SCIENCE APPLIED TO LIVING BEINGS

"What marks the beginning of another age, around 1670, is the widespread and rapid diffusion of new intellectual values. A rejection of the authority of the Ancients, a contempt for erudition, and a search for evidence in reasoning and certitude in facts were the cardinal values."

Jacques ROGER
The Life Sciences in 18th Century French Thought - 1971

The period that starts from the Renaissance and covers the 17th century was rich in inventions and discoveries. Nevertheless, prejudices and old-fashioned ideas remained in the most erudite circles. The **theory of the four elements, earth, water, air and fire**, was still extant. The same is true of the **theory of the three types of spirits, natural, vital and animal**, that were supposed to animate living organisms.

While remarkable progress was being made in physics, thanks to innovative techniques and a mathematical rationalization of experimental results, progress in the Life Sciences was considerably slower. The experimental method was only at its beginnings, particularly with respect to the study of the functioning of the "animal-machine". An understanding of the functioning of an organ in an animal or in Man obviously necessitates an understanding of the animal's or Man's anatomy, as well as an understanding of that organ's topographical relationship with other organs. Daring souls such as the Dutchman Volcher COITER (1534 - 1600), who was a pupil of FALLOPIUS, and others who were inspired by the teachings of the Italian masters of anatomy, helped, in the West, to impose a revision of the teaching of human anatomy, which up to that point had been content to simply follow GALEN to the letter. The teaching of anatomy went from being book-based to being practical. What is more, there was a growth in **comparative** anatomy and in **microscopic** anatomy, both of which were indispensable supports for the budding science of experimental physiology.

3.1. FROM HUMAN ANATOMY TO COMPARATIVE ANATOMY SYSTEMATICS

From the 16th century onwards, the free access to the dissection of human cadavers led to anatomical discoveries that would prove essential to the development of **physiology**, which was considered to be "**live anatomy**". The Italian school continued to be particularly active in this domain. Twenty centuries after the Greek physician ALCMEON (Chapter I), EUSTACHIUS rediscovered a tube-shaped communicating channel between the middle ear or tympanic cavity and the pharynx, which would be mentioned in treatises of anatomy as the EUSTACHIAN tube. FALLOPIUS identified the ovarian tubes in the female reproductive organs (FALLOPIAN tubes). In addition to his discovery of the venous valves, FABRICIUS

D'ACQUAPENDENTE made many observations concerning the digestive, sight and hearing organs, as well as the renal apparatus and the pulmonary apparatus. He gave detailed descriptions of the muscular stomach in birds and the four divisons of the ruminant stomach, as well as different types of placenta and the embryonic appendices in ruminants and rodents. He was also interested in the morphological modifications that occur in the first stages of embryogenesis in the chicken. These pioneers inspired the anatomists of the 17th century. In England, Francis GLISSON (1597 - 1677) described the fibrous envelope of the liver, which would be called GLISSON's capsule, Thomas WHARTON (1610 - 1673) discovered the excretory duct of the submaxillary gland that carries his name and Thomas WILLIS (1621 - 1675) characterized an arterial network of internal carotids, at the base of the brain, mentioned in treatises of anatomy as WILLIS' circle or hexagon. In 1656, WHARTON gave the name thyroid, (from the Greek θυρεός, shield) to a twin-lobed glandular mass that occupied the upper part of the trachea. The function of this gland would only be discovered two centuries later. The Dutchman Franz DE LA BOE, known as SYLVIUS, (1614 - 1672), from the Leyden School of Medicine, described the communicating channel (aqueduct of SYLVIUS) between the third and fourth ventricles of the brain. DE LA BOE was a convinced iatrochemist. He vigorously promoted an alliance between chemistry and medicine, and defended the idea that animal physiology could be explained by chemical reactions. An admirer of HARVEY, he spread the latter's ideas through his teaching. Niels STENSEN (STENON, 1638 - 1687), who was of Danish origin, carried out interesting research on the lacrymal and salivary glands, and showed, by introducing a stylet, that there is a duct (STENON's duct) from the parotid gland and that it opens into the mouth. He observed that the parotid glands secrete saliva. In 1684, the Frenchman Raymond VIEUSSENS (1641 - 1715) published a treatise, *Neurologia universalis*, that dealt with the anatomy of the brain, of the spinal chord and the nerves. The Swiss anatomist Hans Conrad PEYER (1653 - 1712) identified follicular formations of lymphoid tissue in the mucosa of the small intestine. These became known as PEYER's patches.

Physiological studies based on animal vivisection involved a comparison of anatomical structures in Man and animals. In 1622, Gaspare ASELLI (1581 - 1626), a professor of anatomy at Pavia, showed "lacteal veins", which appeared to be small white chords, in the mesentery of a dog "that had recently eaten". At first confusing these with nerves, he quickly recognized that these were vessels that, if they were incised, let a lactescent liquid escape, the chyle. ASELLI had discovered the **chyliferous vessels** that would be found in Man. In 1651, the French physician, Jean PECQUET (1622 - 1674) observed that the **thoracic duct** carried the chyle to the left subclavian artery. At the base of this thoracic duct, PECQUET observed a dilated portion which was named the PECQUET cistern. PECQUET's discovery was questioned by RIOLAN, but confirmed by the Dane Thomas BARTHOLIN (1616 - 1680). In 1650, in Leyden, Olav RUDBECK (1630 - 1702), who was of Swedish origin, discovered, while dissecting a calf, vessels coming from the intestinal wall that were filled with a transparent liquid. These were the **lymphatic vessels**. BARTHOLIN demonstrated that lymph circulates in these vessels. In 1672, the Dutchman Reinier DE GRAAF (1641 - 1673) identified yellowish vesicles that were visible to the naked eye on the ovaries of female rabbits, dogs, sheep, cows and of women. Called the DE GRAAF **follicles**, they were at first considered to be ovules. It was only two centuries later that Karl Ernst VON BAER (1792 - 1876) recognized that the **ovule** corresponds to a microscopic structure contained in the DE GRAAF follicle. DE GRAAF produced an excellent description of the transformations of the ovary and the uterus of female rabbits during the maturation of the fertilized egg and gestation. The English physician Nehemiah GREW (1641 - 1712) was interested in the morphological differences of the stomach and intestine in many fish, birds, and mammals, including Man, and in the significance of these differences according to the types of food eaten. Conscious that a new discipline was emerging in the Natural Sciences, he created the term "**comparative anatomy**" in 1675. The Dutch anatomist Jan SWAMMERDAM (1637 - 1680) is credited with achieving previously unequaled precision in the dissection of three thousand species of insect, and with the analysis of their metamorphoses *via* their anatomical modifications (Figure II.4).

Not only was dissection and observation carried out, but experiments were carried out on anatomical parts. The injection of colored liquids and antiseptics and sometimes of melted wax into blood and lymph vessels revealed the complexity of the vascular network, and, above all, preserved the anatomical parts for demonstrations. This **anatomical injection technique** came into general use in the second half of the 17th century. The Dutchman Frederik RUYSCH (1638 - 1731) acquired a well-deserved renown in this art. He is credited with the discovery of the valves of the lymph vessels.

In **zoology and biology, systematics** was developed on the basis of the knowledge acquired from comparative anatomy. By opening the eyes of the world to the fantastic diversity of living beings, this became a subject for reflection, which would later lead to the idea of evolution. Systematics did not involve experimentation, but it was nonetheless a powerful stimulus. It was a prerequisite for physiological studies carried out on judiciously chosen models. The formidable complexity of the living world, with its infinite variety of morphological and functional differences, made it necessary to have a consistent classification of animal and plant organisms. In Ancient Greece, the zoologist ARISTOTLE and the botanist THEOPHRASTES had begun to make classifications that took their inspiration not only from anatomy, but also from the physiology of animal and plant species (Chapter I-1.2). During the following centuries, almost until the Renaissance, systematics was abandoned.

At the beginning of the 16th century, botanical literature once again began to include detailed descriptions, the value of which was increased by the presence of an abundant iconography. The amount of interest given to particular plants, particularly in the schools of medicine, was related to their medicinal properties. Some illustrious names in this domain in Germany were Otto BRUNFELS (1489 - 1534), Jerome BOCK (1498 - 1554) and Leonard FUCHS (1501 - 1566).

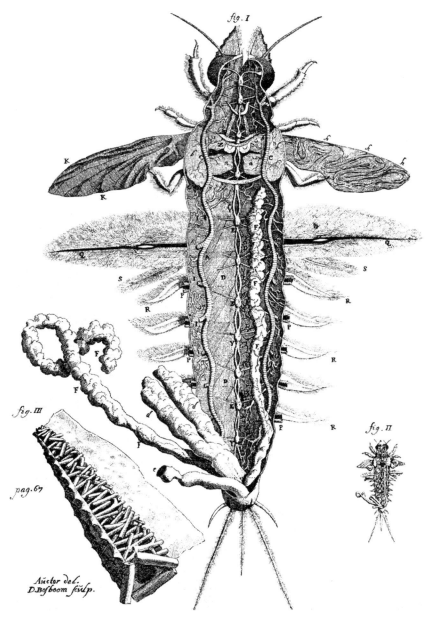

Here we note the extreme precision of detail in the drawing of the different organs of the insect, shown by dissection.

Figure II.4 - Mayfly larva (7 millimeters) dissected by the Dutch anatomist Jan SWAMMERDAM (from Jan SWAMMERDAM - Ephemeri vita, 1675)

By means of detailed, accurate study, it was possible to distinguish essential characteristics from secondary characteristics, to note resemblances as well as divergences, that is to say, it was possible to draw up a knowledge framework that allowed coherent systematics to be constructed. In *On Plants*, CESALPINO, while making reference to THEOPHRASTES and his division of plants into trees, shrubs and grasses, offered a classification that took into account the morphology of flowers and fruits, proceeding by analogy, principles which, in France, provided inspiration for Jean BAUHIN (1541 - 1613) and his brother Gaspard (1560 - 1624), as well as Joseph PITTON DE TOURNEFORT (1656 - 1708) and, in England, for John RAY. Between 1686 and 1704, the latter published the monumental *Historia plantarum*. RAY is credited with the first definition of a **species**: "all plants that arise from the same seed and are reproduced by sowing that seed belong to the same species." These precursors were succeeded by Carl LINNAEUS (1707 - 1778), who wrote the famous *Systema Naturae*, with its **binomial classification** of plants based on the characteristics of the pistil and the number and arrangement of the stamens, and by the DE JUSSIEUs, the most famous of whom, Bernard DE JUSSIEU (1699 - 1777), helped by Michel ADANSON (1727 - 1806), worked on setting up a **natural classification system**. This system took into account the characteristics that were present or absent in neighboring plants, with the species being laid out and associated in such a way that the number of dissimilar characteristics was as small as possible.

Systematics established itself later in Zoology than in Botany for many reasons, including the ease of observation of plants, the interest accorded to medicinal plants because of their therapeutic properties, and the fact that classification of animals seemed to arise from a strictly academic interest. In his *Historia animalium*, published between 1551 and 1558, the Swiss doctor Konrad GESNER (1516 - 1563) sketched out a classification that was reminiscent of that of ARISTOTLE. At this time, zoological literature was being enriched by the publication of detailed monographs concerning fish, birds and marine animals by the French naturalists Pierre BELON (1517 - 1564) and Guillaume RONDELET (1507 - 1566). BELON even went so far as to compare the skeleton of a man and that of a bird. Taking his inspiration from ARISTOTLE, John RAY set out to order the different classes of animals according to the morphology and function of their vital organs. Starting from a preliminary division between animals "having red blood" and animals that do not, he showed that it is possible to divide the first group according to whether respiration is carried out by the lungs or by gills and that another subdivision can be made according to whether the heart has two ventricles or only one.

The binomial classification proposed by LINNAEUS for the plant kingdom was adopted rapidly in Sweden, England and Germany. In France, it came up against the criticisms of BUFFON and DAUBENTON (1716 - 1800). Whatever the principles of classification under consideration, they were underpinned by two key ideas, that of a **continuum** and that of a **hierarchy**. The theory of **transformism** was already on the drawing board. Nevertheless, as the famous paleontologist Georges CUVIER remarked, "**with systematics, we had the vocabulary but not the language.**" It was therefore necessary to go forward, not only in the analysis of structures, but also in the recognition of the relationships between organs, in other words, to adapt the experimental method to a more in-depth exploration of the functions living beings.

3.2. FROM MICROSCOPIC TISSULAR ANATOMY TO THE MORPHOLOGICAL DESCRIPTION OF ANIMALCULES

At the turn of the 17th century, the arrival of optical microscopy provided access to the invisible, revealing a mysterious, unimagined world and turning the ideas that naturalists had had about the living world upside down. Strictly speaking, microscopic observation was not part of the experimental method. Nevertheless, it led to reflection upon the micro-organization of living matter beyond the level which, up until then, had been perceptible by the naked eye. Later, a mastery of microscopy and micromanipulation would open the way for experimentation on the infinitely small.

The first **compound microscopes** appeared in Holland and in Italy at the beginning of the 17th century. They comprised a copper tube at the ends of which two glass lenses were fixed, the objective and the eyepiece (Figure II.5). Tradition has it that the compound microscope was discovered by two Dutch lens grinders, Hans JANSEN and his son Zacharias (1580 – 1638). Other sources affirm that the father of this discovery was GALILEO. It was, in fact, in Italy that the term *microscope* was proposed by Giovanni FABER (1574 - 1629) and that the first drawings of enlarged insect structures were published, from 1630, by Francesco STELLATI (1577 - 1653). *Micrographia*, published by Robert HOOKE (1635 - 1703) in 1665, brought together a large number of engravings showing, with a high degree of accuracy, the morphological details of small animals (insects) or the compartmentalized structures in cross-sections of plant tissues. The structure of cork was compared to that of a honeycomb (Figure II.6). At this point, the term **cell** (from the Latin *cella*, small room) came into being.

M. MALPIGHI
(1628 - 1694)

In the final decades of the 17th century:, with the microscopic examination of plant and animal tissues, Marcello MALPIGHI and Nehemiah GREW opened up a new domain in the morphological sciences, that of **comparative ultrastructural anatomy**. Using cross-sections of the stems of plants, they discovered tubular structures in the form of vessels or even compartmentalized corpuscular structures that would be named utricles (from the Latin *uter*, other) (Figure II.6). The concept of the "cell", the structural and functional unit of the living world, would only be established on undisputed bases two centuries later.

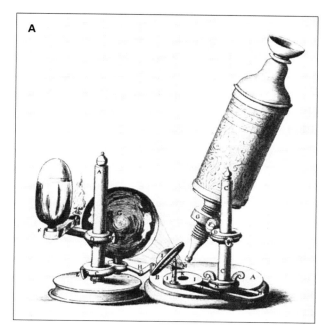

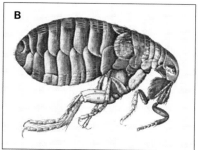

A - Hooke's compound microscope, using the light from a small oil lamp reflected by a water-filled glass globe, then passing through a convex lens (from Robert Hooke - *Micrographia*, 1665).

B - Drawing of a flea by Hooke, observed through his microscope. The original image had a dimension of around 40 centimeters and showed tiny morphological details (from Robert Hooke - *Micrographia*, 1665).

C - Van Leeuwenhoek's simple microscope (magnifying glass). An almost spherical polished glass lens is inserted into a hole cut into a 7 cm long copper plate. The biological sample is fixed onto the point of a rod, the height of which can be adjusted (from Giuseppe Penso - *La conquista del mondo invisibile*, Giangiacomo Feltrinelli Editore, Milano ; French edition, *La conquête du monde invisible*, Les Editions Roger Dacosta, Paris, 1981, all rights reserved).

Figure II.5 - Microscopes used in the 17th century

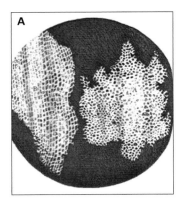

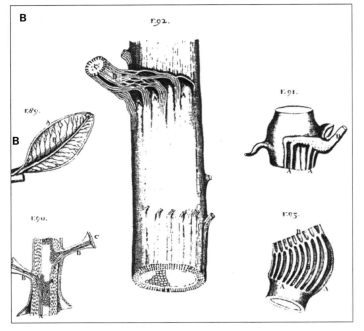

A - Drawing of pores observed under the microscope by HOOKE in a thin cross-section of cork (from Robert HOOKE - *Micrographia*, 1665).

B - Drawing of vessels in plant tissue observed under the microscope by GREW (from Nehemiah GREW - *Anatomy of plants*, 1682).

Figure II.6 - The first analyses of the macroscopic structure of biological objects

MALPIGHI noted the existence of pores in leaves. These were stomata, through which the exchange of gases takes place. Dissecting a silk worm under the microscope, he identified a complex network of tubes, the tracheas, which opened onto the surface of the insect's body and acted as the respiratory apparatus by conducting air to the organs. The apparent analogy between these tracheas and the tubular structures of plants was a source of confusion for a some time. The use of the microscope led MALPIGHI to look at the development of the chick embryo. He described the different stages of its development (Figure II.7). Because of this, he may be considered as a pioneer of embryology. Using an ingenious perfusion technique, MALPIGHI highlighted the network of capillaries that irrigates the alveoli in the frog lung. Water injected into the pulmonary artery came out *via* the pulmonary vein. "Washed" in this way, the lung showed up very fine vessels, the capillaries. The presence of capillary vessels that provided anastomoses between arteries and veins supplied the missing "building block" for HARVEY's theory about the circulation of the blood (Chapter II-1).

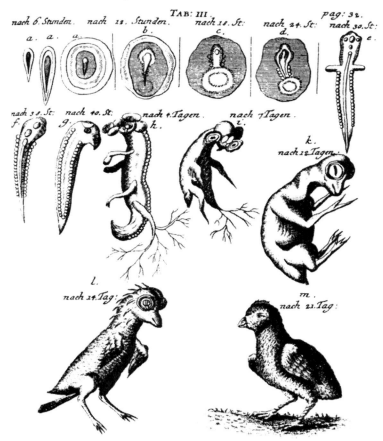

Drawing of the different stages of development of the chick embryo from the fertilized egg to the chick.

Figure II.7 - Development of the chick embryo
(from Marcello MALPIGHI - *Opera omnia : de formatione pulli in ovo*, 1672)

While the use of the microscope was growing, detractors of the new instrument, in order to discourage its use, emphasized the fact that there could be artifacts (irisation phenomena, deformation of images) due to chromatic aberration and aberrations in sphericity. With the aim of correcting **chromatic aberration**, the English optician John DOLLOND (1706 - 1761) came up with the idea of combining two lenses made of silicates having different refraction indices, for example crown glass (potassium silicate and lime) of index 1.52 and flint glass (potassium silicate and lead oxide) of index 1.59. **Aberrations in sphericity**, due to the unequal refraction undergone by light in the different regions of a lens, was corrected by means of a diaphragm that only allowed the light rays to pass through the center of the lens.

A. VAN LEEUWENHOEK
(1632 - 1723)

Anton VAN LEEUWENHOEK, a self-taught draper who lived in Delft, Holland, was fascinated by the world of the infinitely small, and, in a few years, had become a recognized pioneer in this domain. Keeping to the **simple microscope** (see Figure II.5), he perfected its optics to such an extent that enlargements of around 200 times could be obtained. The simple microscope was reduced to a biconvex lens acting as a magnifying glass, inserted into a metal plate and placed opposite and at a small distance from the object to be observed (see Figure II.5). VAN LEEUWENHOEK is said to have manufactured nearly 500 simple microscopes in 50 years. He sent the Royal Society of London more than 400 letters in which he noted down observations ranging from the flow of the blood globules in the capillary vessels to a description of spermatozoids and single-cell organisms. Having no humanist background, but keen to understand the secrets of Nature, VAN LEEUWENHOEK may be considered as one of the founding fathers of **cytology**.

In October 1676 the Royal Society of London received several letters from VAN LEEUWENHOEK describing unicellular microscopic beings that could be found in ponds or even in water in which leaves or seeds had been seeped (infusions). These unicellular beings were **protozoans**. It was impossible to find out what their function was at this time. However, the proof of their existence would have an effect on debates concerning the spontaneous generation of microscopic beings. In 1683, VAN LEEUWENHOEK discovered even smaller, highly mobile "creatures" in dental plaque. No doubt these were **bacteria**. He was aware of the **quantitative aspect** of his findings. In order to give an idea of the size of a particular microorganism, he used the thickness of a hair, seen under the microscope, as a reference. The significance of the bacterial world would only be evaluated two centuries later when a correlation would be established between bacteria and infectious illnesses. From 1677 onward, VAN LEEUWENHOEK informed the Royal Society of a new finding, the presence of tadpole-shaped mobile **animalcules** in the sperm of a varied sample of mammals (Man, rat, dog, sheep), birds, amphibians, fish, mollusks and arthropods.

A controversy then developed concerning the formative structure of the embryo. In 1651, HARVEY had stated that sperm acted on the ovule from a distance, by means of a volatile, fertilizing principle. The thousands of animalcules that VAN LEEUWENHOEK discovered in sperm provided physical support for the idea of the fertilizing principle for the ovule. At this time, the preformation theory, according to which the adult individual was already present in miniature form in the germ, was still believed. This is where the famous quarrel between **animalculists and ovists** began. The former postulated that the human sperm was able to develop, on its own, into a *homunculus*, generator of the future human body. The ovists, in contrast, considered that the ovule was the point of departure for the future embryo. In order to support their thesis, they put forward the **parthenogenesis of aphids**, discovered by the Swiss naturalist Charles BONNET (1720 - 1793).

L. SPALLANZANI
(1729 - 1799)

This quarrel could have come to an end after the celebrated **"artificial fertilization"** experiments carried out on the frog by an Italian priest, Lazzaro SPALLANZANI (1729 - 1799), at the beginning of the 1770s. The experimental protocol was simple. SPALLANZANI made watertight pants with which he clothed his male frogs. He then collected their sperm during pseudo-couplings with females, finding that the latter remained sterile. Nevertheless, if the male frog semen was poured over virgin eggs removed from the bellies of female frogs, this gave rise to tadpoles that developed into frogs. Thus, artificial insemination had been invented. It would be reproduced in the dog in 1780. The theory of **preformation** was soon opposed by the theory of **epigenesis**, which the German embryologist Kaspar Frederik WOLF (1733 - 1794) was one of the first to support, stating that the embryo is formed gradually from an undifferentiated mass.

The revelation of structures that were invisible to the naked eye, by means of the microscope, had unsuspected repercussions within the scholarly world, repercussions that COURNOT writes about in eloquent terms in his *Considerations on the Progress of Ideas and Events in Modern Times*: "for naturalists, the discovery of the microscope had the same effect as the discovery of the telescope for astronomers. It opened up a new world, the strangenesses, or even the wonders of which certainly deserved to attract general attention [...]. These two infinities that Man probed almost at the same time, with instruments of the same type although of different sorts, and between which he appeared to be suspended, seemed to philosophers to be the counterparts of one another."

3.3. THE BIRTH OF PHYSIOLOGY: SEEKING THE EXPERIMENTAL METHOD

Compared to the spectacular growth of anatomy in the 17th century, that of physiology remained modest. The limiting factors were the small number of measurement instruments available, such as balances, thermometers and barometers, and the absence of a methodology that was able to verify the chemical parameters of the metabolism. Given this still rudimentary technological context, the metabolic experiment carried out in 1629 by Santorio SANCTORIUS (1561 - 1636), who held a chair of medicine at the University of Padua (1611 - 1624), has even more of a symbolic value. SANCTORIUS had been a pupil of FABRICIUS D'ACQUAPENDENTE. He was interested in the perspiration, a form of respiration through the pores of the skin, which was therefore different to pulmonary respiration. He developed a **weighing system** for himself, using a balance that he had had specially made for this purpose and which measured his weight before and after eating, after exercise, then at rest and during sleep (Figure II.8).

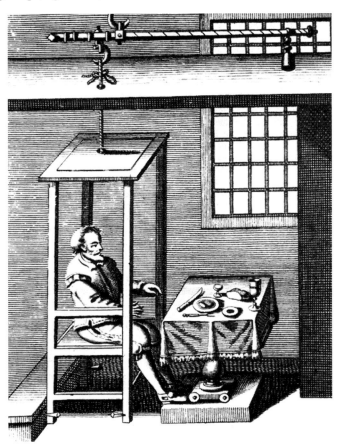

The weighing of a human being allows a quantitative approach to vital phenomena: the absorption of food and drink, excretion, perspiration and respiration. The experiment carried out by SANCTORIUS (SANTORIO) marked the birth of metabolic physiology.

Figure II.8 - The balance in physiology
(from *De medicina statica aphorismis*, 1641, published after the death of SANCTORIUS)

The measurements were made several times a day, for several days in a row, making it possible to evaluate what had been absorbed, what had been lost, and what had been perspired. SANTORIO noted that the weight of the excreta was lower than that of the food ingested. He concluded from this that a share of the products of digestion was lost during perspiration. It would be learnt later that respiration, with the emission of CO_2, is another form of loss of carbonaceous material. SANTORIO's balance was one of the first examples of a **measurement instrument used for a carefully programmed biological experiment**.

The effect of heat on the expansion of air or water had been known for a long time. HERO of Alexandria had created the name **thermoscope** to designate an instrument that measured the variations in the volume of air contained in a flask and subjected to hot or to cold (Chapter I-1.3). Nevertheless, no serious quantification had been carried out. At the end of the 16th century, GALILEO rediscovered the air thermoscope. This was a tube that was half-filled with water and then turned upside down in water-filled basin. As the pressure of the air remaining in the tube varied with temperature, the column of water in the tube rose or fell as the temperature changed. In fact, in this apparatus, the liquid in the basin was in contact with the atmospheric air, which exerted pressure upon it. The apparatus therefore operated not only as a thermometer but also as a barometer. Once TORRICELLI had demonstrated that air has weight (Chapter II-4.3), it was understood that in order to avoid interference by atmospheric pressure, it would be necessary for the thermometer to be a closed tube.

Modern thermometers began to appear at the end of the 17th century and were improved upon during the 18th century. Some of these were real works of art (Figure II.9). Others were simple closed glass tubes with one bulbous end that acted as a reservoir. At first, the liquid used in the tube to indicate temperature changes was ethanol, which at this time was referred to as spirits of wine. In 1717, the German physicist, Daniel FAHRENHEIT (1686 - 1736) replaced alcohol with mercury. In 1724, FAHRENHEIT proposed a new thermometric scale that took as its reference points the value of 0° for the temperature of a mixture of crushed ice and of an ammonium salt and the value of 96° for the temperature of the healthy human body. On this scale, the freezing point of water was 32° and the boiling point of water was 212°. In 1730, René Antoine FERCHAULT DE RÉAUMUR (1683 - 1757), physicist and naturalist, presented an alcohol thermometer that was calibrated from 0° to 80°, to the Academy of Sciences in Paris. The alcohol was colored red with a tincture of sorrel, making it easier to read the level reached by the alcohol in the glass tube. In 1741, the Swedish physicist Anders CELSIUS (1701 - 1744) used the values of 0° and 100° as reference points, the latter referring to the melting point of ice, and the former to the boiling point of water. In 1750, for reasons of ease of reading, the values were reversed, so that zero degrees became the temperature of melting ice and one hundred degrees became the temperature of boiling water. This revised CELSIUS scale was adopted for good in France in 1794, while the FAHRENHEIT scale continued to be favored in Anglo-Saxon countries.

The Italian thermometer represented in this figure was made up of a hollow glass tube that was closed and coiled around in a graduated spiral, with swellings at both ends. The lower swelling, which acted as a reservoir, and part of the spiral were filled with liquid (alcohol, known as spirits of wine). The rise or fall of the column of liquid in the spiraled portion made it possible to identify variations in temperature. This piece of scientific apparatus was bequeathed to posterity as an *objet d'art*. In addition to spiral thermometers, straight thermometers were manufactured for utilitarian purposes.

Figure II.9 - Spiral thermometer manufactured in Italy in the 17th century
(Photo Franca PRINCIPE, IMSS - Florence, with permission)

It took very little time for the advantages of using the **thermometer as an instrument to measure the temperature of living beings** to be appreciated. The first applications and discoveries soon appeared. The thermometer made it possible to demonstrate that the body temperature of Man and of mammals is maintained at a constant value that is independent of the outside temperature, which gave rise to the important idea of **thermal regulation** being a fundamental characteristic of life. The Italian doctor, mathematician, physiologist and astronomer, Giovanni Alfonso BORELLI (1608 - 1679) used vivisection experiments to prove that the heart is at the same temperature as the rest of the body of an animal, thus refuting the theory put forward by DESCARTES that the heart was a kind of kettle distributing its heat to the rest of the body. BORELLI had MALPIGHI as a pupil, and the latter was to become one of the pioneers of microscopic anatomy (Chapter II-3.2).

In 1727, the English chemist and naturalist Stephen HALES (1677 - 1761) described a **very simple system for collecting gas**. This was a graduated tube that was filled with water and upended into a water-filled basin. A tube carrying gas was attached to the base of the graduated tube. The gas, being lighter than water, rose in the graduated tube and took the place of the water at the top. A few years later, the water was replaced with mercury. The principle of the **eudiometer**, which is used to measure gas, was derived from this system. The eudiometer is a graduated glass tube, the upper part of which has two platinum wires running through it. An electrical spark between these wires triggers a chemical reaction between gases present in the tube. If these gases are hydrogen and oxygen, they combine to form water. If one of the gases is provided in excess, the volume that remains of this gas, once determined, gives an indication of the **stoichiometry of the reaction**.

Using his **vacuum pump that was able to produce a vacuum (air pump)** (Chapter II-4.3), BOYLE observed that mammals such as mice were unable to survive without air. BOYLE interpreted this by postulating the presence in air of a "vital quintessence" that animated the "vital spirits". This vital quintessence would later be identified as oxygen.

If we look at research carried out on the mechanism of **digestion**, a process that is relatively accessible to experimentation, it is interesting to see how it evolved from the beginning of the 17th century. At this time, the Belgian doctor Jan Baptist VAN HELMONT, an enthusiast of the ideas of PARACELSUS, considered that in addition to a vital principle made up of a "major archeus" upon which "secondary archei" at the level of each organ were dependent, there was a **chemistry that was specific to living beings**. This theory was illustrated by the mechanism of digestion. VAN HELMONT distinguished six phases in this mechanism. The first was identified by analysis of a bolus regurgitated after various periods of time following the absorption of food. The initial phase takes place in the stomach, and involves the dissolving of the food, which is compared to a "fermentation" caused by acid juices. During the second phase, which takes place in the duodenum, the bolus loses its acidity and is transformed into chyle. The third phase corresponds to the passage of the chyle into the mesenteric veins. During the fourth phase, the liver sends a fermenting agent into the mesenteric veins and the chyle is transformed into blood. The fifth and sixth phases continue in the heart. There, in the left ventricle, the blood combines with air from the lungs to produce the vital spirits, a leftover from GALEN's theory of spirits. However whimsical the ideas involved in some of these phases of the digestive process may seem today, overall, VAN HELMONT's theory has the merit of looking for a mechanism comprising a progression of reactions and describing the logic involved. In 1673, Thomas BARTHOLIN made a fistula in the pancreatic duct of a dog, in order to collect the secretions from it. This experimental innovation, which would be put into common practice and perfected two centuries later by Claude BERNARD (1813 - 1878), did not lead to any tangible results at this time, the understanding of chemistry being too rudimentary.

SYLVIUS (DE LA BOE) was an ardent defendant of the ideas of **iatrochemistry**, for which the paradigmatic model was the digestion, considered to be a set of fermentations, as opposed to the ideas of **iatromechanics** that partly arose from the image of the **animal-machine** formulated by DESCARTES. A fervent partisan of iatromechanics, BORELLI explained that the digestion is mainly carried out by a trituration of the food resulting from the contractions of the stomach. This theory received the support of a celebrated doctor from the University of Leyden, Hermann BOERHAAVE (1668 - 1738), the author of *Institutiones medicae* (1708). At the beginning of the 18th century, faced with the powerful influence of the doctrines of the iatromechanics, a more realistic outlook began to appear. In 1714, Jean ASTRUC (1684 - 1766), from Montpellier in France, wrote a treatise on *The Cause of Digestion: in which we refute the new system of trituration and grinding*. He showed that saliva, bile and pancreatic juices are agents that "dissolve the food with which they are mixed." Nevertheless, it was not until the time of RÉAUMUR and SPALLANZANI that the experimental method was applied to the mechanism of the digestion in a lucid manner.

In his treatise on digestion in birds, which appeared in 1752, RÉAUMUR emphasized differences in the muscle power of the gizzard, in order to explain differences in food transformation mechanisms. In certain birds (turkey, chicken, duck), the gizzard is able to grind glass beads with its powerful contractions. However, in birds of prey such as the buzzard, whose not very muscular stomach resembles that of mammals, food is digested less by trituration than by dissolution due the action of ferments. RÉAUMUR made these birds swallow small, highly-resistant tubes made of lead or brass, which were open at both ends and contained pieces of meat held in place by a wire. After a few hours, pressure was applied to make the birds regurgitate the tubes. It was observed that the pieces of meat that had been in the tubes had been dissolved without having been ground, which proved that there was a "dissolving agent" in the stomach of the buzzard. In this case, mechanical trituration was not necessary to digestion.

Thirty years later, SPALLANZANI repeated RÉAUMUR's experiments, but varied a larger number of experimental parameters, such as the nature of the stomach (muscular, membranous), the type of food (whole meat pieces, ground meat, bread...), and the temperature. These experiments, described to the Academia del Cimento in 1777 and published in Geneva in 1783 in a work that was translated into French by the Swiss naturalist Jean SENEBIER (1742 - 1809), under the title *Experiments on the digestion of Man and of different animal species*, confirmed RÉAUMUR's results. However, it was necessary to go further and prove the chemical nature of digestion. Most of all, in order to refute the hypothesis of a vital force that was responsible for digestion, it was necessary to show that digestion could be achieved outside the animal. With this in mind, SPALLANZANI put small pieces of sponge in metal tubes, made turkeys swallow the tubes, retrieved the tubes after a few hours, and squeezed the pieces of sponge to express the liquid they were impregnated with. He then put this liquid into glass bulbs, in the presence of ground meat. He observed that after 2 to 3 days at 37°C, the meat had been completely liquefied. This was the first example of **artificial digestion**. Artificial digestion was also carried out with the gastric juices of the crow. SPALLANZANI wrote that "the ease with which it is possible to make crows vomit in order to obtain a large quantity of gastric juices enables me to carry out a much larger number of experiments with gastric juices than is possible with the gastric juices of gallinaceous birds, which can only be obtained by killing the birds." Not content with proof obtained from experimentation on animals, SPALLANZANI widened his research on digestion to include Man, using himself as an experimental subject. He swallowed small cloth sacks filled with raw or cooked meat, and regurgitated them after a few hours. He concluded that "Man can digest membranes, tendons, cartilage, and even bones that are not hard, no matter what has been said by physiologists and doctors led astray by dubious experiments that were not carried out with sufficient care […]. Food, therefore, is digested in Man's stomach, as in the stomachs of animals, only by the action of the gastric juices." In the 19th century, the process of digestion would be analyzed in detail and would turn out to be a complex phenomenon involving not only the stomach but also other regions of the digestive tract. Its mechanism would be made clear by studies of artificial digestion using tissue homogenates (pancreas, small intestine) and the juices obtained from glands such as the pancreas, collected by means of artificial fistulae.

The discovery made by HARVEY in the 17th century, concerning the mechanics of blood circulation, and the discoveries made by RÉAUMUR and SPALLANZANI in the 18th century, concerning chemical modifications to food during digestion, were convincing examples of how powerful the experimental method could be when applied to the study of living beings. However, due to a lack of appropriate techniques, much more modest progress was made in the understanding of the mechanisms involved in **relational functions** such as muscular contraction, reflex phenomena and nerve impulses. Such explanations as were given remained brief. In the movements of the arms and legs, resulting from muscular contractions, BORELLI saw a simple mechanism that followed the same principles as a lever in experimental physics (Figure II.10). From 1654 onwards, the English anatomist GLISSON used vivisection in animals to show that the expulsion of bile into the intestine was associated with contraction of the bile duct. From this he concluded that the bile duct contracted because it was irritable.

This idea of **irritability** was taken up by Albrecht VON HALLER (1708 - 1777) in his famous treatise *Elementa physiologiae corporis humani*, the publication of which was spread out over approximately 10 years (1756 - 1766). With VON HALLER, traditional anatomy became more animated, thanks to objective reflection concerning the vital properties specific to each organ. During the 17th century, and at the beginning of the 18th century, the idea of DESCARTES (Chapter II-3.4.3), taken up again by BOERHAAVE, that there was a subtle fluid carrying "animal spirits" inside the nerves, had remained in vogue.

A. VON HALLER
(1708 - 1777)

VON HALLER characterized the irritability of the muscle fiber that undergoes excitation as a shortening movement followed by a return to the original length. He distinguished **irritability** from **sensitivity**, a process instrumentalized by the nerves, on the basis of the creation of experimental lesions in specific nerve territories. It would take the experiments of Luigi GALVANI (1737 - 1798), a physicist and physiologist from Bologna, to show that a muscle contracts when the nerve that innervates it is excited, and to give rise to the suspicion that there is an "innate animal electricity", and then the experiments of a physicist from Padua, Alessandro VOLTA (1745 - 1827), to make it possible to admit that a nerve can be stimulated by the application of an electrical current. Later, experiments by the Berlin physiologist Emil DU BOIS-REYMOND (1818 - 1896) demonstrated convincingly that the excitation of a nerve is followed by a change in its **electrical state**. DESCARTES' "animal spirits" gave way to electrical charges.

H. BOERHAAVE
(1668 - 1738)

In 1761, Giovanni Battista MORGAGNI (1682 - 1771) published *De Sedibus et causis morborum per anatomen indigatis*, a work that exposes the fundamentals of pathological anatomy, i.e., the *post mortem* study of the lesional modifications to organs that are caused by an illness. With **pathological anatomy**, a new exploratory pathway was opened up for medicine.

In the area of **plant physiology**, VAN HELMONT's well-known willow experiment is a striking example of an expressed desire to understand the causes of mechanisms, using experimentation, even if the means employed were still rudimentary. VAN HELMONT wrote "I took an earthen pot and in it placed 200 pounds of earth which had been dried out in an oven. This I moistened with rain water, and in it planted a shoot of willow which weighed five pounds. When five years had passed the tree which grew from it weighed 169 pounds." During the five years, the tree was watered regularly with rainwater or distilled water. The tree's increase in weight, i.e., 164 pounds, could not have come from the earth, as the weight of the latter "had decreased by less than 2 ounces." Here was an experiment that had been carried out with care. The conclusion, that the willow tree's increase in weight could not be explained by the absorption of earth, was sufficient. VAN HELMONT's error was to push his interpretation too far, and to postulate, without formal proof, that the willow tree's wood was manufactured exclusively from water. The idea that plant growth is a consequence of the assimilation of carbon, or, in other words, the synthesis of organic molecules from the carbon dioxide in the air, and also of the assimilation of nitrogen, would only be demonstrated much later (Chapter III-6.2.3).

II - THE BIRTH OF THE EXPERIMENTAL METHOD

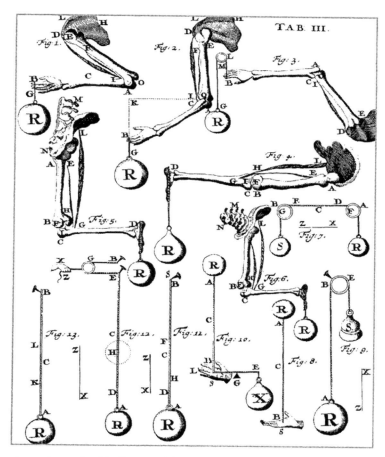

The drawings shown in this figure, which are designed to show an analogy with devices found in industrial mechanics (winch, pulley, lever), were in agreement with the Man-machine philosophy that was in vogue at this time.

Figure II.10 - A mechanical vision of the movements of the arm
(from Giovanni Alfonso BORELLI - *De motu animalium*, Rome, 1680.
Liège University Library, with permission)

The story of VAN HELMONT's willow experiment is typical of the inescapable fit that occurs between the formation of new ideas and the spirit of the times in which they are formed. Because of this, the experimental method progresses as a function of the chopping and changing of the boundary between the originality and prowess of an experiment, on the one hand, and the anchoring of modes of thought that are embedded in tradition, on the other hand. Only four centuries ago, the theory of the four elements, water, earth, air and fire, a theory dating from the time of the natural philosophers of Ancient Greece, was still extant. In order to interpret experimental results relating to the growth of the willow tree, VAN HELMONT could only really have recourse to the theory of the four elements that was accepted and taught in his time.

3.4. FIRST CONCEPTUAL CONTROVERSIES CONCERNING THE USE OF THE EXPERIMENTAL METHOD IN LIVING BEINGS

"Experimentation involves a methodical interrogation of Nature, which presupposes a language to formulate the questions and a dictionary to read and interpret the answers."

Alexandre KOYRÉ
Studies in the History of Scientific Thought - 1966

From its beginnings, the experimental method led to discoveries that disturbed traditional beliefs. REDI's refutation of the theory of spontaneous generation, TREMBLEY's demonstration of the regeneration of the hydra and the refutation of the role of the *pneuma* as the cause of muscle contraction are typical examples of discoveries that fed impassioned controversies concerning the very nature of the experimentation carried out.

3.4.1. Refuting the theory of spontaneous generation

Up until Renaissance, the idea of **spontaneous generation** was accepted without question. It was believed that maggots and flies arose spontaneously from the rotting of meat. This belief was so well-anchored that it took years for people to be convinced that it was untrue. By means of patient observations, Ulisse ALDROVANDI (1522 - 1605), who was a doctor and zoologist at the University of Bologna, noted that maggots developed on meat on which flies had landed.

F. REDI
(1626 - 1698)

The final death-blow for the theory of the spontaneous generation of flies was given by the Italian naturalist Francisco REDI (1626 - 1698). Like ALDROVANDI, REDI observed that a piece of meat left in the open air attracted flies and that after a few days the surface of the meat was crawling with maggots. Some time later a swarm of flies escaped from the meat. This suggested that there was a sequence of interconnected events; flies laid eggs on the meat and these eggs developed into maggots that then metamorphosed into flies. In order to be sure, REDI moved on to experimentation. In two parallel series of jars he put a grass snake, some fish, four eels and a piece of veal. The jars in the first series were closed with parchment. The others were left open. In a few days, the open jars were filled with maggots that gave rise to flies. The closed jars remained intact. An objection was made that the parchment that covered the closed jars to prevent flies from entering had prevented the entry of air. REDI replaced the parchment with very finely-woven cloth, known as "Naples cloth". Here again, no maggots or flies appeared. This experiment, which involved a **comparative control test**, acting here as a **cross-check**, is a typical example of how the experimental method, in all its elegant simplicity, could be applied to the solution of a problem, in this case the spontaneous generation of insects, a problem that had for centuries been the subject of sterile discussion.

The quarrel concerning spontaneous generation would come to the fore again in the 18th century. If it could be accepted that a fly was born of a fly and not from a piece of meat, was it not possible to imagine, on the other hand, that microscopic beings with mysterious functions that could not be observed directly, but only with the use of a microscope, might arise spontaneously from broths of meat or infusions of hay? The possibility that microscopic agents might be the cause of infectious diseases had been put forward by Girolamo FRASCATORO (1478 - 1553), a doctor in Bologna, in a treatise on infectious diseases. However, at this time, such a concept was pure hypothesis. In the middle of the 18th century, Georges BUFFON put forward the idea that animals and plants were made up of an assembly of thousands of infinitely small organic components that serve as molds for new components. The question arose as to whether the very small, mobile figures that the microscopes of this period were able to distinguish were living creatures or assemblies of organic molecules with the potential to become living beings. The English priest and naturalist John Turbeville NEEDHAM (1713 - 1781), encouraged by BUFFON, worked on this problem. NEEDHAM prepared a "very hot" mutton broth that was poured into a flask that had previously been heated with hot cinders. The flask was closed with a "well-sealed" cork stopper. After four days, the flask was full of microscopic organisms. This experiment was repeated with around sixty different decoctions of animal and plant material. In every case, after a few days, the decoction was swarming with microorganisms. NEEDHAM became convinced that plant matter had generated creatures of the animal world by means of a vegetative force. Thus, right in the middle of the 18th century, the idea of spontaneous generation was experiencing a rebirth. The Swiss physiologist Charles BONNET cast doubts on the methodological rigor of NEEDHAM's experiments. Following advice from BONNET, SPALLANZANI repeated these experiments in 1765, taking precautions with respect to heat sterilization that NEEDHAM had no doubt failed to take. "After several days," wrote SPALLANZANI, "I did not find the slightest sign of any animalcules." Nevertheless, he did note that certain microorganisms are killed by the heat in just a few minutes, while, for others, boiling for nearly an hour was necessary. The argument would bounce back again in the 19th century. Louis PASTEUR (1822 - 1895) put an end to it (Chapter III-2.2.2). With the progress made by bacteriology in the 19th century, it came to be recognized that the stumbling block for experimentation into the spontaneous generation of microorganisms was a lack of knowledge of the fact that, during a bacterial cycle, encapsulated living particles, endospores, which are resistant to being heated to 100°C, may appear. This discovery, which was made by the German botanist and microbiologist Ferdinand COHN (1828 - 1898), led to a modification of the process of sterilization by heat. Instead of sterilizing a medium by heating it once up to 100°C, the Irish physicist John TYNDALL (1820 - 1893) recommended the intermittent heating of organic media, a technique that was given the name **Tyndallisation**. After a few minutes at 100°C, which would kill thermosensitive bacteria, heating was interrupted. After the medium had cooled to room temperature, thermoresistant bacterial spores would germinate, giving rise to thermosensitive bacteria that could be destroyed by heating the medium again to 100°C.

3.4.2. The enigma of the regeneration of the hydra

In 1740 the Swiss naturalist Abraham TREMBLEY (or TREMBLAY) (1700 - 1784) made an accidental discovery concerning the freshwater hydra and its unsuspected ability to regenerate. TREMBLEY was interested in aquatic plants and grew them in a bowl filled with pond water. He observed that green multitentacled structures developed on these plants. These structures were mobile, with arms that stretched and then retracted, and that they were morphologically similar to the roots of plants (Figure II.11). He likened these structures to plant material, and to confirm this idea, he tried to make them reproduce by taking cuttings, a procedure for reproducing plants that was well-known at this time. He cut a hydra along its length into two halves, which he placed in a flask filled with water. Two weeks later, buds "resembling horns" appeared where the cuts were made. These buds grew and formed arms. In March 1741, TREMBLEY sought the advice of RÉAUMUR. In agreement with Bernard DE JUSSIEU, RÉAUMUR thought that these mobile organisms were part of the animal kingdom. He thought that they should be called polyps "because their horns appear to be analogous to the arms of the sea animal that bears this name." These small-sized polyps were later named **hydra**. TREMBLEY continued his experimental work in a systematic manner; single or multiple transverse sections of the hydras, longitudinal sections, cutting into small pieces. In his *Memoire to Serve as a History of a Type of Fresh-Water Polyp with Horn-shaped Arms* (1744), he wrote that the small pieces, both those that had arms and those that did not, became perfect polyps. After this came grafting experiments, which involved bringing together either two halves of the same hydra (autograft), or bringing together the head of one hydra and the posterior part of another (homograft). In all cases, these experiments were successful. This regeneration of the hydra was not an isolated case. It was reproduced by Charles BONNET with multiple sections of earthworms. In 1768, SPALLANZANI showed that a vertebrate, the newt, also had regenerative capabilities. After amputation, the tail, the feet and even the jaws **regenerated**, to become just like the original part that had been removed. Although these tests gave spectacular results, they remained anecdotal up until the discovery, in the 20th century, of **stem cells**, and the demonstration of their potential for developing into specific tissues (Chapter IV-2.3.1). We now know that after amputation of the foot of a newt, cells located under the epithelium, close to the wound, divide and differentiate into all the types of cell involved in making a new foot.

The mechanism for this phenomenon remained an enigma for many years. Current studies concerning the operation of stem cells have opened up promising prospects within the framework of regenerative medicine.

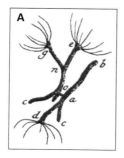
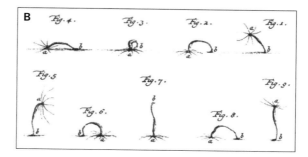
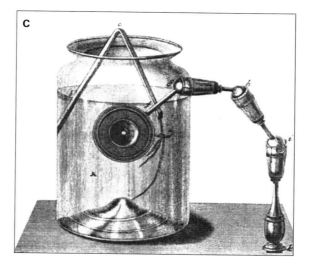

A - Artificial production of hydras obtained by transverse section or longitudinal cutting of the body of the hydra or by grafting. The monstrous hydra obtain after sectioning or cutting, represented here, has seven heads.

B - Movement of a hydra on a solid surface.

C - Apparatus with which TREMBLEY carried out his observations on the hydra. The hydras contained in a bowl filled with water were examined through the glass side of the bowl using a magnifying glass mounted on an articulated arm. It was under these conditions that the sections were carried out on the hydras and that the regeneration of the sectioned hydras was observed.

Two centuries later it would be known that the phenomenon of regeneration is determined by the differentiation of stem cells.

Figure II.11 - The regeneration of hydras
(from Abraham TREMBLEY - *Memoires concerning the History of Polyps*, 1744)

3.4.3. Refusing the idea of the pneuma as the agent of muscle contraction - Birth of the idea of the reflex

In keeping with experiments that had led to the refutation of the theory of the spontaneous generation of flies, and that had demonstrated the phenomenon of tissue regeneration in small animals, at around the same time, other experiments were carried out to investigate the mechanism of muscle contraction and that of nerve impulses. As the number of experimentally determined facts grew, erroneous beliefs were gradually eliminated, over the years, until a rational explanation could be put forward. GALEN had shown that ligaturing a nerve paralyzes the muscle that it innervates. Since then, people had been taught that a nerve, when excited, brought a fluid, the *pneuma*, to the muscle that it innervated, making the muscle contract. In his *Treatise on Man* (1664), DESCARTES likened the *pneuma* to "animal spirits contained in the cavities of the brain, which escape *via* pores. These spirits," he said, "have the force to change the figures of the muscles into which they are inserted, and, in this way, to mke the limbs move." DESCARTES compared muscles activated by these "**animal spirits**" with "musical fountains of the Royal gardens, where the force of the water is sufficient to make the different machines move and play the instruments, depending on the arrangement of the different pipes that carry it." Thus, the "spirits" that circulate inside the nerves are compared to the water that flows inside the pipes: the muscles that contract have their equivalent in the machines of the fountains. This reasoning led to the erroneous conclusion that a muscle increases in volume during its contraction, due to the entry of "animal spirits" into its structure.

The term "**reflex**" does not exist in DESCARTES' *Treatise on Man*. Nevertheless, it is an obvious fact that when someone approaches his or her foot to a flame, he or she feels a sensation of heat and rapidly pulls the foot away from the flame (Figure II.12). In order to explain this sudden removal, DESCARTES imagined a nerve net that carried information from the foot to the brain, the latter ordering the "animal spirits" that it held to return along the same nerve net in order to "swell" the muscles of the leg and allow the leg to move away from the fire.

A very simple experiment carried out by the Dutch naturalist Jan SWAMMERDAM, concerning the contraction of an isolated frog muscle in response to stimulation of the nerve that innervates it, was enough to invalidate the theory of "animal spirits" that supposedly cause the muscle to swell. SWAMMERDAM immersed an isolated frog muscle into a small bowl filled with water, the level of which he measured. When it was excited, the muscle changed shape, but not volume, this result being checked by the absence of any variation in the level of water in the bowl. During the same period, WILLIS used the metaphor of an explosion to account for the spread of nerve impulses from an irritated area, and also the metaphor of the "recoil" after a bullet is fired in order to explain the return of nerve impulses to the same zone. He used the term *"Motus reflexus"* for the first time. The brain was still considered to be the giver of orders. During the 18th century, experiments were carried out on decapitated animals, frogs, rabbits and dogs. The Briton Robert WHYTT (1714 - 1766) and the German August UNZER (1727 - 1799) demonstrated that, in addition to intentional mechanisms that involved cerebral activity, there were **automatic mechanisms** that were independent of the brain. Thus, little by little, the idea took shape that, situated in the spinal chord, and independent of the brain, there were nerve centers that were responsible for automatic actions, which were later named **reflex centers and actions**. In 1784, an anatomist at the University of Prague, Georg PROCHASKA (1749 - 1820) suggested that there are, in fact, structures in the spinal chord that, having received an impulse brought from the periphery by the sense nerves, reflected this impulse *via* remote motor nerves, the muscles responding to this impulse by contracting. This is how the concept of the reflex center came into being. This concept would be established experimentally during the 19th century following experiments carried out by Charles BELL and François MAGENDIE, which would determine the motor and sense functions of the anterior and posterior roots of the nerves of the spinal column (Chapter III-1).

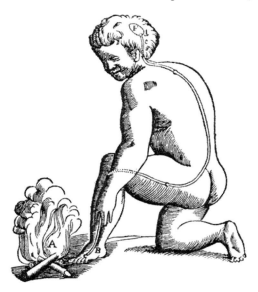

This image from the *Treatise on Man* describes how DESCARTES explains the reflex movement that causes an individual to remove his foot (B) when it is exposed to a flame (A). The skin of the foot perceives a sensation of heat, which is communicated to a hollowed out nerve, through which it is assumed that "spirits" circulate. This is followed by a stretching out of the nerve, which leads to the opening of a pore situated at its end, in the brain (DESCARTES compares this arrangement with the ringing of a bell that is suspended from a cord that is pulled. The opening of the pore allows the animal spirits contained in space (F) in the brain to rush into the nerve and to stimulate muscles at the other end, the contraction of these muscles causing the foot to be moved away from the fire.

Figure II.12 - Theory of the reflex movement according to DESCARTES
(from René DESCARTES - *Treatise on Man*, 1664)

4. THE EXPERIMENTAL METHOD AND ITS IMPACT ON THE PHYSICAL SCIENCES IN THE 17TH CENTURY

"The Galilean and Cartesian revolution was prepared for by a long effort of thought. There is nothing more interesting, more instructive or more fascinating than the history of this effort, the history of human thought obstinately wrestling with the same eternal problems, encountering the same difficulties, struggling, without a respite, against the same obstacles, and slowly but surely forging the instruments and tools, the new concepts, the new ways of thinking that would enable it to overcome them all."

Alexandre KOYRÉ
Studies in the History of Scientific Thought - 1966

In the 7th century, both the **physical sciences** and astronomy developed in a manner that was almost parallel to that of the development of the **life sciences**, the latter being exemplified in particular by the discovery of the circulation of the blood, but also by an increasingly critical questioning of beliefs inherited from medieval scholasticism, such as spontaneous generation or the theory of natural, vital and animal spirits. Such parallel development is not surprising because, at this time, ideas circulated easily between the different areas of science. Knowledge had not yet been compartmentalized. A naturalist was able to have pertinent discussions with a mathematician, a physicist or a chemist. Naturalists and physicists analyzed the phenomena of living beings and inanimate matter according to the same logic, based on the simple principles of mechanics. If DESCARTES referred to fountains, clocks and levers to explain the circulation of the blood, heartbeats or the movement of arms and legs, this was because the mechanics of inanimate matter had begun to be understood and even mathematized, and it was highly tempting to compare its operation with that of living beings.

The new physics looked again at questions asked by the philosophers of Ancient Greece concerning the structure of the Universe, the nature of movement, and the existence of the vacuum. Bringing together a reasoning that was based on a logic stripped of any subjectivity and an experimentation that was made all the more rigorous by the use of accurate measurement instruments, 17th century science would revolutionize many of the concepts that had up until then been taught as a credo. This revolution began with a series of discoveries that were rapidly formalized into laws: KEPLER's (1571 - 1630) laws concerning the elliptical orbits of the planets, which refuted PTOLEMY's geocentrism, GALILEO's laws concerning movement, which opposed ARISTOTLE's theories, and which forecast the principle of inertia clarified some time later by DESCARTES and NEWTON (1642 - 1727), BOYLE-MARIOTTE's law concerning the relationship between gas volume and pressure, and the demonstration of the existence of the vacuum by BOYLE and PASCAL.

4.1. A NEW THEORY OF THE COSMOS

In 1543, the year he died, COPERNICUS' major work *De revolutionibus orbitum cælestium* appeared. At this point, medieval astronomy could be summed up according to the philosophy inherited from ARISTOTLE, a doctrine that, in agreement with common sense, claimed that the Earth remained immobile while the Sun followed a circular orbit around it in a sky filled with fixed stars and moving planets. After COPERNICUS, the terrestrial globe was no longer at the center of the Universe. The Sun took the place of the Earth, which rotated on its own axis and around the Sun, with a one-day-long complete rotation on its axis, and a one-year-long rotation in a circle around the Sun. The Earth was no more than a simple planet, the same as Mars, Venus, Mercury, Jupiter and Saturn, which also had rotational movements. As for the stars, which ARISTOTLE and PTOLEMY had considered to be fixed, COPERNICUS recognized slow variations in the positions of the stars, which he explained as a slow conical oscillation of the Earth. For Copernicus, circular movement, which both ARISTOTLE and PTOLEMY had considered to be the perfect movement, remained the basis for the movement of the planets. The Universe was still a closed space, centered on the Sun. However, was this Universe, on which theological belief imposed limits, an illusion of the senses and of thought? A preacher, philosopher and theologian, Filippo BRUNO (brother GIORDANO) (1548 - 1600), while agreeing with the heliocentric system put forward by COPERNICUS, denied its finiteness. In *De l'Infinito* (1584), BRUNO rejected the idea of the sphere of fixed astral bodies, the stars, and pushed to infinity the limits of a Universe for which God remained the guarantor of unity. The Aristotlian points of reference (up and down, front and back, etc.) no longer made any sense in a cosmic immensity containing an infinity of stars that were as many suns around which planets rotated. Warned of this audacity, which made a mockery of the Scriptures, the Inquisition took action in 1592. BRUNO did not recant. On February 17th, 1600, he was burned alive, in Rome, on the orders of the Holy Office.

More than half a century after the death of COPERNICUS, Johannes KEPLER wrote of his revolutionary ideas concerning the elliptical orbits of the planets in *Astronomia nova* (1609). In 1600, KEPLER had been lucky enough to be associated with the work of Tycho BRAHE (1546 - 1601), a Danish astronomer, who had just settled in Prague. At the beginning of his career in Denmark, Tycho BRAHE had been helped by the support of King FREDERIK II (1534 - 1588), who had an observatory constructed for him on the island of Hveen. This observatory was equipped, at great expense, with the most highly developed instruments of the period, including armillary spheres, quadrants, sextants and a theodolite. Over approximately twenty years, using observations made with the naked eye, Tycho BRAHE drew up a detailed catalogue of nearly one thousand stars, and recorded the movements of the planets with an astonishing accuracy. Nevertheless, PTOLEMY's geocentric system remained in vogue. When FREDERIK II died, Tycho BRAHE's financial difficulties began, and he was forced to exile himself to Prague, in Bohemia, under the protection of the emperor RUDOLF II. It was here that KEPLER joined him. Tycho BRAHE had been unequaled as an observer, with respect to the accuracy of his innumerable measurements. Unfortunately, he did not manage to produce the expected theoretical fruits of his labor.

Making use of the accumulated observations in the laboratory workbooks that Tycho BRAHE had left to him, KEPLER analyzed the movement of the planet Mars around the Sun, within the framework of Copernican theory. His calculations led him to the conclusion that **the orbit described is an ellipse** for which the Sun is a focus, and that the area of the ellipse that is swept through by the vector radius of a planet in motion is proportional to the time taken for it to sweep through. This gave rise to KEPLER's first two laws, which were extended to include the other planets. According to the third law, the ratio of the squares of the periods of revolution of the various planets around the Sun to the cubes of their average distances from the Sun is the same for every one of the planets. A few decades later, Isaac NEWTON would show that attraction is a universal phenomenon, and if there is attraction between the Sun and the planets, there is also attraction between the planets.

The eclecticism of the scholars of this epoch is illustrated by KEPLER's work on the eye and on sight. KEPLER understood that the eye operates according to the laws of physical optics, the crystalline lens causing light rays to converge on the retina, ideas that were developed further by DESCARTES in his *Treatise on Dioptrics*.

In 1609, date of the publication of KEPLER's first two laws, GALILEO pointed his astronomical telescope towards the heavens. He observed that there are craters on the moon as there are on Earth, and he noted that there are satellite planets revolving around Jupiter. In 1610, GALILEO mentioned what he referred to as the phases of Venus, corresponding to variations in brilliancy that could be explained by the positions occupied by the planet during its trajectory around the Sun. In 1613, he observed dark marks on the Sun. Moreover, these marks moved from day to day, a movement that would later be explained by the slow revolution of the sun on its axis, but which at the time was considered to be a challenge to Aristotlian dogma concerning the absolute perfection of celestial bodies. GALILEO observed the amazing multitude of stars in the Milky Way. In 1632, he published a *Dialogue Concerning the Two Greatest Systems in the World*, i.e., the geocentrism of PTOLEMY and the heliocentrism of COPERNICUS, showing his support, of course, for COPERNICUS. This work was written in the form of a Dialogue between three accomplices, SALVIATI [1], in the role of COPERNICUS, who acts as spokesperson for GALILEO, SIMPLICIO, a defender of ARISTOTLE, and SAGREDO [1], an enlightened amateur who wishes to increase his knowledge. The dialogue occurs over four days. The first day is

---

1 Two of the people involved in GALILEO's discussion, SALVIATI and SAGREDO, were members of his circle. Giovanni Francesco SAGREDO (1571 - 1620) had been a consul of Venice and Filippo SALVIATI (1582 - 1614) had been GALILEO's pupil at the University of Padua, before going on to a diplomatic career.

devoted to refuting the Aristotlian division between the terrestrial and celestial worlds. The second day is devoted to an analysis of the fall of an object from the top of a tower, an argument used by SIMPLICIO in favor of the notion of the Earth's immobility, but an argument considered by SALVIATI to be fallacious because the stone that is dropped is following the same movement as that being carried out by the tower, that is to say, a circular movement around the center of the Earth. The third and fourth days are devoted to many proofs of an astronomical nature that favor the existence of a heliocentric system. This *Dialogue* aroused the reprobation of pope URBAN VIII (1568 - 1644). It unleashed the furies of the Inquisition and its sale was banned. The reassuring ideas of the thinkers ancient times, concerning a closed, spherical Universe, had been shaken up by COPERNICUS, Giordano BRUNO had dared to look towards the infinite, and KEPLER had striven to mathematize the heliocentric system, but nothing irremediable had been achieved. However, GALILEO's astronomical telescope had provided irrefutable proof that the Earth is a planet, lost, with others, in the immensity of the cosmos, and that the Sun is far from being a perfect astral being. On June 22nd, 1633, having been ordered by the Inquisition to recant, GALILEO admitted that in his writings, through sheer vanity, he had allowed himself to formulate hypotheses that were too bold and unconvincing. This renunciation prevented him from being burnt at the stake, but it did not prevent him from being imprisoned for life, although this took the milder form of his living under house arrest near Florence.

4.2. A NEW THEORY OF MOVEMENT

"What required greater genius was to look for and discover in the most vulgar phenomena, the falling of a stone, the swinging of a suspended lamp, that which so many philosophers, so many doctors, so many thinkers about divine and human things, had had in front of their eyes for thousands of years, without even dreaming that there was something there to look for and to discover."

Antoine Augustin COURNOT
Considerations on the Progress of Ideas and Events in Modern Times - 1872

GALILEO
(1564 - 1642)

In the home where GALILEO lived after his trial in 1633, he looked once more at his old notes, in which he had recorded the results of experiments on movement. He had carried out these experiments in Pisa from 1589 to 1591, and then in Padua for more than ten years, starting in 1592, well before he became interested in astronomy. He drew conclusions and postulated theories that would be set out in the *Discorsi* published in Leyden in 1638: *Discourses and Demonstrations on Two New Sciences*. These sciences referred to the resistance of materials and to uniform rectilinear and uniformly accelerated movements. The theories concerning movement were revolutionary, and they made GALILEO the founder of a new mechanical science, the **science of movement**. Here again, the break with ARISTOTLE is obvious. ARISTOTLE had noted that a projectile propelled through the air by a motive agent continued to move after being separated from it. Trusting to his perception of natural phenomena *via* the senses, he deduced that to move necessitated being driven, and thus the continuous presence of a motor. The argument continued with the postulate that the air in which the projectile is moving takes the place of the projectile and acts as this motor (Chapter I-1.2). With GALILEO, an object that was projected into the air and continued its movement is in a state that is indifferent to the movement: there is no need to conceive a virtual motor in order to explain the movement.

When thinking about a body falling from a certain height, ARISTOTLE had observed that the fall became more and more rapid. He said that this was because the body was being called by its **natural place**, i.e., the center of the Earth (Chapter I-1.2). GALILEO analyzed the **accelerated movement** of the fall of heavy bodies in **mathematical terms** and, for the first time, he established a **quantitative relationship** between time and the position of the heavy body in its fall. The experimental model used by GALILEO was the **inclined plane**, a useful model that made it possible to measure the kinetic properties of the slowed movement of a small bronze ball. The hypothesis was made that the laws governing the movement of the ball along the inclined plane were the same as those governing its movement in vertical free fall towards the ground, the latter movement being much more difficult to analyze. The inclined plane comprised a piece of wooden molding in which a channel had been cut, the channel being lined with tanned and polished parchment. The bronze ball was released at the upper end of this channel. The resistance offered by the friction of the ball against the parchment was considered to be negligible. The time taken by the ball to travel a certain distance was measured using a water clock known as a **clepsydra**. This was a receptacle that had a small hole drilled in its lower half, this hole being closed with a stopper. The receptacle was filled with water. The hole was opened at the moment when the ball was released. The water that flowed out was collected and weighed. The weights measured were assumed to be proportional to the time taken by the ball to travel through spaces of well-determined lengths, which could, given certain approximations, be considered as an acceptable criterion. GALILEO showed that "the speeds acquired by a 'weight' traveling down planes that have been subjected to different degrees of inclination are equal when the heights of the drop with respect to the vertical are equal." The acceleration of a mobile body moving along an inclined plane decreases with the angle of inclination of the plane, which makes measurement easier. Based on these experiments, GALILEO put forward his **law of uniformly accelerated movement**, i.e., "the spaces described by a body falling from rest with a uniformly accelerated motion are to each other as the squares of the time-intervals employed in traversing these distances." Certain science historians, particularly Alexandre KOYRÉ (1902 - 1964), have doubted the authenticity of the results relating to the inclined plane, finding their accuracy astonishing given the rustic nature of the methodology that was used. Others have credited GALILEO with good faith. What emerges from this confrontation of opinions is that experiments were carried out by GALILEO and a law was deduced from a reasonable approximation of the results. What is interesting in the inclined plane experiment is the thought processes that preceded it, with the design of **original apparatus** that could be handled easily and was able to provide the key to an enigma that the simple observation of the free fall of a body, which occurs much too quickly, was unable to solve.

Another fundamental aspect of the *Discorsi* of 1638 was the discussion of the movement of a mobile body thrown along a horizontal plane, in the absence of any resistance. GALILEO postulated that this body would continue its uniform movement if the area of the plane was infinite. The word **"inertia"** was not pronounced, but the concept is inherent to the postulate. What happens to the mobile body if the plane is interrupted? GALILEO showed that its fall described a parabolic trajectory that is the resultant of two movements, a horizontal movement linked to the residual inertia inherent to the ball, which gradually disappears, and a vertical movement linked to the attraction of gravity. Such a parabolic movement for the trajectory of a projectile was suggested during the same period by LEONARDO DA VINCI (1452 - 1519) and the mathematician Niccolò TARTAGLIA (1499 - 1557). There is a noticeable difference between these ideas and those of ARISTOTLE, for whom a projectile, once propelled, traveled first along a rectilinear trajectory that was suddenly interrupted by its vertical drop (Chapter I-1.2). These considerations would be taken up again by René DESCARTES, and then by Christian HUYGENS (1629 - 1695). Isaac NEWTON synthesized the work done on the physics of movement by GALILEO, DESCARTES and HUYGENS in his *Philosophiae naturalis principia mathematica* (1687), a treatise on rational mechanics, the laws of which mathematized Nature and the Universe. NEWTON stated clearly that in the absence of an outside agent, a body animated with a uniform rectilinear movement will maintain this state of movement. This is the **principle of inertia**. Scholars in China had given a premonition of this principle from the 5th century BC. In the *Mo Jing* written by MO ZI (480 - 390 BC), it was said that the movement of a mobile body ceases if an opposing force is applied, but if there is no opposing force, the mobile body continues along its path without changing trajectory. The Newtonian universe would manage to impose its mathematical logic with the law of **universal gravitation**, according to which bodies attract one another with a force that is proportional to the product of their masses and inversely proportional to the square of the distance between them. The simplicity of this law and its universality – forces of attraction operate everywhere – swept away a whole range of superfluous hypotheses.

4.3. PROOF OF THE EXIXTENCE OF THE VACUUM

Renaissance hydraulic engineers, who, in the gardens of Florence, collected water from deep sources by means of suction pumps and pipes, observed that below a depth of 32 to 33 feet (around ten meters) the water would no longer reach the surface. In order to explain this maximum depth limit for the lifting of water, various conjectures were made, in which the old concept of "abhorrence of a vacuum" involved in the theory of plenism, once again raised its head. Evangelista TORRICELLI (1608 - 1647), a disciple of GALILEO, tried to explain the phenomenon by postulating an equilibrium between the column of water sucked up by the pump and the mass of air that was outside the column of water and which exerted a pressure at the base of this column. He thought that the lifting of the water reached its limit when the weight of the column of water was equal to the weight of atmospheric air. It was almost impossible to check this hypothesis in the field. Because the hypothesis implied a relationship between the weight of water and that of the air, TORRICELLI conceived an experiment in which water was replaced by a much heavier body of liquid. He chose mercury (which at this time was called quicksilver), the density of which is approximately thirteen and one half times that of water. The experiment was carried out in 1643, with the help of his assistant Vincenzo VIVIANI (1622 - 1703). A one-meter long glass tube, closed at one end, was filled with mercury, and then upended into a basin that was also filled with mercury. The column of mercury in the glass tube dropped a little, and stabilized at approximately 76 cm above the level of mercury present in the basin, leaving an unfilled space at the top of the glass tube, a result that was in agreement with the starting hypothesis. The experiment was repeated with tubes of more than one meter in length. In every case, the column of mercury stabilized at a height of 76 cm. Why didn't the column of mercury in the upended tube flow out completely into the basin? As TORRICELLI had predicted, this was because the weight of the air on the surface of the mercury in the basin counterbalanced the weight of the mercury in the glass tube. However, what was contained in the space that was left free above the column of mercury in the glass tube? A vacuum? Mercury vapor? Or even air that might have been able to infiltrate the space through the column of mercury? There was even the suggestion that there might be a "funicule" connecting the column of mercury to the top of the tube. As illustrious a thinker as DESCARTES raised doubts about the existence of a vacuum, postulating that bodies were made up of a network of particles between which circulated a "subtle matter". These different opinions sparked passionate debate between partisans of the vacuum and their opponents. TORRICELLI, a prudent man, did not take part in these arguments. His piece of apparatus, which was afterwards adapted to measure air pressure, was named the **barometer**.

In 1644, the Minim friar Marin MERSENNE (1588 - 1648), who was a physicist and divulger of the science of his time, recounted TORRICELLI's experiment in France, arousing curiosity and interest and causing the debate between **vacuists** and **plenists** to come to the fore once again.

In 1646, Blaise PASCAL, who was 23 years old, and a physicist, Pierre PETIT, reproduced TORRICELLI's experiment. They then repeated it with numerous variations.

*B. Pascal
(1623 - 1662)*

At the end of 1647, PASCAL asked Florin PÉRIER, his brother-in-law, to organize the famous Puy-de-Dôme experiment, which was carried out on September 19th 1648. The results were published the following month in an *Account of the Great Experiment on the Equilibrium of Liquors*. In PASCAL's mind, the aim was to check the effect of altitude and, as a consequence, of the weight of air, which is necessarily lower at higher altitudes. Comparative measurements carried out at the summit and at the foot of the Puy-de-Dôme volcano in France showed that the level of mercury in TORRICELLI's apparatus, at the summit, was slightly, but significantly, lower than the level observed at the foot of the volcano, a result in accordance with the idea that air has weight, but less weight at the summit of a mountain than on a plain below. That which PASCAL had predicted in a letter sent to PÉRIER was confirmed: "If it is found that the height of the quicksilver (mercury) is less at the top than at the bottom of the mountain," he wrote, "it will necessarily follow that the weight and pressure of the air is the sole cause of the suspension of the quicksilver (in TORRICELLI's apparatus), and not the horror of a vacuum, because it is certain that there is far more air that weighs down on the foot of a mountain than on its summit, but it is not possible to say that nature abhors a vacuum more at the foot of a mountain than at its summit." From the Puy-de-Dôme experiment, it was possible to deduce that atmospheric air exerts pressure on all terrestrial bodies.

In his *Treatises on the Equilibrium of Liquors and the Weight of the Air*, PASCAL gave a simple explanation of how air enters the lungs during respiration: "when we breathe, air enters the lungs [...] because [...], pushed by the weight of its whole mass, it enters and falls due to the natural and necessary action of its weight, which is so intelligible, so easy and so naive an explanation that it is strange that anyone went looking for the horror of the vacuum, supernatural qualities and other such distant or chimerical causes to account for it." In short, the living organism reacts to the weight of the air in the same way as the column of mercury in TORRICELLI's apparatus, which is an example of the mechanistic idea of the physiology of living beings that was born at around this time.

In 1654, the German physicist and engineer Otto VON GUERICKE (1602 - 1686), who was mayor of Magdeburg, developed a **vacuum pump that was able to pump the air out of an enclosed space**, i.e., to create a vacuum in this space. He obtained the enclosed space by sticking together two hollow bronze hemispheres, each with a diameter of 24 cm, in order to obtain a hollow sphere. The inside of the sphere was connected to the vacuum pump by means of a pipe. After the air had been extracted, it was observed that it had become impossible to separate the two hemispheres unless a considerable force was applied, requiring the combined pulling force of teams of more than a dozen horses. This experiment, which was carried out at Regensburg, confirmed the theory that air has weight, and demonstrated the power of air pressure on objects at ground level. Nevertheless, VON GUERICKE's experiment did not provide an answer to the question of whether or not a vacuum modifies physical phenomena that are habitually observed under normal conditions of atmospheric pressure. The answer to this question was provided by the Irish physicist Robert BOYLE. He was assisted by Robert HOOKE, who would later become illustrious for his microscopy experiments. In 1659, BOYLE had a spherical glass receptacle of around thirty liters in volume made. This receptacle had openings that could be closed with valves. It was connected to a highly-perfected vacuum pump (air pump), which was equipped with a crank-operated rack system for pumping the air (Figure II.13).

The vacuum pump included a glass flask with a capacity of approximately thirty liters, the upper part of which could be sealed, connected by its lower part to the body of the pump. A rack system made it possible to pump out the air contained in the glass flask and to create a vacuum.

Figure II.13 - BOYLE's first air pump (1658) (from Robert BOYLE - *New Experiments Physico-Mechanicall, touching the Spring of the Air, and its Effects*, Oxford, 1660)

Experiments were carried out over approximately ten years. BOYLE checked what became of the column of mercury of a barometer-type instrument when it was placed in the vacuum pump and a vacuum was formed: the level of mercury in the glass column dropped, showing that the pressure of atmospheric air was no longer being exerted. Thus, the space left free in the glass column, which was in equilibrium with the vacuum of the vacuum pump, could only be empty. This "**vacuum within the vacuum**" experiment should have brought an end to the objections of the plenists, but this does not take into account the polemical relentlessness of the philosopher Thomas HOBBES (1588 - 1679), who, in his *Dialogus Physicus de Natura Aeris* (1661) remarked that the reliability of BOYLE's experimental devices was open to dispute and that the conclusions that were drawn from the experiments were based on hypotheses that could be called into question. HOBBES' criticisms, and the plenist doctrine that he supported, finally collapsed in the face of experimental evidence, particularly when BOYLE's results were confirmed by the Dutch physicist Christian HUYGENS. BOYLE's experiments on the vacuum (43 in all) where published from 1660.

Some of these 43 experiments touched on **animal physiology**. BOYLE observed that animals (mice, birds) that were enclosed in the body of the vacuum machine died of suffocation within a few minutes of when the vacuum was created. He noted that a lit candle goes out in a vacuum, and he asked himself whether there might be a relationship between respiration and combustion, an idea that would be taken up at a later date by LAVOISIER (Chapter III-5.2). From all of these results BOYLE concluded that part of the air is essential to respiration and to life. It would not be until the development of the chemistry of gases in the 18th century that the fraction of atmospheric air that is essential to respiration would be identified as oxygen. BOYLE was not only a great physicist; he was also an inspired chemist. He was responsible for the preparation of a large number of mineral compounds such as copper chloride, copper sulfate and tin chloride, as well as the discovery of natural colored indicators such as cochineal carmine, sunflower dye and campeachy tree wood, the colors of which are modified by the addition of an acid or a base.

Proving the existence of the vacuum by the fact that it could be created brought an end to the quarrel between vacuists and plenists. This gave rise to a new outlook on physical phenomena beyond the terrestrial atmosphere. The applications to animal physiology were a prelude to research on "breathable air" and the identification of its chemical nature.

4.4. TOWARDS OTHER REVELATIONS CONCERNING THE INANIMATE WORLD

Since ancient times, it had been known that if amber is rubbed with wool it acquires the power to attract certain light bodies such as wisps of straw. William GILBERT (1540 - 1603), who was physician to England's queen ELIZABETH I (1533 - 1603), became interested in this phenomenon and qualified the force of attraction as an **electrical force** (from the Greek ἤλεκτρον = amber). His experiments on the "lodestone", which spontaneously attracts iron filings, are a reminder of those of an inspired genius, Pierre de MARICOURT, in the 13th century. They confirmed that this was a phenomenon that differed from the production of electricity obtained by rubbing amber. GILBERT went further. He saw the terrestrial globe as a vast magnet, the magnetic field produced by this magnet being responsible for the movement of the needle of a compass. The idea of **terrestrial magnetism** is clearly brought up. Nevertheless, the physics of materials was still in its infancy, and the difference between the animate and the inanimate was so slender that GILBERT likened terrestrial magnetism to the soul of the Earth.

The **first electrical machine** capable of producing electricity by simple friction was developed by Otto VON GUERICKE, the same person who had invented the vacuum pump (Chapter II-4.3). VON GUERICKE's electrical machine consisted of a sulfur sphere fixed to an axis that was rotated by hand while the other hand pressed down on the sphere, acting as a friction strip. This first friction electrical machine was followed by others that were much more developed. At the beginning of the 18th century, Stephen GRAY (1670 - 1736) in England and Charles François DU FAY (1698 - 1739) in France studied the properties of electricity generated by friction. DU FAY perfected a small instrument for measuring the electrification of a body, a forerunner of the electroscope, a type of small pendulum made up of a light conducting ball suspended from a wire. Later, it was realized that chemical reactions accompanied electrical phenomena. In 1785, Charles-Augustin DE COULOMB (1736 - 1806) stated the laws of electrical and magnetic attraction and repulsion after research of a high degree of precision, which made him one of the founders of electrical science. In 1799, VOLTA's battery would give rise to the first use of electrical phenomena as a source of electrification.

In 1630, Jean REY, a French doctor from the Périgord area who was interested in chemistry, published an essay on *A Search for the Reasons why Tin and Lead Increase in Weight when they are Calcinated*. He postulated that atmospheric air plays a part in this phenomenon, and prophesized the discovery that would be made by the Frenchman Antoine Laurent LAVOISIER (1743 - 1794) one hundred and fifty years later, concerning the role of oxygen in the formation of **metal oxides**.

The phenomenon of the rainbow had occupied people's minds since Ancient Times. In the 13th century, Theodoric VON FREIBERG (1250 - 1310) had published a theory, based on a hypothesis made by Roger BACON, which talked of the refraction of sunlight on drops of rainwater. It was necessary to give an explanatory meaning to this phenomenon. In *Philosophical Transactions* (1672), NEWTON described a crucial experiment on the **breaking down** and **reconstitution** of **white light**. The experimental material comprised two glass prisms and a piece of wood with a hole drilled in it. White light that was shone on the first prism was broken down into a set of light rays of different colors, which had been deviated by different amounts. If the prism was rotated, these rays could be selectively sent through the hole drilled in the piece of wood. A ray of a certain color, having been subject to certain refraction, and selected by the hole in the piece of wood, would fall on a second prism. This ray would be defracted at the same angle as when it left the first prism, and when it left the second prism, its color remained the same. **The light rays that make up white light therefore have specific characteristics of color and refractive power.** Continuing his experimentation, NEWTON sent light rays of different colors, arising from the diffraction of white light by a prism, though a biconvex lens. White light came out the other side. This very simple experiment, carried out with simple equipment, showed the complex nature of white light. The mystery-filled rainbows of rainy days were shown to be no more than the banal result of the breaking down of sunlight by rain droplets acting as many tiny prisms.

In just a few decades of the 17th century, GALILEO's astronomical telescope showed the immensity of the cosmos and supported COPERNICUS' heliocentric theory, the rolling of a ball down an inclined plane led GALILEO to formulate the laws of uniform accelerated movement and BOYLE's vacuum pump helped to establish the validity of the idea of a vacuum. NEWTON demonstrated universal attraction, formulated the principles of mechanics, based on the idea of inertia, and described a theory concerning the composition of light. These discoveries were added to those of the circulation of the blood and other, equally remarkable, discoveries concerning spontaneous generation, digestion, and so on. In all these cases, the **experimental method** showed its efficacy. The growth of the experimental method contributed to the invention and perfecting of many instruments, such as the microscope, the astronomical telescope, the barometer, the calculating machine, the air pump, the precision balance, the pendulum clock and the thermometer. The mystical spirit of the Middle Ages, which had considered Nature to be sacred, from the forest to Man, faded away bit by bit as it came up against a thought process that was audacious and that sought to give a **simple, cause-and-effect explanation** for the phenomena of a manipulated Nature. A new scientific era, which was distinguishing itself from medieval scholasticism, had begun.

What was remarkable in this new way of investigating Nature in the 17th century, this experimental method, was an increasing awareness of its investigative power, as well as its development in both the sciences of inanimate matter and those of living beings. At this point in history, there was no frontier between these two domains. Each was comprehensible to the other. Both were distinguishable for the sobriety and the pertinence of their experimental protocols, which led to irrefutable demonstrations and showed inspired thought. HARVEY's tourniquet experiment had nothing to envy GALILEO's inclined plane experiment in terms of simplicity and ingenuity. In both cases, rigorous thought and flawless experimental practice led to major discoveries.

Having resolved the controversies that arose from Aristotlian concepts, physics, armed with the principles of the experimental method and new mathematical tools such as differential calculus, would, in the 18th century, continue its conquest of unexplored territory, particularly the areas of electricity, magnetism and light. "The eighteen century was the Enlightenment, the Age of Reason," wrote Allen G. DEBUS (b. 1926) in *Man and Nature in the Renaissance* (1978). "Its science was 'Newtonian' in that it was experimental science characterized by quantification and the use of mathematical abstraction in the description and clarification of a science that rejected and vilified the mysticism and magic so common to the Renaissance." The event that was going to throw science into the modern world, at the end of the 18th century, was the collapse of an esoteric alchemy and its replacement by a quantitative chemistry that was able to use simple equations to explain reactions between molecular species.

5. OPENING UP CHEMISTRY TO QUANTITATIVE EXPERIMENTATION IN THE 18TH CENTURY

"Compared with physicists, mechanics and surveyors, chemists appear as the real inventors of the art of experimentation. If they were the last to form theories, this is because their task was much harder."

Jean-Baptiste DUMAS
Lessons in Chemical Philosophy - 1836

During the 17th century, there had been an accumulation of discoveries both in the area of experimental physics and in that of the life sciences. In the 18th century, it was Chemistry's turn to open up to objective experimentation, breaking away from medieval alchemy, with its secret cabinets and its esoteric formulae. As emphasized by Jean-Baptiste DUMAS (1800 - 1884), 18th century chemistry became an experimental science, with its paraphernalia of crucibles, stills and ovens, and a vast area of use of natural products (Figure II.14). It tried hard to **resolve the molecular complexities of living beings by extracting, separating and purifying**. To this essentially empirical experimentation, it later added **a theoretical structure**.

G.E. STAHL
(1660 - 1734)

Until the middle of the 18th century, it was believed that air, when it is breathed in, penetrates directly into the blood, either to cool it, or to provide it with some invigorating principle. At the beginning of the 18th century, Georg Ernst STAHL formulated his phlogiston theory which postulated that fire, which was still considered as an element, an inheritance from the philosophers of Ancient Greece, exists in two states, either combined or free, and that the combination of a body with fire makes this body combustible.

Distillation of acids in the laboratory of an "assayer", according to Lazarus ERCKER. Comparison with Figure I.5 shows that over the space of three centuries the basic apparatus had remained more or less the same as in the Middle Ages. The progress that has been made was mainly in the isolation of numerous organic and mineral substances.

Figure II.14 - Chemical technology at the turn of the 18th century
(from Lazarus ERCKER - *Aula subterranea*, Frankfurt, 1701.
Liège University Library, with permission)

Georg STAHL flouted an observation that was nevertheless well known, that a metal heated in the presence of air increases in weight. For STAHL, a metal heated slightly in the presence of air is "calcinated": it is transformed into a "metallic lime", and at the same time an invigorating material that is enclosed in the metal, **phlogiston**, is given off in the form of a flame. A vigorous heating of metallic lime, in the presence of charcoal that is heated until it is red hot, regenerates the metal in its native form. STAHL believed that the incandescent charcoal, which is rich in phlogiston, gives off enough of it to convert the metallic lime into metal. If metallurgical iron becomes covered with rust, this is because it has lost phlogiston. Rusted iron that is heated to red hot refixes the phlogiston and so is reinvigorated. There was a certain logic to STAHL's system, and this was the first known chemical theory. Its sole, but nevertheless formidable, inconsistency, was that it paradoxically explained the increase in weight of the metal in "metallic lime" by the loss of a material, phlogiston.

LAVOISIER's genius was shown in his understanding that **iron increases in weight when it rusts because it fixes an element in the air, oxygen**, and that this notion is valid for all oxidizable metals. However, curious as this may seem, the negative weight of phlogiston did not seem paradoxical to such eminent chemists as Antoine DE FOURCROY (1755 - 1809), Claude BERTHOLLET (1748 - 1822) and Louis Bernard GUYTON DE MORVEAU (1737 - 1816) in France and Richard KIRWAN (1733 - 1813) in England. It was only when faced with the evidence for LAVOISIER's ideas, in the absence of arguments against, and perhaps under the influence of a bandwagon effect, that, one by one, they came around to these ideas. Such a degree of error and such a level of hesitation illustrate how far an erroneous theory that has become embedded in tradition can withstand the reality of the facts.

5.1. THE BIRTH OF PNEUMATIC CHEMISTRY
AND THE STUDY OF THE EXCHANGE OF GASES IN LIVING BEINGS

"The idea of respiration remained completely unsuspected until the last quarter of the 18th century. Interpretation was made difficult by ignorance, at this point in time, concerning the nature of the oxygen, nitrogen and carbon dioxide gases that entered into the composition of the atmosphere."

Emile GUYÉNOT
The Life Sciences in the 17th and 18th centuries - 1941

The **chemistry of gases**, or **pneumatic chemistry**, mainly developed in the second half of the 18th century. Its success can be explained by the fact that, over the same period of time, the rigorous experimentation and bold ideas of talented chemists, who were both attentive observers and skilled technicians, were concerned with finding the solution to the same problems, those of the composition of air and the transformation of air during respiration. Among these chemists were Joseph BLACK (1728 - 1799), Joseph PRIESTLEY (1733 - 1804) and Henry CAVENDISH (1731 - 1810), who were British, the Dutchman Jan INGENHOUSZ (1730 - 1799), Jean SENEBIER from Switzerland, Carl Wilhelm SCHEELE (1742 - 1786), who was a Swede of German origin and the Frenchman Antoine Laurent DE LAVOISIER. A **scientific quorum** (Chapter II-7.3) had been achieved. This manifested itself in an explosion of discoveries in the domain of gaseous exchange in living beings. Physiology had opened up to a new area that had previously remained a mystery, the mechanism of respiration.

At the beginning of the 17th century, VAN HELMONT had observed that in certain grottos, an aeriform principle that was heavier than air was given off, and that it stagnated at ground level. This principle was also found during the combustion of wood and during fermentation in barrels in breweries. This became known as the **gas sylvester**. The term "gas" was a neologism that VAN HELMONT had created to designate invisible "fluids" that were expandable but not breathable, unlike atmospheric air. It was recognized that a property of this gas sylvester was to make lime water cloudy, which gave rise to the proposed name of **chalky air**. After this, BLACK called it **arial air** or **fixed air**. This was carbon dioxide or CO_2. Unlike atmospheric air, which was called **common air**, fixed air was unable to support life. BLACK, who had noted this, showed that fixed air is given off during respiration. In 1762, Henry CAVENDISH was responsible for the discovery of an **easily-inflammable** gas that was lighter than air, unsuitable for respiration, and easy to prepare using the action of an acid (hydrochloric acid, for example) on a metal such as iron. This gas was **hydrogen**. CAVENDISH also showed another form of air, called **mephitic** or **foul** air, which is obtained by passing atmospheric air over incandescent charcoal. This gas was **nitrogen**. In 1774, PRIESTLEY and SCHEELE discovered **vital** or **breathable air**, which was later called **oxygen** by LAVOISIER. In the same year, SCHEELE prepared **chlorine gas** and PRIESTLEY, **nitrous air or nitrous oxide**. CAVENDISH obtained water when he burned hydrogen in the presence of air. In 1785, LAVOISIER showed that water is not a single element, but a compound of hydrogen and oxygen.

5.2. GASEOUS EXCHANGES IN LIVING BEINGS

J. PRIESTLEY
(1733 - 1804)

During the summer of 1771, PRIESTLEY experimented on the "transformation" of atmospheric air by green plants. The latter were kept in a closed atmosphere inside a glass jar. After one week, the air in the jar was collected. This air, which enabled a mouse to survive, made the flame of a candle burn brighter. In a completely different experiment, in 1774, PRIESTLEY observed that when he concentrated the sun's rays, *via* an enormous lens, onto a "mercury precipitate" (mercury oxide) contained in a glass receptacle, the heat caused a gas to be given off, and this gas was also able to re-light the flame of a candle that was about to go out and to make it burn brightly at the same time as metallic mercury accumulated. In 1777, SCHEELE described a similar gas in his *Chemical Treatise on Air and Fire*. He called it **fire air**. In keeping with the spirit of phlogistics, PRIESTLEY named this "fire air" **dephlogisticated air**.

In the middle of the 1770s, SCHEELE tried to reproduce PRIESTLEY's experiments concerning the improvement of atmospheric air by green plants. The result was negative. Confused by SCHEELE's failure, PRIESTLEY decided to re-explore the behavior of green plants exposed to light. The experiments were repeated, but this time, they were carried out with all the rigor of the experimental method; the proof being corroborated by the counter-proof. The results are reported in a work that became a classic, *Experiments and Observations relating to various branches of natural Philosophy with continuation of the Observations on Air* (1779). In this work, PRIESTLEY describes how breathable air is given off by strawberry or mint plants put in a trough of water under a glass jar. It is necessary for the jar to be exposed to the sun. If the jar is covered with "black sealing wax" in order to make it opaque, the plants no longer give off breathable air. PRIESTLEY insisted on the fact that it is not the heat of the sun, but its light that is responsible for breathable air being given off. In addition, he noted that exposure to the sun encouraged the formation of "green matter", which would later be identified as chlorophyll. He postulated that this "green matter" is indispensable for the production of breathable air.

When INGENHOUSZ repeated PRIESTLEY's experimentation in 1779, the controversy between PRIESTLEY and SCHEELE concerning breathable air did not die down. INGENHOUSZ confirmed that the giving off of dephlogisticated air (oxygen) by green plants is only observed if the plants are under light, and that the release of the gas is more-or-less immediate. "I observed," he writes in his treatise, *Experiments on Plants* (1780), "that the plants not only have the faculty to correct impure air within the space of six days or more, as Mr PRIESTLEY's experiments seem to indicate, but also that they carry out this important task in only a few hours, and in a most complete manner [...], this operation is by no means continual, but begins only some time after the sun has risen above the horizon [...], this operation in plants is more or less vigorous, depending on the clearness of the day, the position of the plant, which may be more or less within range of the influence of this heavenly body, and all plants in general corrupt the surrounding air during the night and even in the middle of the day, in the shade." Using aquatic plants held in water in a jar exposed to the sun, INGENHOUSZ describes the formation of bubbles on the surface of the leaves, which come together in the end. "If we examine the air that makes up these bubbles," he notes, "we are soon convinced that is far from being common air. We find that it is of a quality far superior to the best air of the atmosphere. It is truly dephlogisticated. An animal lives in it much longer than in the purest known air, it greatly increases the volume of a candle flame, which acquires a brilliance that dazzles the eye, and a candle that has gone out re-lights, if there is a tiny spark left." INGENHOUSZ prophesized that "this ethereal fluid, which the leaves give off in abundance, must naturally contribute a great deal to purify the atmosphere." One and a half centuries would go by before the enigma of the production of oxygen by water photolysis, when green plants are illuminated (Chapter III-6.2.3), would be resolved, in 1941, by Samuel RUBEN (1913 - 1943) and Martin KAMEN (1913 - 2002). RUBEN and KAMEN's experiments, which provided the proof of the existence of photolysis, used water marked with ^{18}O, the heavy oxygen isotope, and involved physical analysis by mass spectrometry. Thanks to a fantastic technological leap forward, Mankind passed from a summary understanding of chemistry to the isolation and quantitative analysis of isotopes of natural elements, in this particular case, ^{18}O. Nevertheless, the fact remains that despite the most limited technical facilities, a very simple, ingenious, rigorously-carried-out experiment, with counter-proof, had led PRIESTLEY and INGENHOUSZ to discover the fundamental role of light in the production of oxygen by green plants.

A.L. DE LAVOISIER
(1743 - 1794)

It was at the time of PRIESTLEY that LAVOISIER carried out some remarkable experiments that invalidated the theory of phlogistics. The quantification of the measurements that were made, both by weighing and by the evaluation of gas volumes, was carried out with an accuracy that had never before been achieved. It should be noted that, in his laboratory, LAVOISIER had a remarkable range of instruments available to him, particularly **high-precision balances**, one of which was able to weigh 4 g with an accuracy of 0.1 mg and the other, a weight of around 600 g with an accuracy of 5 mg, a **gasometer** and a **calorimeter** (Figure II.15), the sophisticated design of which required a consummate technical know-how and skill. In November 1772, LAVOISIER sent the French Academy of Sciences a sealed envelope. Opened in May of the following year, this letter provided the proof that the combustion in air of sulfur and of phosphorus results not in a loss of weight, as the phlogistics theory would lead us to expect, but, on the contrary, in a weight gain. This was a specific fact, the explanation of which would keep LAVOISIER occupied for years, culminating in the discovery of the **phenomenon of oxidation** by oxygen in the air. In the same way, if tin is heated in a closed, air-filled vessel, the weight of the whole assembly does not change, but that of the tin increases noticeably. However, the volume of the air decreases. To sum up, here we have a demonstration that the metal has combined with a principle contained in the air.

In an article published in May 1777, LAVOISIER describes two basic experiments concerning the nature of atmospheric air.

In the first, mercury is heated slowly in a sealed enclosure filled with atmospheric air. A red mercury precipitate is formed by a reaction known as calcination, a substance that at the time was called "calx of mercury". When all the mercury has been "calcined", the volume of air has decreased by one fifth. "The remaining air," writes LAVOISIER, "does not precipitate lime water. It puts out lights and quickly kills any animals that we put into it." LAVOISIER concludes that "when it is calcined, mercury absorbs the most breathable part of the air, only leaving the unbreathable or mephitic part." If the calx of mercury (mercury oxide) is now heated with a very hot flame, in a retort the outlet tube of which goes to a bell jar upended over a mercury-filled basin (Figure II.16), it is possible to collect the same breathable gas as that described by PRIESTLEY and SCHEELE. The "calx of mercury" therefore results from the combining of mercury and of the breathable gas.

The increasing sophistication of chemical technology in the 18th century, and the rich variety of apparatus available (here we see the instruments in LAVOISIER's laboratory at the end of the 18th century) are in contrast with the rusticity of the equipment of the preceding centuries.

Figure II.15 - Gasometer and calorimeter used by LAVOISIER
(from Antoine LAVOISIER - *Elementary Treatise on Chemistry*, 2nd ed., Cuchet Libraire, 1793)

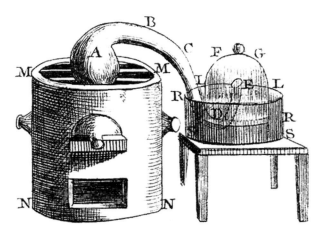

The intense heating of "calx of mercury" (mercury oxide) leads to oxygen being given off. This experiment provides the proof that oxygen, combined with mercury to form an oxide, can be released from the oxide by intense heat.

Figure II.16 - Experiment on the release of oxygen by heating mercury oxide in a retort, carried out by LAVOISIER
(from Antoine LAVOISIER - *Elementary Treatise on Chemistry*, 2nd ed., Cuchet Libraire, 1793)

The increase in weight of the "calx of mercury", compared with the starting weight of the mercury, is thus explained in a logical way. LAVOISIER's results provided a simple explanation for the observations made by Jean REY who, one hundred and fifty years earlier, had postulated that a "thickened part of the air" became fixed to metals such as lead and tin, increasing their weight. In addition, he showed that heating mercury moderately, in contact with air, produced a new substance (that would be called mercury oxide) and that this substance, when heated fiercely, broke down to regenerate mercury and liberate the breathable gas (which would be called oxygen). Proof was therefore given that the "new substance" was a "compound substance" resulting from the combining of two chemical species, mercury and breathable air (oxygen).

The second experiment, described by LAVOISIER in 1777, involved putting a sparrow inside a closed receptacle containing a very limited volume of air. "After about ten minutes, the bird's respiration became difficult, and death followed in an hour." Curiously, the volume of the air did not decrease significantly. However, the air was foul, in that it no longer supported life. It put out the flame of a candle and, a crucial point, it turned lime water milky. This air was therefore chalky air or fixed air (later called carbon dioxide, CO_2). This experiment showed that respiration removes the breathable (vital) part of the air (later called oxygen, O_2) and produces carbon dioxide. The fact that **respiration** in Man also involves the **consumption of breathable air (O_2)** and the **production of chalky air (CO_2)** was demonstrated by experiments carried out by LAVOISIER in 1790 on his colleague Armand SÉGUIN (1767 - 1885) (Figure II.17).

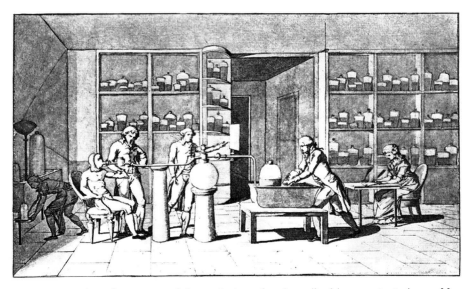

The consumption of oxygen and the emission of carbon dioxide were tested on a Man who had not eaten. SÉGUIN, the subject, was a chemist and a colleague of LAVOISIER. Here, he is breathing oxygen through a mask. Madame LAVOISIER is taking notes.

Figure II.17 - Experiment concerning Man's oxygen requirements in LAVOISIER's laboratory
(from Edouard GRIMAUX - *LAVOISIER 1743-1794*, Félix Alcan, Paris, 1888)

SCHEELE and PRIESTLEY had concluded from their experiments that animal respiration produced very little fixed air (carbon dioxide) and a lot of foul air (nitrogen). These erroneous conclusions were contradicted by LAVOISIER's results; only fixed air (carbon dioxide) and not foul air (nitrogen) is generated by animal respiration. No doubt impressed by the fact that the combining of vital air or breathable air with sulfur or with phosphorus produces acids, LAVOISIER decided to give the name oxygen (from the Greek ὀξύς = acid and γένος = birth) to vital air. Later, it would be discovered that there are exceptions to this definition. Thus, hydrochloric acid (HCl) does not contain oxygen.

In his *Reflections on Phlogistics*, published by the Academy of Sciences in 1783, LAVOISIER wrote: "combustion is none other than the effect that occurs in the moment at which the oxygen principle becomes engaged in a new combination [...]. This doctrine is diametrically opposed to that of STAHL and his disciples, according to which it is in combustible bodies that the heat matter is located, the combined fire, and the phlogiston that escapes at the moment of combustion. In contrast, I suggest, and believe that I have demonstrated, that the air and the fuel each contribute." In fact, phlogistics had arisen from unsound reasoning. Considered as a principle of lightness, its loss from a metal made the latter, paradoxically, heavier. The rebuttal of phlogistics opened the way for modern chemistry, based on the idea of simple substances, compound substances and combinations between chemical species, and on the application of these notions to living beings. The nature of respiration had remained unsuspected up until the end of the 18th century because of a lack of knowledge concerning the composition of the gases making up the atmosphere. By giving identities to these gases and quantifying their appearance and disappearance, pneumatic chemistry highlighted the elegant simplicity of **gaseous exchanges in the animal and plant worlds.**

Experiments on "winey" fermentation, i.e., the fermentation of grape must by yeast, which were carried out by LAVOISIER from 1787 - 1788 (but, sadly, not continued), ushered in the great era of yeast, a model micro-organism that would leave its mark on the cell biology of the 19th and 20th centuries. Apparatus designed by LAVOISIER (Figure II.18) made it possible to collect both the carbon dioxide given off by "winey" fermentation of grape must, using a trap comprising an alkaline solution, and water, another product of fermentation, by means of another trap comprising a deliquescent salt. The increase in weight of both traps, the decrease in the weight of the fermented must, and the production of alcohol by the "winey liquor" were correlated, and used as indices for monitoring the progress of the fermentation.

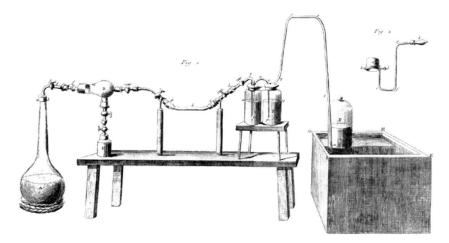

The fermentation apparatus produced by LAVOISIER includes a flask, A, filled with grape juice and yeast. Receptacles B and C are used to collect the scum that overflows from flask A during fermentation. The horizontal tube, h, contains a deliquescent salt that traps water given off during fermentation. Receptacles D and E contain an alkaline solution that traps the carbon dioxide produced by the fermentation. Finally, receptacle F is a safety trap, but in fact no gas appears here. Despite its relative rusticity, this instrument may be considered to be the ancestor of REGNAULT and REISET's respirometer (see Figure III.7)

Figure II.18 - Apparatus designed by LAVOISIER for analysis of the gases given off during "winey" fermentation
(from Antoine LAVOISIER - *Elementary Treatise on Chemistry*, 2nd ed., Cuchet Libraire, 1793)

It was when he was doing this work that LAVOISIER set forth the principle of the transformation of organic matter: "nothing is created, neither in the operations of the art, nor in those of nature, and one can assume in principle that in every operation there is an equal quantity of matter before and after the operation, the quality and the quantity of the principles are the same, and that there are only modifications. The entire art of carrying out experiments in chemistry is based on this principle. For all of them one is obliged to assume a true equality of equation between the principles of the bodies being examined and those that are drawn from them through analysis. Thus, since the must of grapes produces carbonic acid gas and alcohol, I can say that grape must = carbonic acid + alcohol." Having observed that the product of fermentation of grape must, called spirits of wine, is the same as the product of fermentation of apple juice, LAVOISIER proposed to replace the term "spirits of wine" with the term "alcohol", which was inherited from Arab alchemists. Thus, wine fermentation becomes alcohol fermentation.

Because he considered that it is the sugary material in grape must that is fermented, LAVOISIER decided to replace grape must with a sugar solution, to which he added some beer yeast to initiate fermentation. He had therefore moved from a complex system (grape must) to a simplified system (sugar solution), making chemical analyses easier. These analyses showed that the sugar is transformed into carbon dioxide and alcohol, as is the case with grape must, and in proportions that were found again 25 years later by GAY-LUSSAC.

Of equal importance, with respect to his revolutionary experimental practice, LAVOISIER introduced a new language and a rejuvenated nomenclature into his *Elementary Treatise on Chemistry* (1789). Strictly speaking, this semantic revolution had been prepared for by GUYTON DE MORVEAU. The latter, who was a magistrate in the Dijon parliament, was a passionate enthusiast of chemistry and mineralogy. In 1783, he resigned from his job as a lawyer to devote himself to the study of chemistry. His initiative, a reform of the chemical nomenclature of the day, which was a particularly confusing one, full of synonyms, was welcomed with enthusiasm not only by LAVOISIER, but also by BERTHOLLET and DE FOURCROY. In the *Elementary Treatise on Chemistry*, the term oxygen replaces all synonyms: vital air, dephlogisticated air, empyreal air and so on. The simplest compounds of oxygen are named acids and oxides. The term associated with acid or oxide defines the chemical species that enters into the combination, for example sulfur for sulfuric acid, lead for lead oxide. In order to designate the degree of oxidation, an appropriate prefix or suffix is associated with the name of the chemical species. Thus we talk of hyposulfurous, sulfurous and sulfuric acid in order to show the increasing richness in oxygen of acids arising from sulfur. The old terms that refer to the origin of the products, or to immediately-perceptible properties, are replaced with terms having a chemical sound. Thus sorrel salt becomes oxalic acid, sal volatile succini (amber salt) becomes succinic acid and the astringent principle of plants, gallic acid. Taking his inspiration from the principles laid down by the Abbot

Etienne BONNOT DE CONDILLAC (1715 - 1780), according to which "we can only think with the help of words, and languages are true analytical methods," LAVOISIER declared that "the impossibility of separating the nomenclature of a science from the science itself is due to the fact that every branch of physical science must comprise three things; the series of facts that make up the science, the ideas that represent these facts, and the words in which these ideas are expressed. **Like three impressions of the same seal, the word ought to produce the idea, and the idea to be a picture of the fact.** As ideas are preserved and communicated by means of words, it necessarily follows that we cannot improve the language of any science without at the same time improving the science itself; neither can we improve a science without improving the language or nomenclature that belongs to it. However certain the facts of any science may be, and however just the ideas that may have been formed from these facts, only false impressions will be communicated to others as long as we lack the words by which these may be properly expressed." This is an excellent lesson on the value of using the right words to express an idea in scientific language!

5.3. MEASUREMENT OF ANIMAL HEAT AND THE BIRTH OF BIOENERGETICS

In 1780, LAVOISIER and LAPLACE (1749 - 1827) presented the French Academy of Sciences with a *Memoire on Heat*, in which they described a double-walled calorimeter that they had designed. The space between the two walls was filled with crushed ice. This piece of apparatus enabled them to make high quality calorimetric measurements. The authors explained that this was a simple and rational method for measuring the quantity of heat given off by a substance during a chemical reaction: "If we bring a mass of ice into an atmosphere that is at a temperature above zero on the thermometer, all parts will be subject to the action of the heat of the atmosphere, until their temperature has reached zero. In the latter state, the heat of the atmosphere will stop at the surface of the ice, without being able to penetrate it. It will be employed only in melting the first layer of ice, which will absorb it when resolving into water. A thermometer placed in this layer will remain at the same temperature, and the only observable effect of the heat will be the change of state of the ice from solid to liquid (water)." The quantity of water collected gave an idea of the amount of heat given off by an animal (a guinea pig) placed inside the chamber of the calorimeter (see Figure II.15). In order to provide correlative data, the guinea pig's respiration, i.e., the quantity of oxygen consumed and the quantity of carbon dioxide given off was measured, by means of absorption using a "caustic alkali". In parallel, LAVOISIER and LAPLACE carried out the same type of experimentation for the combustion of charcoal. They noted that there was a good correlation between the amounts of heat given off by the charcoal and by the guinea pig, with respect to the volumes of oxygen consumed and of carbon dioxide given off, their ratios being close to one, leading to the conclusion that "**respiration is thus a combustion, a very slow one to be sure, but nevertheless** very similar to that of charcoal," a magic formula that laid down the bases of **bioenergetics**. LAVOISIER and LAPLACE hypothesized that this combustion phenomenon took place in the lungs, which is a perfectly understandable error given the absence of knowledge concerning cell physiology at this time. It would only be at the end of the 19th century that it would be recognized that cells respire, i.e., that they oxidize metabolites derived from the products of digestion (Chapter III-2.2.1). Two other sources of error, of which LAVOISIER seems to have been conscious, were recognized: not all of the oxygen consumed was found in the carbon dioxide given off (some was present in water), and, in addition, the animal placed in the calorimeter was subject to heat loss that increased with the duration of the experiment. In fact, these few corrections had very little weight in the face of the power of a prospective suggestion that invoked the **concept of an analogy between respiration and combustion**, in terms of energetics.

The 18th century came to an end with the death of LAVOISIER and the birth of modern chemistry. The phlogistics theory had been abandoned. The notion of air as one of the four elements of Nature, as imagined by EMPEDOCLES, could not survive when faced with the proof of the diversity of gases. It had been shown that atmospheric air is a mixture of oxygen and nitrogen. The oxygen cycle in Nature became obvious. Respiration consumes the oxygen that is given off by green plants exposed to light. An understanding of the nitrogen cycle would have to wait for the 19th century. The mechanism of carbon assimilation (carbon dioxide taken up by green plants in light) would only be deciphered in the 20th century.

While the chemistry of gases took priority position in the birth of modern chemistry, the **purification of natural products** corresponding to **organic molecules**, should not be forgotten. SCHEELE, in particular, deserves credit for having purified and crystallized tartaric acid from grape must, lactic acid from fermented milk, citric acid from lemons and many other organic acids, including malic, uric and oxalic acids. SCHEELE also isolated glycerol as a product of the breakdown of fats. Physiologists of the 19th century would find these substances in cell metabolism. Twentieth century biochemists would assign them a place in reaction chains or cycles catalyzed by enzymes within living cells.

6. *EXPERIMENTAL SCIENCE AS SEEN BY THE PHILOSOPHERS OF THE 17TH AND 18TH CENTURIES*

"Do not believe something based simply on hearsay. Do not believe based on faith in traditions, just because they have been honored for many generations. Do not believe something because general opinion holds it to be true, or because people talk about it a lot. Do not believe something based on the evidence of one or other of the wise men of Ancient Times. Do not believe something because the probabilities favor it, or lengthy acquaintance makes

you inclined to think it true. Do not believe in what you have imagined to yourself, thinking that a superior power has revealed it to you. Do not believe anything based on the sole authority of your masters or priests. Only that which you have experienced for yourself, experimented on and recognized for true, that which is in agreement with your well-being and that of others; believe in that and behave accordingly."

BUDDHA
Anguttara Nikaya - 6th century BC

At the beginning of the 17th century, the break with medieval scholasticism that had begun with the Renaissance, grew. Experimental approaches in the physical and natural sciences, often encouraged by political patronage, showed their efficacy and opened up often-unexpected perspectives. What is more, a new way of questioning Nature came to the fore. The objective, realistic **how** of experimenters was substituted for the metaphysical **why** of medieval philosophy, dominated by theology. It was in this context that the philosophers Francis BACON, Robert BOYLE and René DESCARTES formulated rules for putting into practice the experimental method, or, in other terms , for codifying scientific research.

6.1. *FRANCIS BACON AND INDUCTION IN SCIENTIFIC REASONING*

F. BACON
(1561 - 1626)

Francis BACON (1561 - 1626), the son of the Lord Keeper of the Seal under England's Queen ELIZABETH I, is considered to be the father of **experimental philosophy**. After studying law at Cambridge, he undertook a law career which led him to hold various political offices within the environment of Royal power. BACON was not an experimenter, and no discovery has been attributed to him. Nevertheless, his prophetic vision of the role of the scholar in society, as well as his political aura, made him one of the most influential promoters of science at the turn of the 17th century. He was persuaded that scientific knowledge was a necessary condition of the power of the State, and wished those who governed to consider themselves responsible for the advancement of science.

The scientific basis of Francis BACON's doctrine was the **inductive method**, which involves, for a given phenomenon that is subject to experimentation, observing the largest possible number of specific cases, under well-defined conditions, collecting and comparing the observations and drawing conclusions by means of *a posteriori* reasoning concerning the validity of hypotheses that were formulated *a priori*. In the Natural Sciences, he said, only the inductive method is valid. The deductive method, which goes from general, *a priori*, principles towards the specific, passes by the facts.

In his essay, *On the Dignity and Advancement of Learning* (1605), BACON explains that, for science to advance, it is necessary to **quantify results**. "Many parts of Nature," he wrote, "can neither be invented with sufficient subtlety, nor demonstrated with sufficient perspicuity, nor accommodated unto use with sufficient dexterity, without the aid and intervening of the mathematics." Neither a blind empiricism nor dogmatic rationalism, but a judicious middle way between these extremes; that is the secret of the experimental method. BACON wrote that "the most empirical of philosophers resemble ants, who simply gather and fill their reserves, while rationalists resemble spiders who weave everything inside their bodies." He wished to be shown "a philosopher who, like the bee, possesses a disposition that is in between, who gathers things from far away, but who absorbs what he gathers by his own means and processes it afterwards."

In 1620, Francis BACON published his major work, *Novum Organum*. The title is self-explanatory. This work revises, and even abandons, the concept of science that ARISTOTLE proposed in his *Organon*. To fight against medieval scholasticism, which had helped to sterilize the imagination and creativity, the *Novum Organum* proposes action and the establishment of a new order. The allegorical engraving of the frontispiece of the work (Figure II.19) represents the columns of HERCULES that ships pass by in order to go explore unknown lands and return loaded with merchandise. The legend *"Multi pertransibunt et augebitur scientia"* ("Many will go to and fro, and knowledge will be increased") is significant: if one dares to affront difficult, unexplained problems, then the harvest will be rich in discoveries. Of things that were already known, Francis Bacon wrote that before they were known it would have been difficult to suspect their existence. "If a man had been thinking of the war engines and battering-rams of the ancients, though he had done it with all his might and spent his whole life in it, yet he would never have lighted on the discovery of cannon acting by means of gunpowder." Even ten big catapults would be less efficient than a cannonball. Rather than be content with reproducing what has been learnt in past, on a larger scale, it is necessary to **innovate**. This is the secret of progress. In addition, it is necessary to force oneself to follow the rule of objectivity, and thus reject **four types of idols of the human mind** which, being handed down from medieval scholasticism, obstruct the mind and block any progress in scientific understanding. **These are the illusions for which our senses are responsible, current prejudices that tend to last, the routineness of our habits, and erroneous concepts spread by Aristotlian philosophy**. Later, John LOCKE (1632 - 1704) would take his inspiration from this philosophy. In an essay on *Human Understanding*, he would use his theory of the blank slate or *tabula rasa* to demonstrate that it is the experiment, and the experiment alone, which generates ideas. In *Novum Organum*, BACON introduced the term **crucial experiment** for the first time, in order to denote an experiment that could be used to settle whether or not other experiments were valid. Many scientists boasted of having carried out the crucial experiment. Later, the word "crucial" was disputed and its scope redefined (Chapter III-8.2).

The illustration on the cover of *Novum Organum* represents ships returning to port with their loads of previously-unknown exotic products. This allegory concerns a new science, which was founded on the experimental method and acquisition of wealth through discovery.

Figure II.19 - Frontispiece of Francis BACON's philosophical work concerning a new concept of science, *Novum Organum Scientiarum* (1620), Dutch edition of 1650

In *De Augmentis Scientarium* (1623), Francis BACON emphasized certain experimental criteria; varying an experiment by varying the materials, varying conditions, reproducing situations observed in Nature, exploring the effect that is the opposite of what was expected, and finally, not neglecting applications. All of this corresponds to what BACON called the "Hunt of Pan", a hunt guided by reasoning, implementing experiments carried out in a certain order and a certain direction.

BACON added that there are still experimental hazards, and that nothing is more foolish than to attempt an experiment not because you are led to do so by reason or some other fact, but because no-one else has tried to do it. It is possible, however, that beneath this extravagance is hidden something that is really big, if one only had the courage to move all the stones in Nature. All the great secrets in Nature are found off the beaten path and outside the realm of our understanding. This is an idea that can be found sketched out by Paul FEYERABEND (1924 - 1994) in his *Against method*, four centuries later, in which a strong role is given to the effects of serendipity.

The *New Atlantis* (1627), published a year after Francis BACON's death, speaks of an ideal research community, SOLOMON's house, which is governed by wise men according to a hierarchy, and is provided by the State with laboratories that have been equipped with the most modern instruments, botanical gardens, libraries and museums containing collections of animals or plants from around the whole world. A similar utopia had already been proposed by Tommaso CAMPANELLA (1568 - 1639) in his *City of the Sun* (1602). The model that had been Alexandria, in Ancient Greece (Chapter I-1.3), was a distant memory.

For Francis BACON, who was a politician as well as a philosopher, **knowledge is power**, and the aim of any research is a better understanding of Nature in a quest for progress. The idea that the **applications** of a discovery should not be neglected is one of the themes that are developed in *De Augmentis scientarum*. These principles would influence Thomas HOBBES, one of the founding fathers of **political science**, in his ideas concerning the role of the State as a regulator of scientific progress.

In his *Sketch for a Historical Picture of the Progress of the Human Mind* (1790) CONDORCET (1743 - 1794), in a few sentences, sums up the contribution made by Francis BACON to scientific understanding: "BACON revealed the true way to study Nature, and to use the three instruments that she has given us so that we might penetrate her secrets, observation, experimentation and calculation. He wishes the philosopher, thrown into the middle of the Universe, to begin by renouncing all the ideas that he has received and even all the ideas that he has formed, so that he can create a kind of new understanding, in which he can only accept precise ideas, correct notions, and truths for which the degree of certainty or of probability has been rigorously weighed up."

6.2. ROBERT BOYLE AND REQUIREMENTS OF EXPERIMENTAL PRACTICE

In contrast to Francis BACON, Robert BOYLE was an experimenter, both in physics and in chemistry. It was only incidentally, no doubt following a controversy with Thomas HOBBES, that Robert BOYLE came to lay down his views concerning experimentation. His demonstration of the existence of the vacuum, made possible by the pneumatic pump, had been sharply criticized by HOBBES, one of the latter's arguments being that one cannot expect anything other than a **probability** from natural causes. Faced with HOBBES' allegation that Nature offered enough directly-observable, convincing and interpretable experiments, BOYLE's defense was an apologia for experimental science.

R. BOYLE
(1626 - 1691)

In the *Sceptical chymist, or, Chymico-Physical Doubts and Paradoxes* (1661) BOYLE attacked the concept of the four elements that had been inherited from Ancient Times and spread by medieval scholasticism. Chemistry had become a science of matter that could be distinguished from alchemy, which was still very prevalent, with its illusions about the spontaneous generation of life from the inert and the transmutation of base metals into gold. Being an objective experimenter, and careful about instrumental details, BOYLE reasoned more like a practitioner than a theorist or philosopher. He considered science from at least three different angles, **instrumentation**, which generates results, the **collection and circulation of these results**, and their **impact on society**. BOYLE related these three aspects to three "technologies", **material, literary and social**. The idea of error comes through in this discussion. How far is an instrument reliable? Are there any criteria for confirming the veracity of the results of an experiment? In restricting the role of subjectivity due to direct observation by the sense organs, the instrument allows **objective verification** of the progress of an experiment and **precise quantification** of the results.

In BOYLE's time, the number of scientific instruments was still limited, e.g., barometers, thermometers, balances, pneumatic machines, microscopes and astronomical telescopes, but judicious use of those that were available allowed interesting discoveries to be made concerning living beings, such as the absence of life in a vacuum, the same heat being radiated in the heart and in the rest of the body (the heart was therefore not a kettle) and the revelation of the existence of microscopic beings. Later, in the 18th century18th century, calorimetric experiments would allow a distinction to be made between the notions of temperature and of quantity of heat, which would lead to the definition of specific heat. The success encountered by scientific instruments had the effect of stimulating technical advances in experimentation and making the latter more credible.

For BOYLE, any experimentation should take place according to a **precise protocol, in front of witnesses**. Working hypotheses, and the way in which an experiment is carried out, with its hazards as well as its results, should be reported in **registers** signed by the experimenter and the witnesses. Newly created Academies should participate in the publication of these results and in their circulation. This admission of the public into the preparation and carrying out of experiments was in total opposition to the secrecy with which the alchemists had surrounded themselves, and their impenetrable practices. Not only were the public invited in as spectators, they were also able to discuss the experimental protocol and even refute the conclusions that were made. In short, BOYLE made experimentation a stimulus for progress in the sciences and **prohibited any pompous discussion or trivial speculation**, leading the Royal Society to choose as its motto *"Nullius in verba"* (on the words of no one). This way of doing science would continue in British scholarly circles, and would be marked by a concern for detail, the attention paid to performance and the reliability of instruments, and the precision shown in drafting experimental protocols.

6.3. RENÉ DESCARTES AND THE CARDINAL PRINCIPLES OF SCIENTIFIC RESEARCH

R. DESCARTES
(1596 - 1650)

René DESCARTES, a man of the laboratory, experienced in the practices of the physical sciences of the period, discussed experimental science from quite a different angle than Francis BACON. In his *Discourse on Method* (1637), which was a preface to three essays covering the topics of Dioptrics, Meteors and Geometry, DESCARTES stated and developed four principles for "rightly conducting one's reason and seeking truth in the sciences." These principles have not aged. They remain the plinth on which all experimental science stands. The first principle is **doubt**: "never accept anything as true which I cannot accept as obviously true." The second principle emphasizes the necessity of sticking to **limited questions**: "divide each of the problems I am examining into as many parts as I can, as many as shall be necessary to best resolve them." The third principle shows how to **move from the simple to the complex**: "begin with the simplest and easiest to understand matters, in order to reach, little by little, as if by degrees, the most complex knowledge." Finally, the fourth and last principle refers to the **ability to summarize**: "make my enumerations so complete and my reviews so general that I can be assured that I have not omitted anything." These were simple ideas that would be universally recognized and would have a considerable influence, both in the immediate and long after, due to their rational foundation.

The objective philosophy of DESCARTES was inspired by mathematical rigor as well as by his understanding of the laws of mechanics and of human anatomy. HARVEY's discovery of the circulation of the blood, the first mechanical representation of a physiological function, would lead DESCARTES, reasoning by analogy, to formulate the concept of the **animal-machine** that is found in the *Treatise on Man* (1664): "all living beings," he wrote, "are machines, automata, mechanisms. Man is different because he also has a soul." He also wrote: "I suppose that the body is nothing other than a statue or an earthly machine that God has formed deliberately to make it more similar to us than is possible." In his *Principles of Philosophy* (1644), he states: "I do not see any difference between the machines made by artisans and the various bodies that Nature alone composes, except that the effects of the machines only depend upon the arrangement of certain tubes, or springs [...]." The animal-machine is compared to a clock or a watch with its cogs leading to a precise movement of the hands, the purpose of the positioning of these hands being to assign a temporal value to each moment. A fundamental point in DESCARTES' philosophy is that Nature operates according to "laws that God has decreed." God does not interfere in the operations that obey these laws and which proceed according to the principles of Physics. Within the Universe, only the human spirit is capable of spontaneous activity. *Discourse on Method* was published anonymously by DESCARTES. *Treatise on Man* was published posthumously in 1664. These facts speak for themselves. GALILEO's trial had had such repercussions that despite an intellectual rebellion that was tacitly tolerated, the fear of severe repression by ecclesiastical authorities stifled any attempt to oppose the reigning religious orthodoxy.

The ideas of BACON, DESCARTES, and BOYLE inspired new forms of communication for scientific thought. There is no doubt that they had a great deal to do with the formation of learned societies, and the academies, and with the circulation of ideas by means of scientific journals.

The Cartesian doctrine, which was based on the notion of the primacy of an innate intelligence, was the subject of impassioned discussions, and indeed, arguments, on the part of eminent philosophers such as GASSENDI, LOCKE and LEIBNIZ. Although he was far from biological experimentation, Father Pierre GASSENDI, who was a philosopher, mathematician and astronomer, supported the theory of the circulation of the blood put forward by HARVEY. GASSENDI argued that a search for the causes of a phenomenon is indissociable from a certain teleological point of view. In *Syntagma philosophicum*, which became known when published posthumously in *Opera omnia* (1658), GASSENDI underlined the *a priori* inaccessibility of material, efficient causes able to explain the production of anatomical structures and of physiological mechanisms. He tried to reconcile a materialist philosophy inherited from DEMOCRITUS and EPICURUS with the teachings of the Christian tradition relating to the immortality of the soul and the transcendence of God. The soul was considered to be an igneous principle, a flame that was lit during life, and went out upon death.

J. LOCKE
(1632 - 1704)

John LOCKE, a rationalist empiricist, postulated that two sources feed our understanding. The impression of the senses is the source of simple ideas that are generally of a quantitative kind, such as size, shape and movement, while the thought that is based on these simple ideas leads to understanding. In short, there is nothing in the understanding that has not previously been in the perception of the senses. At the turn of the 18th century, the German mathematician and philosopher Gottfried Wilhelm LEIBNIZ (1646 - 1716) developed the idea that comprehension of the mechanisms of life must arise from the **analytical deciphering** of the living substance, while admitting that "nothing exists without a sufficient reason allowing it to do so." The term **organism** appeared as a new concept in LEIBNIZ. The organism was considered to be a machine of divine fabrication that, once built, operates on its own. The monad theory postulated that the smallest parts of this machine are interdependent force centers, endowed with their own operating ability, and subject to laws, some mathematical in nature and others metaphysical. Other illustrious philosophers of the time, Baruch SPINOZA (1632 - 1677), Nicolas MALEBRANCHE (1638 - 1715), Georg STAHL (1660 - 1734), and George BERKELEY (1685 - 1753), continued to ask questions about the idea of life, the criteria that define the living state, thought and the transcendence of the soul, the relationship between experience and knowledge, and ideas and concepts and their formulation in words and in language.

6.4. CONTRADICTORY CURRENTS IN THE PHILOSOPHY OF THE SCIENCES IN THE 18TH CENTURY

D. HUME
(1711 - 1776)

Two eminent 18th century philosophers, David HUME (1711 - 1776) and Immanuel KANT (1724 - 1804), stood out because of their reserved attitude with respect to the inductive reasoning extolled by Francis BACON, as well as their critical attitude towards the experimental fact, in contrast to BACON's enthusiasm for a science that must "go to the things themselves." Here, in fact, was a noticeable crossroads in a scientific revolution whose new instruments, microscopes, telescopes and air pumps, were revealing **what had never been seen** and demonstrating **what had never been proven**. It was necessary to be convinced of the credibility of the information received by means of these instruments. HUME's empirical skepticism reflected that of a period that was witnessing the flowering of a scientific technology overflowing with audacity. This new technology would be accepted, but with caution. The inductive method of reasoning would be used, but it would be subject to safeguards. In his *Treatise of Human Nature* (1739) HUME wrote: "there is nothing in any object, considered in itself, which can afford us a reason for drawing a conclusion beyond it and even after the observation of the frequent or constant conjunction of objects, we have no reason to draw any inference concerning any object beyond those of which we have had experience."

In agreement with HUME, and perhaps in a more marked fashion, KANT denied that induction was a means to penetrate the secrets of the phenomena of Nature. In his *Critique of Pure Reason* (1781), KANT emphasized the difference between **analytical judgment**, which brings nothing new to the understanding (e.g., a triangle has three sides) and **synthetic judgment**, which widens the understanding, is deduced from an experience and leads to questioning (e.g., the tiger is a dangerous animal). The **phenomenon** is contrasted to the **nomenon** ("thing in itself").

Phenomena are accessible to our senses and are included in the domain of objective reason, that is, the domain of positive science, while nomena, which are beyond any possible demonstration, as is the case with the existence of God, belong to metaphysics, "a science at the limit of pure reason." POPPER would later refer to this limitation of scientific knowledge in terms of a "demarcation problem."

I. KANT
(1724 - 1804)

The part played by subjectivity in the accomplishment of understanding is one of KANT's major lines of thought: "experience," he wrote, "is a mode of knowledge that requires understanding, the rules for which I must presuppose in myself before even being confronted with objects, and thus, as a consequence, in an *a priori* manner." He also wrote that "reason only penetrates absolutely that which it produces itself, according to its own plans, and it must advance with the principles of its judgment, according to unchanging laws, and constrain nature to respond to its questions, not letting itself be driven by nature as with reins, because otherwise accidental observations that are made, without a pre-established plan, will never converge towards a necessary law that is the only thing that reason searches for and demands." Jean HAMBURGER (1909 - 1992) formulated these ideas nicely in the Preface to his *Philosophy of the Sciences Today* (1986): "Man looks at the world through the spectacles of his spirit and he cannot see with eyes other than his own."

For KANT, since pure reason cannot decide upon the existence of God, Man must forge his own ethical links based on principles that are accessible to reason. This gives rise to the famous maxims of the *Critique of Practical Reason* (1788): "act only from the maxim that makes you wish it to become a universal law, act in such a way that you treat humanity as well in your person as in the person of someone else, always as an end and never as a means." Continuing with his idea, in his *Critique of Judgment* (1790), KANT distinguishes two forces that drive the living being, a **motive force**, comparable to that of a machine, and a **formative force** belonging only to living things, which is capable of forming something from inert materials with self-reproducing structures operating in an autonomous manner.

The Cartesian concept of the animal-machine extrapolated to Man was radicalized in the 18th century by Julien OFFROY DE LA METTRIE (1709 - 1751) in his 1748 publication, *Man a Machine*. In it he says: "the human body is a machine that winds its own springs, the living image of perpetual movement [...]. Vital, animal, natural and automatic movements are only carried out by the action of these springs." The conclusion was brutal: "Man is a machine and in the Universe there is only one substance that is variously modified." In contrast to the dualism of DESCARTES, with the soul distinct from the body in Man, the monism of LA METTRIE claimed that thought, reasoning and feelings are in fact only the expression of the specialized operation of the brain. The period in which LA METTRIE lived was also the period of automata which, by means of sophisticated mechanisms, mimicked human behavior. An illusion was created using Jacques DE VAUCANSON's three famous automata (1736 - 1737), the flute-player, a 1.5 meter tall android that was able to play twelve musical melodies by means of the movements of its lips, which could open, close, move forward, or move backward; the tambourine player and the famous "digesting duck which flapped its wings, ate and gave back digested grain." Although it was a simple simulation, VAUCANSON's astonishing demonstration impressed those who saw it. The operation of the living being became comparable to that of a factory, with its transmission systems, cogs, pulleys and belts. It was during this same period that Denis DIDEROT (1713 - 1784) published his famous *Letter to the Blind for Those Who Can See* (1749), a work that would put him in prison at Vincennes for a few months. In this *Letter* he showed a surprisingly atheistic materialism. In it, he conjectured that Nature proceeds in its creations by means of trial and error, and that only successful trials are viable. This was an amazing thought, well in advance of his time.

For a certain time in the middle of the 18th century, the **mechanistic philosophy of living beings** was very much in vogue. It was supported by some very well-known people, such as the famous Dutch physician Herman BOERHAAVE, and the Swiss physiologist Albrecht VON HALLER, who was the author of the celebrated *Elementa Physiologiae Corporis Humani*, which was published in Lausanne between 1759 and 1766, and was known for his theory of the fiber, which was considered to be a structural and functional unit of living organisms. Nevertheless, it quite quickly became obvious that a completely mechanical organization based on cables, springs and pulleys, no matter how sophisticated it might be, could not account for the complexity of the operation of animal or plant structures. The mechanistic trend in the explanation of the functions of living beings was in opposition to the ideas of **animism**, for which Georg STAHL, in Germany, had been the standard bearer at the beginning of the 18th century. STAHL's animism was based on two principles, i.e., finalism and the absence of corruption of organs. An organizing entity that is spiritual in nature, the soul, integrates a certain number of physico-chemical events, adapts them to the operation of living organisms, making them a fundamental principle of life. STAHL's animism was succeeded, in France, by **vitalism**, with such thinkers as Théophile DE BORDEU (1722 - 1776), Paul Joseph BARTHEZ (1734 - 1806) and Xavier BICHAT (1771 - 1802). Johannes MÜLLER, who was well-known for his work on the sense organs, and was considered to be one of the founders of German physiology, was also a convinced vitalist. Although it was an inheritance of the theory of animism, vitalism could be distinguished from it because it disregarded the idea of the soul. For Ernst MAYR (1904 - 2005) (*This is Biology*, 1997), vitalism was clearly an opposition movement that was characterized by its rebellion against the mechanistic philosophy of the Scientific Revolution and physicalism from GALILEO to NEWTON. He vigorously contested the doctrine that stated that an animal is nothing other than a machine. It is possible that the emergence of a triumphalistic mechanical physics in the 17th century and the beginning of the 18th century, then that of a revolutionary analytical chemistry in the second half of the 18th century, rubbed off on the philosophical vision of the mechanisms of living beings, and that there was thus a transition between the physicalistic concept of the Man-Machine and the deliberately undetermined one of the vital principle.

On the side-lines of animism, but joined to it by the importance given to the concept of the soul, Abbé Etienne DE CONDILLAC's *Treatise on Sensations* appeared in 1754. His doctrine of sensualism dismisses innate ideas and implies that the material and ability allow thought operations that result from stimuli from the environment. CONDILLAC wrote that "if we consider that recollecting, comparing, judging, discerning, imagining, being astonished, having abstract ideas, having them of number and of duration, knowing general and specific truths, are only different ways of being attentive, and that passions, loving, hating, hoping, fearing and wanting, are only different ways of desiring, and that, in the end, the origins of being attentive and desiring are only sensing, we can conclude that sensation covers all of the faculties of the soul." LAVOISIER made CONDILLAC's philosophy his own, rejecting *a priori* ideas, innate ideas, and always coming back to Francis BACON's principle, according to which all that we know derives from the experiment.

The taxonomists of the 18th century, Carl LINNAEUS, Charles BONNET and Antoine (1686 - 1758), Bernard (1699 - 1777) and Antoine Laurent (1748 - 1836) DE JUSSIEU, aimed to put some order into the stupefying diversity of Nature's productions, both in the animal kingdom and in the plant kingdom, and to give it some intentional value. From this arises the Kantian idea of the teleonomic organization of the universe, explaining the phenomena that take place in it and a perfect organization of the living world in which each class of animals and plants has its place, Man being placed on the highest rung of the evolutionary ladder. Taking this idea as a basis, and arguing that the human mind can have an intuitive vision of the living being and be capable of penetrating its secrets, the German philosophers Georg Wilhelm Friedrich HEGEL (1770 - 1831), Friedrich Wilhelm Joseph SCHELLING (1775 - 1854), Lorenz OKEN (1799 - 1851) and Johan Wolfgang GOETHE (1749 - 1832) developed a speculative doctrine concerning the organization of the world. This doctrine, known as "**Nature Philosophy**", tried to show that there is a fit between the operations of the mind and the fundamental laws of Nature: these laws, once known, should make it possible to have access to the secret of causes and to solve the questioning of the why. The purpose of "Nature Philosophy" was to construct a system that was able to make sense of Nature, based on a unitary concept encompassing all of the processes that take place within it, from the mineral world to the inanimate and animate organic world, with Nature rising from the inferior towards the superior by means of a system of metamorphoses, and with the human being situated at the summit of the system. This Nature Philosophy, a contemporary of German Romanticism, was to prosper from the last decades of the 18th century to the beginning of the 19th century.

This evolution in ideas in the 18th century was not without its effect on university teaching. Although scientific societies remained the sources of various innovations, universities, thanks to reforms that had become necessary, helped to advance knowledge. The teaching of physics became dissociated from that of philosophy, experimental physics was at the forefront, and chemistry was beginning to emerge. Chemistry was now part of the official teaching program in Paris, both at the Collège Royal (the future College of France) and at the Jardin du Roi (the future Natural History Museum). Everywhere in Europe, exchanges of ideas between scholars became more frequent, the artisanal approach to making scientific instruments produced many successes, and knowledge became more widespread. In this sense, the 18th century can be seen as a century of transition, a sort of trampoline towards the 19th century take-off of experimental science, in all domains, but most markedly in the Life Sciences.

7. IS THERE AN EXPLANATORY LOGIC FOR THE BIRTH OF THE EXPERIMENTAL METHOD?

"A scientific discipline is born as a new way of considering the world, and this new way is structured in tune with the cultural, economic and social conditions of a period."

Gérard FOUREZ
Construction of the Sciences. The Logic of Scientific Inventions - 2002

The discovery of the experimental method in the 17th century in Europe did not arise out of a process of spontaneous generation. It came about through the ideological and cultural chopping and changing that characterized the Renaissance and the 16th century. It associated innovative ideas and reasoning on the basis of logic with observable progress in instrumentation. It gave a concrete form to a marriage between, on the one hand, a **global approach** that was full of intuition, and guided fundamental science in its thirst for discoveries and, on the other hand, a **rational, rather analytical approach**, which was oriented towards the technical, and the inventive. From the point of view of the historian of the sciences, the appearance of this phenomenon is all the more curious in that highly civilized countries such as China, which were well in advance of the West in terms of industrial and agricultural technology, remained ignorant of this method until recent times. Here we have a summary of contextual events, which are either precursory or correlative, and are suspected of having had a triggering influence on the intellectual revolution that made the experimental method into an indispensable tool for exploring phenomena in Nature and discovering laws from them.

7.1. SOCIOPOLITICAL CRISES

During the Renaissance, Francis BACON wrote that the face of the world was changed by three innovations (inherited from China); **printing**, which made it easier to spread ideas that might be controversial and to spread the knowledge that was being acquired; **gunpowder**, and hence the production of fire-arms, which led to the disappearance of chivalry and an upheaval in the system of state control that had been in place for centuries, and, finally, the **compass**, which became the navigation instrument of the great explorers. Medieval civilization, which had been introspective, was now opening up to a new world that welcomed knowledge and was eager to learn. In the *Grammar of Civilizations* (1987), Fernand BRAUDEL (1902 - 1985) talks of the Renaissance as a period of history in which thought became liberated, authority, both religious and scientific, was put into question, and intellectual ability came to be recognized as more important than the privilege of birth; "Europe came out of a period of multisecular Lent." **With the Renaissance, everything became possible.** Italy was the epicenter of this liberating movement that spread rapidly beyond its frontiers. In the Italian city-states of Venice, Genoa and Florence, which had, due to their trade and their arts and crafts, achieved both flourishing economies and their independence from the 13th century onwards, wise **patronage** had allowed the blossoming of **learned societies** that would later become **Academies**, on the sidelines of the conservative universities. This patronage would also give a boost to the artistic and literary surge that, along with humanism, was to be the distinctive mark of the Renaissance.

At the end of the 15th century, unknown lands that were thought to be peopled with legendary monsters were found by intrepid explorers, with the American continent being "discovered" by Christopher COLUMBUS in 1492 and India being reached by Vasco DA GAMA (1496 - 1524) in 1497. The first circumnavigation of the globe was achieved by one of Ferdinand MAGELLAN's ships (1480 - 1521), in 1519. From these faraway lands, samples of plants, some of which had therapeutic properties, were brought back. This was the period in which the first Natural History Cabinets were formed, gathering together vast collections of objects brought back from foreign lands. Herbalists fixed dried plants on glue-covered paper in a herbarium, leading to the birth of a passion for botany and the creation of Botanical Gardens attached to the Universities (Padua 1545, Pisa 1547, Bologna 1567, Leyden 1577, Heidelberg and Montpellier 1593, Paris 1626).

In the 16th century, and in the middle of the religious crises that were overturning the established order, questions began to be asked about the power of the papacy, which had remained uncontested throughout the Middle Ages. This started in 1517 with the circulation of the 95 theses written by Martin LUTHER (1483 - 1546), which were then taken up in dogmatic form by John CALVIN in his *Institutio Christianae religionis* (French translation in 1540). The Protestant Reformation was supported by men of considerable scientific and philosophical renown, including ERASMUS of Rotterdam (1467 - 1536). As it spread, the Protestant Reformation undermined the allegiance to the Pope of certain reigning authorities. In 1534, in England, HENRY VIII (1491 - 1547) issued the Act of Supremacy, which established the king as the only head of the Church of England. In 1559, England's break with Rome was made complete by the promulgation of the Act of Uniformity by ELIZABETH I.

At the beginning of the 17th century, the continent of Europe was ravaged by the Thirty Years War (1618 - 1648), which arose from the antagonism between protestant German princes and imperial catholic authorities. In England, a civil war broke out and a Puritan gentleman, Oliver CROMWELL, took power in 1649. In France, the *Fronde* (1648 - 1652) civil unrest that was triggered against Cardinal MAZARIN (1602 - 1661), finally came to nothing. The hostile attitude of the Gallican Church (the Roman Catholic Church in France) with respect to the papacy led to a declaration being made in 1682, with the support of LOUIS XIV, in which the freedom of the Gallican Church was demanded and the Pope was only to be allowed spiritual power. When this long period of crises came to an end, a new order had established itself in Europe, with the continent now divided between catholic and protestant nations. The political role of the pope, which had been predominant up until the Renaissance, began to fade. The split between theological and political power that had begun at the end of the Middle Ages was almost complete. The repercussions with respect to science were far from negligible. Although it remained subject to the eye of the Church for some time, science had begun its move towards **secularization**. In keeping with the spirit of the times, the English philosopher Thomas HOBBES, while exiled in France, published *Leviathan*. This work is a polemic on political and natural philosophy, which discusses both the power of the State and that of the science that was taught in the Universities. The term "Leviathan" refers to a biblical monster from the book of JOB. This work's aggressive stance with respect to established dogma and traditions became notorious, and its repercussions were in keeping with the degree of polemic that it triggered. In the final chapter, *The Kingdom of Darkness*, HOBBES attacks "Aristotlelity", a philosophy that he says is a vain one, based on a conciliatory compromise between catholic doctrine and a revision of the philosophy of ARISTOTLE. HOBBES countered the inconsistent, theoretical atmosphere that arose from this philosophy with a breath of fresh air that could give rise to teachings based on the science of reason.

To sum up, the period of western history from the 15th to the 17th century was witness to a sociocultural revolution, the main source of which was a revision of the allegiances of States with respect to the power of the Catholic Church. Paradoxically, while tensions, splits and conflicts were growing within the political climate of Europe, the sciences, arts and literature where undergoing an unprecedented degree of growth. There was a flowering of genius, and argumentation against dogmas that previously had been considered to be untouchable, bringing to mind the words of Ernest RENAN (1823 - 1892) in his *History of Science* (1890): "movement, war, alarms; these are the real environments in which humanity develops. Genius only grows powerfully when there is a storm, and all great creations in thought appear during situations of trouble."

7.2. A KNOWLEDGEABLE SOCIETY

The prohibitions that, in the Middle Ages, prevented access to an understanding of living beings, faded away with the Renaissance. In the Montpellier School of Medicine, from the 14th century onwards, dissections of human cadavers were authorized, under supervision, and in 1556 an anatomy amphitheater was built there. In Italy, the Padua School of Medicine saw a succession of highly talented physicians and anatomists. VESALIUS rectified the errors made by GALEN, who had extrapolated observations made on monkey cadavers to Man. The impetus given to human anatomy was helped by the development of a remarkable iconography, thanks to artists of genius such as LEONARDO DA VINCI, who was not only a draftsman, painter, sculptor and anatomist, but also a visionary inventor. From a comparison of the wing movements of bats, birds and butterflies, he deduced a procedure for artificial flight. He appears to have been the first anatomist to have obtained a mold of the ventricles of the brain by injecting liquid wax into the ventricular space.

A. PARÉ
(1509 - 1590)

In the ferment of a society in which there was little hesitancy about leaving the beaten pathways of thought, **men who had escaped the norms of the past** because they had trained on the margins of official teaching and practiced of their art outside what were considered to be orthodox rules, came to the forefront. Ambroise PARÉ and Bernard PALISSY are famous examples. PARÉ, who was the first great reformer of surgery, did not know Latin. Although he was despised by his peers, he was recognized by the court and given the title of surgeon to the king. Bernard PALISSY, who was a simple potter, benefited from the protection of CATHERINE DE MEDICI (1519 - 1589), and taught chemistry at the Tuileries, outside official authority.

Wishing to defend their prerogatives, University authorities, who were in decline but still arrogant, continuously argued against novel ideas that led to progress, occasionally confronting royal power. In 1530, FRANÇOIS I of France (1494 - 1547), having tired of the escapades of the Sorbonne, decided to create another center for the teaching of science, the College Royal, which would become the College of France. He brought in scholars from Italy. This was not an isolated incident. When, a century later, the dean of the Paris School of Medicine prohibited the teaching of the circulation of the blood as discovered by HARVEY, LOUIS XIV decided to create a special chair of anatomy at the Jardin du Roi, in order to get around the School of Medicine's prohibition (Chapter II-1).

Mathematics held a place of honor in the scientific revolution of the 16th and 17th centuries and many mathematicians have handed their names down to posterity, including Niccolò TARTAGLIA and François VIETE (1540 - 1603), promoter of algebra, John NAPIER (1550 - 1617), the inventor of the logarithms, Pierre DE

FERMAT (1601 - 1665), a precursor of analytical geometry with René DESCARTES and the calculation of probabilities with Blaise PASCAL and Abraham MOIVRE (1667 - 1754), Isaac NEWTON and Gottfried Wilhelm LEIBNIZ, who discovered infinitesimal calculus, and Jacques BERNOULLI (1654 - 1705) and his brother Jean BERNOULLI (1667 - 1748) who are known for their work on integral calculus. This forceful entry of mathematics into scientific culture went along with a mechanistic vision of the world. In the physical sciences, this led to an attempt to **mathematize the laws of Nature**. As an indirect consequence, the credibility of an experimental result was increased from the moment at which it became possible to quantify the extent of the error to which it was subject. It was only in the much longer term that the impact of mathematics was to be felt in the Life Sciences.

At the beginning of the 17th century, the Academies came into being, presenting a challenge to Universities in which the scholastic tradition was still accepted, any attempt to reorganize science or even allude to such a reorganization was severely repressed, and sterile verbal jousting paralyzed the spirit of innovation. These included the *Academie dei Lincei* in Rome (1603), the oldest of the Academies, founded by Duke Federico CESI (1583 - 1630), the *Academie del Cimento* in Florence, which was set up in 1648 by FERDINAND II DE MEDICI (1621 - 1670) and built up in 1657[2] under the leadership of the cardinal-Prince LEOPOLD of Tuscany (1747 - 1792), the Royal Society of London (1662) and the Paris *Académie des Sciences* (1666)[3]. These were followed by the Academies of Berlin (1700), Moscow (1725), and Stockholm (1753). Faced with the overwhelming emergence of European Academies, colonial America was not to be left behind. The Boston Philosophical Society was founded in 1683 by Increase MATHER (1639 - 1723), followed in 1769 by the American Philosophical Society under the direction of Benjamin FRANKLIN (1706 - 1790). In 1780, under the impetus of John ADAMS (1735 - 1826) the American Academy of Arts and Sciences was founded.

These Academies were not research centers. They were institutions in which people discussed the science that was coming into being and commented upon the experiments presented by others. In France, the nascent Academy of Sciences benefited from of the generosity of COLBERT (1619 - 1683), who attracted valuable foreign scholars such as HUYGENS and CASSINI (1625 - 1712) and nominated others such as RÉAUMUR to enquire into new techniques developed by engineers and artisans throughout the kingdom. After this prosperous period under COLBERT, who was highly receptive to scientific currents, came the period under LOUVOIS (1639 - 1691) who was less extravagant in terms of work facilities. In France, the Academy of Sciences published its *Journal des Sçavants*, which became the official publication for circulating discoveries and inventions. At the same time, in Eng-

---

2   The *Academie del Cimento* ceased its activity in 1667.
3   From 1635, Father MERSENNE managed to set up a coterie, the *Academia Parisiensis*, which brought together illustrious scholars such as PASCAL, HUYGENS, ROBERVAL...

land, the Royal Society was publishing its *Philosophical Transactions*. Scientific information circulated freely and rapidly. Despite the unrest in Europe at this time, scholars moved around to discuss and compare their results. Thus a network of scientific exchange was set up across Europe, and this network interacted with others of a literary or artistic nature. The time of secretive alchemists' cabinets and esoteric practices was well and truly over.

The Academies often reflected the relationship between a State and its citizens. Thus, in England, the Royal Society was originally an association of people chosen and recruited by Robert BOYLE in Oxford, independently of any state authorities. It was recognized by CHARLES II (1630 - 1685) in two charters that were drawn up in 1662 and 1663. These charters set the statutes of the society, but recognized its independence. Things proceeded differently in France, where the Academy of Sciences was founded according to the will of the State, and remained under its control. This difference in behavior became established. In England, scholars grew closer to the worlds of industry and finance. Writers, artists and artisans were admitted into their ranks, and on occasion, the latter described apparatus of their invention in the pages of the *Philosophical Transactions*. In France, the academic milieu, far from the centers of the economy, would remain dependent on the State, which handed out financing and privileges. The impregnation of English scientific activity with considerations and interests relating to the private economy was to have a far-reaching effect that is shown even today, as biotechnologies make their breakthrough. It is, in fact, in Britain and other countries with an Anglo-Saxon culture that we see a predominance nowadays of the pharmaceutical and agricultural and food industries that come under the heading of the Life Sciences.

7.3. A SCIENTIFIC QUORUM

From the Renaissance to the 17th century, an increasing political consciousness of the role of science and a new spirit of freedom and intellectual emancipation on the part of scholars, with respect to outdated doctrines and teaching, had a great deal to do with the discovery of the experimental method and the judicious application of this method to the study of phenomena in Nature. In order to increase their own influence, city-states such as Padua, Pisa and Florence attracted such scholars, offering them better working conditions and hospitality. Another factor that cannot be ignored, and which was no doubt linked to changes in society, is the way in which the number and quality of the researchers in the different domains of science rose with the increase in potential. "In the 17th century," wrote COURNOT in 1872, "the fabric of scientific history was already so tightly woven, the great discoveries were weighing it down so heavily, that it is easy to feel that they were well ripened."

It is probable that the scientific revolution of the 17th century benefited from a **threshold effect**, or a **quorum**, with respect to the number of scholars who had been converted to the new methods, freed of any dogmatic spirit. Such exchanges of ideas, confrontations of points of view, and even critical discussions could only lead to significant advances in knowledge and sometimes to spectacular discoveries. Anglo-Saxon microbiologists have given the name "**quorum sensing**" to a phenomenon by which a certain threshold cell density arising from the proliferation of bacteria is able to control either the growth of these bacteria, or their virulence and, in terms of pathogenicity, to be one of the factors responsible for the infection of eukaryotic cells. Bacterial "quorum sensing" is due to the emission of molecular signals, which are the means of communication between bacterial cells. They affect specific genes, which are activated. Transposed to a human scale, with particular reference to the world of scholars, the idea of a "**quorum**" means that in a particular location, or in different locations if communication facilities are available, a certain **density of researchers** is necessary for a discovery to be able to be born. Using the analogy of the bacterial world, in which "quorum sensing" involves the secretion and diffusion of molecules from bacterium to bacterium, the degree of activation of a thought in the scientific world is proportional to the intensity of the communication signals carried between researchers. In the same way that we talk of quorum sensing, or a **signaling threshold**, in the bacterial world, we can talk of a communication threshold in the scientific world. Without doubt, it was not just fortuitous that HARVEY's discovery of the circulation of the blood took place following the time he spent in Padua, where the study of anatomy was a tradition that had been well-established since the 16th century by such scholars as VESALIUS, COLOMBO, FALLOPIUS and FABRICIUS D'ACQUAPENDENTE, and where he would have rubbed shoulders with many figureheads of anatomy and physiology. Here we are far from the sterile confidentiality of the secret cabinets of medieval alchemy. From the 17th century onwards, exchanges of ideas were made easier by the creation of Academies where scholars met and discussed, thus generating a scientific quorum that was all the more efficacious in that the eclecticism of this period did not put up barriers between the different scientific domains. Within this context, we should not ignore the role of the increasingly frequent meetings between scholars from different countries, provoked by scientific amateurs who were able to understand and discuss. In France, this was the case for Friar Marin MERSENNE, who was a physicist, a translator of the work of GALILEO [4], a specialist in acoustics, well-informed about the development of this science in China, and open to many other disciplines (mathematics, mechanics, optics, astronomy...), and who corresponded with the greatest thinkers of his time (PASCAL, FERMAT, DESCARTES, GASSENDI, HOBBES...). It was also the case for Nicolas Claude FABRI DE PEIRESC (1580 - 1637), an enthusiast of anatomy and a councilor in the Parliament of Provence, known for his eclecticism, with his passion for astronomy, physiology,

---

4 As Pierre COSTABEL warns readers in a re-publication of MERSENNE's work, *The New Thoughts of GALILEO* (1638), MERSENNE took a number of liberties with the translation of the great Italian thinker's *Discorsi*, which had been published a few months previously, in Leyden.

geology, botany and archeology. With the scientific quorum achieved in the 17th century, criticism of ideas and experiments were more open and more efficient, becoming so many beacons delimiting the pathways of research. The metaphysical **why** of medieval scholasticism had now been replaced by **how**; how can a phenomenon in Nature be reproduced experimentally, how can the results of an experiment be interpreted, how can the conclusions that are drawn from it be used to form innovative ideas for new experiments?

7.4. INSTRUMENTATION,
AN INTEGRAL PART OF THE EXPERIMENTAL METHOD

If the experimental method requires a considerable amount of thought, with the generation of an idea that leads to the design of an experimental protocol, the interpretation of the results of the experiment, and a reasoned generalization of the conclusions that can be drawn, it also necessarily proceeds from instrumentalized observation. An instrument may be of the greatest simplicity. This was the case at the beginnings of the experimental method, with a tourniquet being placed on the arm to compress veins and arteries in HARVEY's experiments (Chapter II-1), or a jar being filled with meat and covered with gauze to keep out flies in the experiments carried out by REDI (Chapter II-3.4). With the accumulation of scientific knowledge, instrumentation became more sophisticated, but also more vulnerable to criticism. BOYLE, who concluded that the vacuum exists because it can be created by means of a vacuum pump, was, for a certain time, the victim of violent sarcasm based on certain technical shortcomings of the machine that he had manufactured (Chapter II-4.3). One of the criticisms was that an artifact had been produced, making any interpretation of the results null and void. Nevertheless, as technical know-how increased, instrumental reliability increased. Thus, at the end of the 18th century, laboratories were already equipped with effective instruments such as high-precision balances, gasometers and eudiometers for accurate analysis of gases, calorimeters, thermometers, etc. A high degree of craftsmanship was needed for the manufacture of these instruments, and a high level of training was required for their use. The experimental method therefore involved a symbiosis between, on the one hand, the art of formulating ideas and translating them into an experimental protocol and, on the other hand, the art of designing instruments and manipulating them with skill and precision. Using an instrument to carry out an experiment implies confidence in that instrument's **performance** and an ability to evaluate the **margin of error** inherent to its operation. In the 17th and 18th centuries, the know-how of artisans and the requirements of experimenters meant that laboratory apparatus became both **more reliable** and **easier** to handle. The measurement and its margin of error became requirements for any experimentation. Taking measurements led to the use of a mathematical language that enabled physicists to theorize the results of their experiments. With accurate measurements of weight, temperature and quantities of gas given off or consumed, as well as with the rudiments of a chemistry that had lately modernized, naturalists were equipped with a technology that foreshadowed the blossoming of the following centuries.

7.5. THE ENIGMA OF THE DISCOVERY OF THE EXPERIMENTAL METHOD AND OF ITS DEVELOPMENT IN THE WEST

At the end of the 17th century, China had reached a high degree of technical civilization. It was well in advance of Europe with respect to many inventions that made everyday life easier, such as printing, paper, gunpowder, the magnetic compass and the harness. However, China did not develop the experimental method that was born in the 17th century in a Europe that was still well behind it from a technological point of view.

Epistemologists have tried to find an explanation for this bizarre differential. In *Chinese Science in the West* (1969), the British naturalist Joseph NEEDHAM (1900 - 1995) wrote that "the Chinese had worked out an organic theory of the Universe that encompassed the nature of Man, the Church and the State, as well as all things past and present," a thesis that was close to that of Jacques MONOD (1910 - 1976), who wrote, in *Chance and Necessity* (1970): "I am tempted to think that if this unique event (the experimental method) arose in the Christian West rather than within another civilization, it is perhaps partly due to the fact that the Church recognized a fundamental distinction between the domain of the sacred and the domain of the profane."

Political factors have also been said to be involved. At the time of the Renaissance, politico-religious quarrels and civil wars weakened the authority of governments and brought the established order into question, thus helping to free creative thought from a sterilizing control based on tradition, and giving science an imaginative power that it had lost. This turbulent instability in the West can be contrasted to the social and administrative stability of China, which NEEDHAM calls "**homeostatic**" and turned-in-on-itself, as well as to the high level of Chinese state intervention with respect to industry, with its strict control over production and marketing. This state of things would last until the 19th century. No equivalent to the Renaissance took place in China. What is more, the mandarins' practices deprived China of some of its greatest minds due to an interventionist, finicky, all-powerful administration, which was very attached to its prerogatives, carrying out recruitment by means of competitive literary exams. Through the power of the mandarins, the bureaucratic, centralized, imperial State only interested itself in the utilitarian areas of agriculture, irrigation and minor industry, considering that any forward thinking with respect to unknown and hazardous areas of science, which was non-utilitarian in purpose, was superfluous and a waste of time.

With the same idea in mind, i.e., a comparison of China and the West, Fernand BRAUDEL put forward the thesis that the industrialization of the 18th century sealed the scientific fate of the West by giving it an indispensable economic momentum:

II - THE BIRTH OF THE EXPERIMENTAL METHOD

"at this time," he writes in *The Grammar of Civilizations* (1987), "Europe, unlike China, had been subject to capitalist pressures that had enabled it to overcome the obstacle, the incentive to do so having been felt long before, from the rise of the great trading cities of the Middle Ages, and, above all, from the 16th century onwards." This remark goes hand in hand with the recognized fact that there was a resurgence of a liberal, anti-authority spirit in Europe.

For Albert EINSTEIN (1879 - 1955), Ancient Greece had a determining influence on the birth of the experimental method in the West. In a letter sent to a correspondent in California, he wrote that "the development of experimental science was based on two major achievements, the invention of a system of formal logic (Euclidian geometry) by the Greek philosophers, and the discovery that it is possible to find a causal relationship by means of systematic experimentation (The Renaissance)." EINSTEIN concluded that in his opinion "it was not surprising that Chinese wise men had not taken the same step forward [...]," and made the disenchanted comment that "what was surprising was that this discovery (the experimental method) was made at all." It should be added that from the western world of this time, under a papal authority that was recognized, even if was occasionally challenged, had emerged a spiritual unity, which, by an interplay of elite minds, would lead to the same desire to understand the secrets of Nature.

Once the experimental method had been discovered in the West, it still needed to be **cultivated** by researchers and **theorized** by philosophers in order to last. From this point of view, the philosophical ideas of BACON, BOYLE and DESCARTES would have a decisive impact with respect to the taking into account of the experimental method. The **scientific quorum** that had contributed to its birth in Europe was recognized and encouraged by well-advised political powers.

In the 17th century, scientific know-how progressed in a spectacular way in several domains as far apart as astronomy and the natural sciences. The gap between advances made in physics and the stagnation of chemistry in the 17th century was filled in the second half of the 18th century, once appropriate techniques made it possible to recognize the complexity of Aristotlian "air" and to refute the sterile theory of phlogistics. Although the revolution in the chemical sciences took place somewhat later than that in physics, any delay was quickly made up. A tangible consequence was the use of chemical methods to understand the mechanism of metabolic transformations in living beings. The amazing progress made in animal and plant physiology in the 19th century was the logical result of this.

As we have seen, other parameters would play a not-insignificant, underlying role in the development of experimental science. The circulation of books, thanks to **printing**, which had become a flourishing industry, helped make access to knowledge more democratic. Latin, which, until the 17th century, had remained the language used to spread scientific thought, was gradually replaced by **vernacular languages**. It then became necessary to learn several languages in order to communicate, but this restriction was compensated for by the advantage that scientific knowledge could now reach a wider audience. When called upon by experimenters, **artisans and craftsmen** responded by manufacturing increasingly elaborate instruments and scientific tools to meet specific needs. In the 18th century, the encyclopedias of William CHAMBERS (1723 - 1796) in England and Denis DIDEROT in France revealed the technical novelties of the major industries and laboratories to the public at large. At the time of the French Revolution, the desire to have science play a practical role was given concrete form by the creation, in 1795, of the Ecole Polytechnique, or Polytechnical School, the main purpose of which was to train engineers who were specialists in military arts, public works, mining and rural engineering.

Why is it that the Alexandria of the ancient world, where so many philosophers, mathematicians and physicians had succeeded one another, was not the place that sparked off the development of the experimental method? There are doubtless several raisons, including the outlying location of Alexandria, violent confrontations with nascent, conquering religions, first Christianity and then Islam, a docile acceptance of the tradition of the time, and the Socratic tradition of which, after all, Alexandria was the successor. These potential handicaps did not exist in the 17th century. The Church was being questioned, and its authority over society had been shaken by the schisms that broke out, one after the other. Thus the city-states of Italy (Venice, Florence, Pisa), whose flourishing economies allowed the support of sciences and arts by means of an enlightened patronage, became the intellectual hotbeds of Europe. These cities attracted many foreign scholars, and contacts with foreign universities were relatively easy. No doubt another reason was the sharpened critical sense of the physicists and physicians of this period, who did not hesitate to question doctrines that were being taught, even while the promoters of these doctrines, such as ARISTOTLE and GALEN, continued to be admired for their intelligence and their talent. Finally, mathematics, in connection with physics, quantified the error involved in experimentation. All of these ingredients probably led to the scientific quorum that was reached in the 17th century in the West achieving a state of efficiency that had been unknown in other times, resulting in the birth of the experimental method.

8. CONCLUSION - THE MARIAGE OF TECHNIQUES AND IDEAS

"If scientists play a major role in human affairs, this is not only because science has modified and continues to modify Man's fate. It is also because science has greatly modified the idea that Man has of society and the world. Scientists have introduced and made use of a type of logical reasoning that is often cruelly lacking in others. Also, scientists have successfully combated dogmatism, the argument of authority, and also authority itself."

André LWOFF
Games and combats - 1981

II - THE BIRTH OF THE EXPERIMENTAL METHOD

The 17th and 18th centuries witnessed the **birth and establishment of the experimental method in western physical, chemical and life sciences**. The basic principles were defined by the philosophers of science, BACON and BOYLE in England and DESCARTES in France. The beginnings were already noticeable in the 16th century. COPERNICUS' publication concerning the revolution of celestial bodies, in 1543, is often considered to be the event that marks the start of modern science. In the same year, VESALIUS published *De humani corporis fabrica libri septem*, which has caused him to be recognized as the father of modern anatomy. In reaction to the immobilistic tradition of the Universities, learned societies such as the Royal Society of London or the Paris Academy of Sciences were formed, providing scholars with the means necessary to carry out their work, present, discuss and circulate their results, and to find practical purposes for them. **Science became utilitarian** as an echo to the thoughts of BACON, who had proclaimed that knowledge is power, and that with knowledge, Man can dominate Nature.

The 17th century was a century of breaks with dogma, but also a century in which eclecticism dominated. A characteristic trait of this period was that experimental physics and the life sciences were often tackled head on by the same people. Thanks to the many scholars and philosophers who had been attracted to the rationality of the experimental method, there was a birth of a European community of the mind which, indifferent to tensions and political hazards, exchanged ideas and techniques. A reasonable and reasoned approach to the phenomena of Nature replaced an uncritical imagination that was often betrayed by the senses, and often appealed to by the practice of medieval magic. With GALILEO, the rolling of a marble down an inclined plane marked the beginning of a new mechanical science. With TORRICELLI, BOYLE and PASCAL, Nature no longer abhorred the vacuum. With NEWTON, the idea of universal gravitation became established as a general law that regulated the movement of the stars and abolished the dogma of sublunary and celestial worlds. No less surprising were the advances made in the domain of the life sciences. Scholars looked at facts for explanations for phenomena found in living beings that had previously been associated with mysterious causes. It was no longer thought that blood seeped slowly through the body to irrigate the members, as GALEN had written, but instead, that it makes a rapid circuit, as shown by HARVEY, or rather two circuits, that of the **major circulation discovered by HARVEY** and that of the **minor (pulmonary) circulation suggested by SERVET**. The **fable of spontaneous generation** collapsed. This idea, which had had insect larvae being born from rotting material and mice being born from silt, was unable to withstand the demonstrations of REDI and SPALLANZANI. Nevertheless, belief in this idea was so deeply rooted that it would be necessary to await the experiments carried out by PASTEUR in the 19th century before it would be completely eradicated. Animal experimentation revealed astonishing and unexpected phenomena. TREMBLEY discovered **regeneration** in the hydra, SPALLANZANI carried out **artificial fertilization** in the frog, HARVEY tackled **embryogenesis** and made the first stages of the formation of a chick embryo visible to the naked eye, using a fertilized egg. Using a microscope, MALPIGHI was able to show the latter in more detail.

In the 17th and 18th centuries, experimental data were quantified thanks to the use of instruments such as the balance, thermometer, barometer and clock. Other technical advances gave access to that which was not accessible to the senses. The telescope revealed the immensity of the stellar world, while the microscope uncovered the profusion of the infinitely small. Inspired by discoveries in the physical sciences, the mechanistic philosophers of the 17th century tried to use elementary principles of mechanics to explain certain functions of living beings such as muscle contraction or nerve impulses. In mathematising experimental results, the **metaphor of the machine** was looking for logical support to reinforce itself.

It was necessary to await the **chemical revolution at the end of the 18th century** to see the mechanisms of chemical reactions interpreted on a rational basis, in tune with experimental fact, and to see the functions of living beings that, before this period, had all-too-often been mechanized, reconsidered in the context of chemical modifications. The chemical sciences, which had been relegated to the secret cabinets of the alchemists, then made a decisive breakthrough thanks to progress in pneumatic chemistry. The idea of the four elements, air, water, earth and fire, which had been inherited from the philosophers of Ancient Greece and taught as a dogma in medieval scholasticism, was dethroned. Air was recognized to be a mixture of oxygen and nitrogen, and water to be a combination of hydrogen and oxygen. The **rejection of the dogma of phlogistics** by LAVOISIER opened the way for the discoveries of reactions of oxidation by the oxygen in the air and directed the birth of analytical chemistry. PRIESTLEY and INGENHOUSZ wrote the **first chapter of the history of photosynthesis** by discovering that green plants exposed to the sun give off oxygen. With LAVOISIER, the nascent science of **bioenergetics** showed that the oxygen in the air breathed by animals is used to burn nutrients, giving off heat and carbon dioxide. This fundamental discovery contained the seed of the notion of the metabolism, that is to say, the destiny of the products of digestion in the economy of the living being. In coming up with the ideas of simple and compound substances, and of combination and dissociation reactions, the new chemistry contributed to the decline of a strictly mechanical view of living things, which gave way to a chemical vision that was a prelude to physiological chemistry, with its reaction equations and metabolic balances.

Encyclopedias that were printed in France and in England during the 18th century, and circulated to the public at large, helped those who read them to become conscious of the power and diversity of the techniques made available to experimental science in order to discover the secrets of Nature and master their forms of expression. The 18th century was not only the century of the Enlightenment, marked by daring liberties of thought expressed in discussions and writings, but also a century of technological growth, which, by means of an all-conquering industrialization, favored the manufacture of original, high-performance instruments. From this point of view, this century may be considered as a turning point in the evolution of the experimental method.

Being conscious of the consequences of the technical advances that boost the economies of States, Man, who had previously allowed himself to be dominated by Nature and who had considered it to be sacred, realise that he was able to control it. Nature became an object of objective research. The decor was now laid out for the great adventure of physiological experimentation in the 19th century.

Chapter III

THE IMPACT OF DETERMINISM IN THE LIFE SCIENCES OF THE 19TH AND 20TH CENTURIES

"The mountain torrent that can only be navigated by raft finally becomes an estuary that whole squadrons can sail upon: this is the image we have of the scientific current, which first makes its way over tortuous terrain, crossing a thousand obstacles on its way, not without undergoing sudden changes of level, but later turns into a majestic river and finally a sea that spreads farther than the eye can see. Until the 19th century, each science had its history, and it could even be said that the sciences as a whole had theirs, but in the current state of the sciences, we can only imagine the mass of books, memoires, anthologies, annals and scientific journals of all sorts, written in all languages, and published all around the globe."

Antoine Augustin COURNOT
Considerations on the Progress of Ideas and Events in Modern Times - 1872

At the end of the 18th century, the progress made in studying the chemistry of gases and the clarification of the idea of the chemical reaction brought an end to a first stage in the application of the experimental method to the understanding of living beings. The theory of the four elements, earth, air, water and fire, which had been inherited from the philosophers of Ancient Greece, was replaced by a rational classification of known chemical species. At the beginning of the 19th century, the English chemist John DALTON introduced the idea of atoms being the elementary components of chemical species. This atomic theory signaled the start of a series of progressive steps made in the chemical sciences. Physiology was waiting for this progress to be made in chemistry, in order to be able to evaluate, or even to quantify, metabolic modifications that are inherent to vital processes such as digestion, respiration, muscular contraction or nerve impulses.

Organic chemistry, which, up until this point, had been applied to the inanimate world, became the **chemistry of carbon-containing molecular species** present in Nature. In a memoir addressed to the Academy of Sciences in 1829, Eugène CHEVREUL (1788 - 1889) wrote: "Since 1824 I have accepted that the chemical species resides in the molecule, a molecule that is an individual. However, this molecule escapes our senses." It would be necessary to wait for more than a century before the physical reality of macromolecules could be shown by means of electron microscopy and their structure, on an atomic scale, could be determined by X-ray diffraction techniques. Nevertheless, in the 19th century, **synthetic chemistry** underwent a breath-taking expansion (in one century, several tens of thousands of organic molecules were synthesized and their structures analyzed, not only in terms of their elementary composition, but also in terms of functional groups). Laws were formulated and theories put forward to explain the mechanisms of **chemical combinations**. In 1808, John DALTON's law of multiple proportions gave rise to the idea of atoms, and in 1814 the law put forward by Amadeo AVOGADRO (1778 - 1856) stated that, under the same conditions of temperature and pressure, equal volumes of different gases contain the same number of molecules. In 1819, the law concerning specific heats was put forward by Pierre-Louis DULONG (1785 - 1838) and Alexis PETIT (1791 - 1820), and isomorphism was discovered by Eilhard MITSCHERLICH (1794 - 1863). In 1831 William PROUT (1785 - 1850) put forward a law stating that the atomic weights of all elements are integral multiples of that of the hydrogen atom, and in 1834 Jean-Baptiste DUMAS formulated the atom substitution theory. In 1836 Auguste LAURENT (1807 - 1853) put forward the theory of nuclei, or radicals, out of which arose the idea of the chemical function, followed by the theory of residues of Charles-Frederic GERHARDT (1816 - 1856), which explained that very often a chemical reaction corresponds to a combination of radical residues with elimination of a simple mineral molecule (water, for example). 1869 saw the publication of one of the foundation stones of modern chemistry, the periodic classification of chemical elements, by Dimitri MENDELEEV (1834 - 1907). Within one century from the time of LAVOISIER, the number of these elements had practically doubled.

In the first half of the 19th century, eminent chemists, including Joseph GAY-LUSSAC (1778 - 1850), Louis Jacques THENARD (1777 - 1857), Jöns BERZELIUS (1779 - 1848), Jean-Baptiste DUMAS, Jean-Baptiste BOUSSINGAULT (1802 - 1887), and Justus VON LIEBIG (1803 - 1873) laid down the instrumental bases of a highly effective **analytical chemistry** that was able to determine the elementary composition of organic molecules as percentages of carbon, oxygen, hydrogen, nitrogen and finally phosphorous and sulfur. The work carried out on benzene by August KEKULÉ (1829 - 1896) marked the birth of **structural chemistry**, which takes into account the spatial arrangement of atoms in a molecule. In 1827, William PROUT suggested that **biomolecules be classified as proteins, lipids and sugars**. The discovery of the polarization of light by Etienne Louis MALUS (1775 - 1812) and then the analysis of rotational polarization by Jean-Baptiste BIOT (1774 - 1862) opened the way for recognition of the optical properties of so-called dextrorotatory or levorotatory molecules.

Stereochemistry, a new area of chemistry, came to the fore with the work of Louis PASTEUR, and, more importantly, it made its entry into the world of living beings. In 1848, PASTEUR discovered that in sodium and ammonium *para*tartrate crystals, which are "inactive" in polarized light, there is an equal mixture of crystals that deviate polarized light, some to the left (levorotatory, or L) and the others to the right (dextrorotatory, or D). These two forms can be distinguished by their geometry, "according to the orientation of their facets of symmetry." The microscopic mold *Penicillium glaucum*, when sown in a medium containing the *para*tartrate (a mixture of tartrates L and D) feeds exclusively on D-tartrate. In 1874, Joseph LE BEL (1847 - 1930) and Jacobus VAN'T HOFF (1852 - 1911) formulated the theory of **tetravalent carbon**: they associated the asymmetry of molecules of the same atomic composition with the spatial arrangement of atoms or groups of atoms attached to one or more atoms of carbon, and called, for this reason, asymmetric carbon. It was recognized that the products of synthesis in organic chemistry are **racemic** (mixture of L and D forms, called enantiomers), while in the living world the products of synthesis are mainly one of the two **enantiomers** in the same racemic mixture. The invention of the polarimeter made it possible to distinguish each of the two enantiomers in a solution, by means of the specific deviation of polarized light.

In 1875, Germany's Emil FISCHER (1852 - 1919), a pioneer of the structural chemistry of sugars (oses) discovered that they reacted with phenylhydrazine to give a first monosubstituted derivative, phenylhydrazone, then a second disubstituted derivative, osazone. In keeping with PASTEUR, FISCHER found that yeasts inoculated into a culture medium containing the racemic D,L of a sugar (for example, D,L-glucose, D,L-fructose or D,L-mannose) only ferment in the D form, leaving the unaltered L form in the medium. FISCHER used this particularity to prepare pure forms of the crystallized phenylhydrazone and osazone derivatives of L-glucose, L-fructose and L-mannose. He used chemical means to determine the configuration. Impressed by the surprising specificity of the action of yeast "ferments" on the very precise geometric configurations of sugars, he postulated that the ferment and the sugar interact like a "lock and key", a vision that proved to be prophetic, looking forward as it did to the notion of enzyme active sites.

Faced with the vigorous development of chemistry, **physics**, helped along by a powerful mathematical current, was not left behind. Joseph FOURIER (1768 - 1830) published the *Analytical Theory of Heat*, in which he developed a mathematical formalism that led him to describe the law of propagation of heat. André-Marie AMPÈRE (1775 - 1836) discovered the fundamental laws of **electricity**, Nicolas Sadi CARNOT (1796 - 1832) and Rodolf Julius Emmanuel CLAUSIUS (1822 - 1888) founded **thermodynamics** and Michael FARADAY (1791 - 1867) demonstrated electromagnetic induction. James Prescott JOULE (1818 - 1889) determined the **mechanical equivalent of the calorie**. James Clerk MAXWELL (1831 - 1879) laid down the bases of a revolutionary theory of electromagnetism. The research carried out by physicists on electricity would find its counterpart in the development of the electrophysiological exploration of the functions of motor muscles carried out by, among others, the Frenchman Guillaume Benjamin DUCHENNE DE BOULOGNE (1806 - 1875) the German Emil DU BOIS-REYMOND. The **principle of the conservation of energy** in living beings, stated independently by Robert VON MAYER (1814 - 1878) and Hermann VON HELMHOLTZ (1821 - 1894), led to the conclusion that work and heat, products of animal activity, correspond to an expenditure of energy that is only possible if it is met by an equivalent caloric intake, arising from food. The idea that living organisms are transformers of energy opened the way for **bioenergetics**. The optical microscope underwent many refinements that helped overcome the defects that had previously prevented it from being recognized as a reliable research instrument. Joseph Nicéphore NIÉPCE (1765 - 1833) and Louis Jacques DAGUERRE (1787 - 1851) were able to fix the image of an object on a metal plate covered with a photosensitive substance, thus establishing the bases of photography.

Taking advantage of the knowledge acquired from organic chemistry, the identification of many natural molecular species and increasingly rigorous analysis procedures, the **chemistry of living beings** underwent a radical transformation within a few dozen years. It was labeled **physiological chemistry**, and later took the name **biological chemistry**, or **biochemistry**. The first chair of physiological chemistry in Europe was created at Tübingen, and held by Felix HOPPE-SEYLER (1825 - 1895). The dynamics of the transformations under study, i.e., the **metabolism**, was the pivotal point. To physiological chemistry was added the physics of living beings, or **biophysics**, which at that time was based on an understanding of the properties of electrical currents.

At the end of the 19th century, the physical sciences had made an impressive amount of progress, which was to have both the immediate and longer term effects in biology. In 1895, Wilhelm RÖNTGEN (1845 - 1923) showed the existence of invisible radiation that had previously been unrecognized. While studying the operation of a CROOKES (1832 - 1919) tube, that is, a glass bulb filled with a rarified gas, and equipped with a cathode and an anode between which a potential difference of several thousand volts was established, RÖNTGEN was surprised to note that the apparatus emitted rays that were able to excite the fluorescence of a screen coated with barium platino-cyanide. If he placed his hand between the bulb and the screen, he saw the bones of his phalanges on the screen. He fixed the shadow of the bones of his wife's hand on a photographic plate. This image, formed by a previously unknown type of ray, called the X-ray, which was able to penetrate tissue, rapidly made a tour of an amazed scientific community. **X-rays** would be used in the following years in medical radiography for showing the skeleton, and, from the middle of the 20th century onwards, for **structural analysis of macromolecules**. At the end of the 19th century, it was not possible to predict that, fifty years later, X-ray radiation would be used on an atomic scale to decode the structure of macromolecules. What we see here is the very long term application of a fundamental discovery. The light used by the human eye to distinguish objects is situated within a very restricted range of wavelengths, between 400 nanometers (near to ultraviolet) and 800 nanometers (near to infrared). In order to visualize atoms with dimensions of one tenth of a nanometer, it is necessary to use light with a wavelength that is much smaller than visible light. This type of light corresponds to X-ray radiation. In the domain of radiation types, the harvest of discoveries continued to grow. Henri BECQUEREL (1852 - 1908) discovered the **natural radioactivity** of uranium in 1896, while Pierre CURIE (1859 - 1906) and Marie CURIE (1867 - 1934) discovered that of radium in 1898. A few decades later, in 1934, Irène JOLIOT-CURIE (1897 - 1956) and Frédéric JOLIOT (1900 - 1958) discovered **artificial radioactivity**. The use of molecules that have been "labeled" with radioactive atoms, with a view to analyzing metabolism and cell traffic and many other functions of living beings, has made it possible to write whole volumes on biology.

At the turn of the twentieth century, **atomic theory** underwent renewal and adjustment following work carried out by Max PLANCK (1858 - 1947), Niels BOHR (1885 - 1962), Werner HEISENBERG (1901 - 1976), Wolfgang PAULI (1900 - 1958), Jean PERRIN (1870 - 1942), and Ernest RUTHERFORD (1871 - 1937). Gilbert LEWIS (1875 - 1946) put forward the concept of **covalence**. Albert EINSTEIN turned the classical notions of time and space of Newtonian science upside down with his theory of **restricted relativity** (1905) and that of **general relativity** (1915).

The spectacular advances that occurred in chemistry and physics in the 19th century provided the life sciences not only with **instrumental and methodological bases** but also with **food for thought**. Within a few decades, systematics, or the science of classification, which had relied on a comparison of anatomical arrangements between animal and plant species, gave way to a physicochemical analysis of the functions of tissues and organs. A complete physiological science was coming into being. The questions arising from the increasingly numerous discoveries were becoming more specific and deeper.

From the point of view of Science, the 19th century was witness to a double revolution, a **technical** one, which imposed its own rhythm on the experimental method, and a **conceptual** one, which delimited the range of domains that an experimenter could now hope to fully understand, thus leading to the specialization of knowledge. During the Renaissance, and up through the 18th century, the life sciences had been explored and taught by scholars whose training and knowledge could cover many areas, such as physics, chemistry, anatomy, and, occasionally, mathematics and philosophy. In the 17th century, at the age of 34, DESCARTES, a philosopher and physicist, decided to study anatomy and physiology, taking an interest in the development of the chick from the embryo stage, and in the mechanisms of reflexes. The Italian doctor BORELLI, who was familiar with mathematics and physics, was quite at ease with both the translation of APOLLONIUS' *Treatise on Cones* and the writing of *De Motu Animalium*, in which he explained muscular contraction in terms of mechanics. The philosophers of the 18th century, DIDEROT, D'ALEMBERT (1717 - 1783), BUFFON, CONDILLAC, MAUPERTUIS (1698 - 1759) and CONDORCET, had as good an understanding of mathematics and physics as of the so-called natural sciences. In his house in Cirey, VOLTAIRE (1694 - 1778), the talented writer and polemicist, experimented with apparatus that had just been perfected by contemporary physicists and chemists. In 1738, he wrote a voluminous work explaining Newtonian theory in lay terms, The *Elements of Sir Isaac NEWTON's philosophy*, which showed his curiosity and his expertise in a domain that was quite removed from that of novelistic, historical or philosophical literature in which he excelled.

In the 19th century, an increasing recognition of the fantastic complexity of Nature made it necessary to focus on more specific research subjects. Scientific eclecticism gave way to a compartmentalization of science. One sign of this trend was the simultaneous creation of the term **biology** by the Frenchman Jean-Baptiste LAMARCK (1744 - 1829) and the German Gottfried TREVIRANUS (1776 - 1837), in 1802. This term, which replaced that of **natural sciences**, the latter being more vague, delimited a specific domain concerning the analysis of living beings, their organization and their development, with the aim of elucidating the conditions under which the phenomena of life take place, and those under which the causes of the existence of life are produced. In the middle of the 19th century, several major events overturned certain traditional concepts: the laws laid down by MENDEL (1822 - 1884), which formalized the modes of transmission of hereditary characteristics, PASTEUR's refutation of the theory of the spontaneous generation of microorganisms, the theory of evolution formulated by DARWIN, founded on the ideas of natural selection and contingency, and cell theory, which was put forward by SCHLEIDEN (1804 - 1881) and SCHWANN (1810 - 1882), and finally laid out by REMAK (1815 - 1865), and which recognized the cell as being the structural and functional unit of living organisms. The latter, adapted to pathology by VIRCHOW (1821 - 1902), shed light on the processes of illnesses (cancer, degenerative diseases…), the etiology of which had remained a mystery up until that point.

At the end of the 19th century, with the proliferation of discoveries that revealed the complexity of living beings, thought was being given to DARWIN's theory of evolution and the contingency characteristics associated with it. Even if there was a common logic of reasoning for the experimental sciences, and if the principle of causality was valid for all, and if research in biology fed on the techniques of physics and chemistry, it nevertheless seemed obvious that the laws of biology did not show the same inflexible rigor as those of physics and chemistry. The divorce which resulted, in certain locations and for a certain time, is today fading away, insofar as structural biology makes it possible to visualize objects of the living world using precise criteria of size and architecture, on an atomic scale.

1. THE RECOGNITION OF PHYSIOLOGY AS AN EXPERIMENTAL SCIENCE IN THE 19TH CENTURY

"Just a few more years, and physiology, intimately linked with the physical sciences, will no longer be able to take a step without their help: it will acquire the rigor of their method,

the precision of their language and the certitude of their results. By raising itself up in this way, it will find itself out of the range of that ignorant crowd which, always laying blame but never learning, is constantly present, in force, when it is necessary to oppose the progress of Science."

François MAGENDIE
Elementary Compendium on Physiology - 1825

The understanding of human anatomy had made remarkable progress during the 16th and 17th centuries. The 18th century had been one of systematics and comparative anatomy. The 19th century was the century of **physiology**. During the last decades of the 18th century, there had been an original intrusion of analytical chemistry into the quantitative exploration of overall physiological characteristics, such as respiration, in terms of oxygen consumption, and the production of heat in relation to respiration. However, this promising beginning had to await new initiatives. Previously considered as a system for explaining the anatomical particularities of each organ, physiology in the 19th century distinguished itself from anatomy by its **instrumentation** and **methodology**, both of which were still somewhat basic, but were nonetheless efficacious. Physiology became a research-oriented **university subject**, with the creation of professorial chairs.

F. MAGENDIE
(1783 - 1855)

In France, François MAGENDIE (1783 - 1855) actively promoted physiology. At this time, physiology was still under the influence of a vitalist current which had taken over as a reaction to mechanistic naturalists. Opposing the vitalist doctrine that was founded on preconceived, often dogmatic, ideas, MAGENDIE declared that science can only be built upon **facts** and that physiology should set itself the task of working out the **relationships between facts** and searching for **laws governing these relations**. He gave the name "vital action" to a set of phenomena that are inherent to life, such as the formation of bile by the liver or urine by the kidney, phenomena for which, at that time, none of the cellular or molecular bases were known. In contrast to the ideas of vitalism, this vital action was considered to depend only on physicochemical phenomena. "All the phenomena in life," wrote MAGENDIE in his *Elementary Compendium on Physiology* (1825) "can be tied to nutrition and the vital action." Armed with an unshakeable pragmatism, he added: "The hidden movements that make up these two are not obvious, and it is not to them that we should pay attention. We should limit ourselves to studying their results, i.e., the physical properties of the organs, the perceptible effects of vital actions, and discovering how each of them contributes to life." Straight away, MAGENDIE differentiated the functions of nutrition and generation from relationship functions. The latter, mediated by the organs of the senses, locomotion and muscular contraction, bring the individual into a relationship with surrounding objects. MAGENDIE carried out an experiment that confirmed and added to certain results obtained by the English physiologist Charles BELL, providing proof that sectioning the anterior root of a spinal nerve leads to a loss of motility in the innervated region, while sectioning the posterior root leads to a loss of sensitivity. This demonstration definitively established the concept of the reflex arch (Chapter II-3.4.3): a sense neuron transmits information located in a peripheral zone, an irritation, for example, to the spinal cord and the resulting response is transmitted *via* a motor neuron to the muscles of the irritated zone. Many other areas of physiology, concerning the heart, digestion and the action of alkaloids such as strychnine and morphine, were explored by MAGENDIE in a brilliant manner.

2. DETERMINISM, THE PHILOSOPHICAL FOUNDATION STONE OF EXPERIMENTAL PHYSIOLOGY

"Magendie mechanized living beings and considered vitalism as madness. The discovery of internal secretions, the formulation of the concept of the internal medium, the highlighting of a few constancy phenomena and regulation mechanisms in the composition of this medium; these are what allowed Claude Bernard to be a determinist without being mechanistic, and to understand vitalism as an error and not a folly, in other words, to introduce a method for exchanging perspectives in the discussion of physiological theories."

Georges CANGUILHEM
Studies on the History and Philosophy of the Life Sciences - 1994 (7th edition)

François MAGENDIE and his pupil Claude BERNARD, Pierre FLOURENS (1794 - 1867) and Paul BERT (1833 - 1886) a pupil of Claude BERNARD, in France, Johannes MÜLLER (1801 - 1858), Emil DU BOIS-REYMOND (1818 - 1896), Ernst VON BRÜCKE (1819 - 1892), Hermann VON HELMHOLTZ (1821 - 1894) and Carl LUDWIG (1816 - 1895) in Germany, and Charles BELL (1774 - 1842), Marshall HALL (1790 - 1857) and John JACKSON (1834 - 1911) in Great Britain were considered to be the leading lights in a modernized **physiology** that was based on **determinism**, a principle that stated that any phenomenon depends on conditions that precede or are simultaneous to it, or, in other words, there is no effect without a cause. Since the time of the discovery of the experimental method in the 17th century, the determinist outlook had been implicit in all the major advances made in experimental science. Thus, when REDI refuted the existence of spontaneous generation in maggots and flies, his thought processes were resolutely determinist (Chapter II-3.4.1). While determinism finally came to the fore in the 19th century as the basis of biological experimentation, helped by the eloquent defense for it made by Claude BERNARD in his *Introduction to the Study of Experimental Medicine* (1865) (Figure III.1), this was only after long and fierce debate, arising from powerful philosophical currents as varied and contradictory as the mechanistic, animist and vitalist theories.

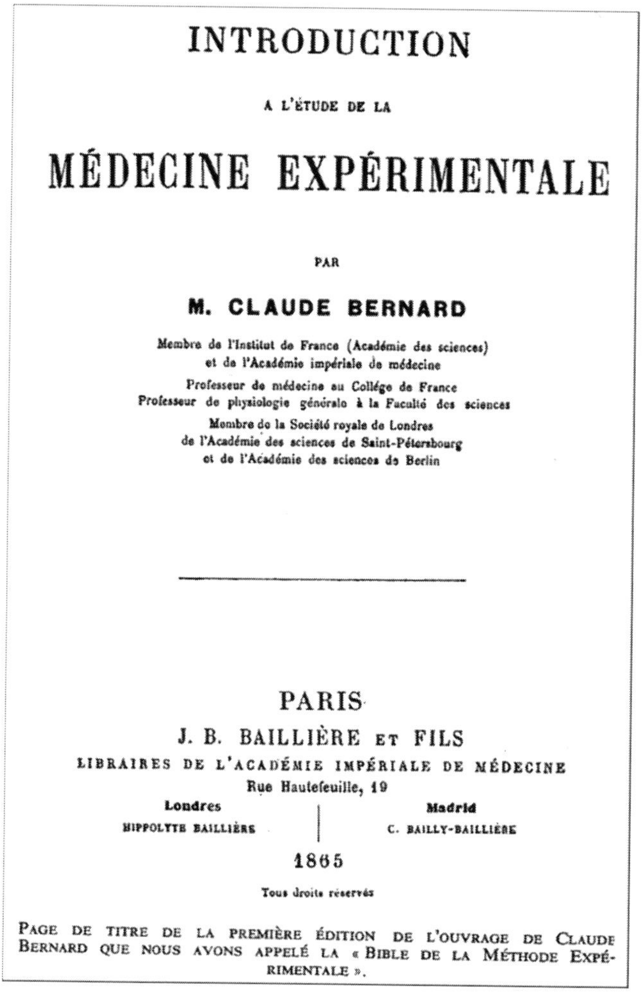

III - THE IMPACT OF DETERMINISM IN THE LIFE SCIENCES IN THE 19ᵀᴴ AND 20ᵀᴴ CENTURIES

> ## INTRODUCTION
>
> A L'ÉTUDE DE LA
>
> # MÉDECINE EXPÉRIMENTALE
>
> PAR
>
> **M. CLAUDE BERNARD**
>
> Membre de l'Institut de France (Académie des sciences)
> et de l'Académie impériale de médecine
> Professeur de médecine au Collège de France
> Professeur de physiologie générale à la Faculté des sciences
> Membre de la Société royale de Londres
> de l'Académie des sciences de Saint-Pétersbourg
> et de l'Académie des sciences de Berlin
>
> PARIS
> J. B. BAILLIÈRE ET FILS
> LIBRAIRES DE L'ACADÉMIE IMPÉRIALE DE MÉDECINE
> Rue Hautefeuille, 19
>
> | Londres | Madrid |
> |---|---|
> | HIPPOLYTE BAILLIÈRE | C. BAILLY-BAILLIÈRE |
>
> 1865
>
> Tous droits réservés
>
> PAGE DE TITRE DE LA PREMIÈRE ÉDITION DE L'OUVRAGE DE CLAUDE BERNARD QUE NOUS AVONS APPELÉ LA « BIBLE DE LA MÉTHODE EXPÉRIMENTALE ».

The impact of the *Introduction to the Study of Experimental Medicine* on the scientific world has been compared by certain authors to that which DESCARTES' *Discourse on Method* had in its time.

Figure III.1 - Title Page (of the first edition, in French, of the *Introduction to the Study of Experimental Medicine* by Claude BERNARD (J.B. Baillière, Paris, 1865)

The new philosophy of experimental research on living beings, which had been spurred on by MAGENDIE, had even more repercussions in countries other than France, as is shown both by the creation, in 1839, of the first Institute of Physiology, by Johann PURKINJE (1787 - 1869), in Breslau, followed by many others, as well as by a fruitful harvest of discoveries in which the German school distinguished itself.

*J. MÜLLER
(1801 - 1858)*

It was during this period that the German physiologist Johannes MÜLLER carried out work on the sense organs and the nervous system that brought together expertise in the areas of physiology, chemistry and pathology. MÜLLER, author of a remarkable treatise on human physiology, was the head of a brilliant school that gathered together talented researchers such as SCHWANN, REMAK and VIRCHOW, who were responsible for cell theory and its circulation, REICHERT (1811 - 1883) who applied cell theory to embryology and DU BOIS-REYMOND who showed the electrical nature of the nerve impulse. The development of animal physiology had repercussions that were beneficial for medicine, which at that time was impregnated with empiricism and subject to contradictory systems of thought. Because of this, medical pathology was gradually transformed into pathological physiology, with causal relationships and laws that tried to tie in with physicochemical principles.

2.1. CLAUDE BERNARD'S DETERMINIST BIBLE

In the 17th century, the credibility of the experimental method was asserted by well-known philosophers such as BACON, DESCARTES and BOYLE, who vindicated it with their proven pedagogical talent. In the same way, in the 19th century, physiology acquired its status as an experimental science on the same grounds as physics and chemistry, deciding to base itself implicitly on the principles of **scientific positivism** professed by Auguste COMTE (1798 - 1857) in his *Course of Positive Philosophy* (1830 - 1842). Although, in the beginning, the determinism of Claude BERNARD seemed to be in line with the positivism of Auguste COMTE, this agreement remained superficial and transitory, COMTE distrusting experimental artifacts and sticking to the fundamental principle of the observation of Nature while BERNARD advocated experimentation in order to unlock its secrets.

Claude BERNARD codified the rules of a determinist doctrine armed with intangible principles, rejecting any recourse to a particular philosophical system, and setting as an essential goal the identification of the direct determining cause of a phenomenon, without worrying about ultimate causes or the essence of life. "In order to find the truth," he wrote in the *Introduction to the Study of Experimental Medicine*, "all the scholar needs to do is to stand before Nature and question it, according to the experimental method, with the help of increasingly perfected investigation tools. I think that, in this case, the best philosophical system is not to have one. In addition," he added, "the great experimenters appeared before the precepts of experimentation," a line of thought that was taken up by Henri BERGSON (1859 - 1941) with his elliptical comment: "Claude BERNARD began by making great discoveries and afterwards asked himself how he should have gone about it."

Claude BERNARD rejected extreme attitudes. He refuted both the vitalist's idea that the essence of vital processes goes beyond human understanding and the materialistic, mechanistic doctrine that explains the functions of living beings in terms of the forces and movements of elementary physics applied to each organ, without taking into account the integrated, interdependent character of these functions. He advocated *a posteriori* reasoning without, however, neglecting the *a priori* idea at the origin of the experiment. To design an experiment, he explained, was first of all to ask a question, with an **independent mind**, without being restrained by the yoke of existing theories. While a preconceived idea is inevitably at the source of an experiment, the researcher must be able to free himself from it once he examines the results of the experiment, and the conclusions he draws must never go beyond the experimental fact.

One safeguard against the erroneous interpretation of an experimental fact is the **control experiment** or **counter-proof**: if a phenomenon is shown when an organism is subjected to a certain condition, the absence of or a change to this condition leads to the disappearance of the phenomenon. The fact that two processes are simultaneous does not mean that they are correlated, so a counter-proof makes it possible to dissociate **simultaneousness** and **correlation**. Bernardian doctrine can be summarized as follows: the researcher observes a phenomenon. Based on this observation, an idea is born in the researcher's mind. This idea is subjected to experimental verification. The results that are obtained are interpreted, and the conclusions that are drawn from them may lead to a new idea, the validity of which is tested by a new experiment. **The experiment is something of a "provoked observation" that is gone through with the fine-tooth comb of the experimental method.** Nevertheless, while the "observer does not reason, he only observes, the experimenter reasons and uses acquired facts as a basis for devising and rationally provoking other facts." It is impossible to overestimate the importance of **the idea** for Claude BERNARD. "The term discovery is given to an understanding of a new fact," he says, "but I believe," he insists, "that it is **the idea that is connected with the discovered fact that is really the discovery.**" While the experimental fact remains at the heart of the Bernardian scientific process, several closely-related facts lead to the proposal of a theory. They provide the framework. The theory gives coherence to these facts, linking them by a chain of cause-and-effect relationships, and making them intelligible.

C. BERNARD
(1813 - 1878)

2.2. THE MANY CONCEPTUAL APPROACHES OF EXPERIMENTAL DETERMINISM IN THE STUDY OF LIVING BEINGS

When acting according to determinist principles, discovery proceeds along pathways whose diversity and originality astonish the layman. Thus, the observation of a phenomenon can give rise to an idea that leads to an experiment being devised and put into practice just to satisfy curiosity. This is **the experiment to see what happens**. A polemic concerning two rival theories can lead to the birth of an experiment that is sufficiently imaginative and well constructed that its results will give preference to one of the theories and render the other null and void. This is the **decisive experiment**. Sometimes, curiosity can lead the researcher to take a detour from his "route" and discover a previously-unsuspected mechanisms that even he or she had not expected. This is the principle of **"serendipity"** [1]. It may also happen that the researcher gets taken up by a first impression, often based on good sense, which then leads in an erroneous direction. Although **analogical reasoning** is often highly useful, it is not without its dangers. Finally, the share of scientific discovery that should be attributed to **luck** should be evaluated correctly: this depends on the perspicacity of the experimenter and the environmental circumstances. In this work, the different approaches outlined above are illustrated by examples taken from the history of the life sciences over the last two centuries.

2.2.1. The experiment to see what happens

"The sort of groping-in-the-dark type experiments that are extremely frequent in physiology, pathology and therapeutics [...] could be called experiments to see what happens, because they are designed to throw up a first previously undetermined, unpredictable observation, the appearance of which could suggest an experimental idea or a research idea."

Claude BERNARD
Introduction to the Study of Experimental Medicine - 1865

Carbon monoxide gas is responsible for fatal poisoning. At the time of Claude BERNARD, in the absence of any hypothesis concerning the cause of the carbon monoxide poisoning, it was necessary to come up with an idea that could be tested by means of an **experiment to see what happens**. Claude BERNARD carried out this experiment on a dog, which he made breathe carbon monoxide. The autopsy showed that both the venous and the arterial blood were crimson in color. Similar observations were made with rabbits, pigeons and frogs. This crimson color is typical of arterial blood which is loaded with oxygen in the lungs. By analogy, Claude BERNARD thought of the presence of oxygen in the venous blood. Having taken some venous blood from the poisoned dog, he passed a current of hydrogen through it in order to displace the oxygen that might be fixed, but the results were negative. His first hypothesis was therefore a false one. Another possibility was that the carbon monoxide had been substituted for oxygen. In order to test this

---

1   This word means finding something that one has not been looking for. It was coined in 1754 by the British writer Horace WALPOLE (1717 - 1797), who was the son of the famous politician Robert WALPOLE, based on a fairy tale entitled *The Princes of Serendip* (an old name for Ceylan, which is today Sri Lanka). In this tale, the Princes discover by accident or sagacity things that they were not looking for.

hypothesis, he took venous blood from a control dog. He placed this blood in a test tube that he upended in a basin filled with mercury, and he passed a current of carbon monoxide through it. After a few minutes, the atmosphere in the tube was analyzed. It contained oxygen. Precise measurements showed that the oxygen in the blood had in fact been displaced by the carbon monoxide. Later, it would be understood that it is in the hemoglobin of the red blood corpuscles that gas exchanges take place. Under normal conditions, the hemoglobin in the arterial blood is loaded with oxygen and becomes oxyhemoglobin. Carbon monoxide displaces the oxygen of the oxyhemoglobin because its affinity for the hemoglobin is greater than that of oxygen.

The study of carbon monoxide poisoning tied in with the problem of blood gases, and the more general question of respiration, a phenomenon that was a subject of considerable study by the physiologists of the 19th century, and which would lead to the concept of cell respiration. Eduard PFLÜGER (1829 - 1910) is credited with having realized, in the 1870s, the significance of cell respiration and the role of hemoglobin as the transporter of oxygen to tissue cells, while Hermann VON HELMHOLTZ is credited with having realized that the production of heat is linked to the respiration of muscle tissue cells. Respiration and heat are two phenomena that are closely linked within the concept of **cellular energetics**. Later on, in the 20th century, it would be discovered that carbon monoxide inhibits cell respiration, not only by substituting for the oxygen in oxyhemoglobin, but also by fixing to an electron transporter of the mitochondrial respiratory chain, the cytochrome oxidase, which reacts directly with the oxygen.

H. VON HELMHOLTZ
(1821 - 1894)

In the 19th century, it was an **experiment to see what happens** carried out .by the Hungarian obstetrician Ignac Fülöp SEMMELWEIS (1818 - 1865), which was responsible for the introduction of antisepsis into medical and surgical practices. In the middle of the 1840s, SEMMELWEIS was an assistant in one of the obstetrics departments in the Vienna General Hospital, under the direction of Professor KLIN. The other department was directed by Professor BARTCH. At this time, there was a high risk of puerperal fever for women who had just given birth, and the number of deaths from this condition exceeded a third of these women. SEMMELWEIS was surprised to observe that percentage of deaths was from two to three times higher in Professor KLIN's department than in Professor BARTCH's. "Cosmic" causes, which were traditionally given, did not make sense. SEMMELWEIS turned his attention to the organization of the work of those who delivered the babies; midwives in BARTCH's department and medical students in Professor KLIN's. SEMMELWEIS observed that the latter, having been dissecting cadavers with their bare hands, then went to the hospital to deliver babies. It was known at this time that an "anatomical" cut, accidentally carried out with a scalpel, often had fatal consequences.

*I.F. SEMMELWEIS
(1818 - 1865)*

Straight away, SEMMELWEIS considered that there might be a relationship between the germs carried on the hands of those tending the women giving birth and the appearance of a virulent septicemia in the parturient. In order **to see what happened**, he came up the idea of having all students wash their hands with a chlorinated lime solution before they entered the labor ward. He went so far as to ask Professor KLIN to do the same. The professor took this an insult and fired SEMMELWEIS on the spot. Taken in by BARTCH, SEMMELWEIS continued his procedure. He asked BARTCH to exchange the midwives of his department with Professor KLIN's students. In the following month, the mortality rate of women who gave birth in BARTCH's department tripled. At this point it was decided that anyone assisting at a birth had to undergo careful washing of the hands. Proof had been provided that hands could carry the germs responsible for infections. SEMMELWEIS's work was a precursor to the idea of antisepsis well before PASTEUR and KOCH, fifty years later, demonstrated irrefutably that there is a relationship between infections and bacteria.

Closer to home, in 1950, a famous **experiment to see what happens** led the virologist André LWOFF (1902 - 1994), at the Institut Pasteur in Paris, to discover the true nature of lysogenesis in bacteria infected with bacteriophages. The bacterium used, *Bacillus megatherium*, carries bacteriophages in the form of prophages. Such bacteria proliferate normally, from time to time liberating bacteriophages by lysis, in low percentages. They are said to be **lysogenic**. Out of curiosity, LWOFF decided to irradiate these bacteria with ultraviolet radiation. This led to the instant lysis of the whole of the bacterial population. From the prophage stage, the virus had switched to the virulent stage. Later, molecular analysis would show that in each lysogenic bacteria cell the expression of the phage genome is repressed. By abolishing this repression, UV radiation allows the expression of the phage genome and the reproduction of hundreds of bacteriophages, thus leading to bacterial lysis. LWOFF's experiment to see what happened gave rise to the fundamental concept of the negative regulation of gene expression, a concept that would be illustrated brilliantly by the work of Jacques MONOD and François JACOB (b. 1920) on the mechanism of the repression of β-galactosidase expression in the enterobacterium, *E. coli*.

Another example of an **experiment to see what happens**, which was in the news recently, is the relationship between the presence of a gastric ulcer and the colonization of the stomach mucosa by a helical shaped Gram-negative bacterium, *Helicobacter pylori*. This bacterium was isolated at the end of the 19th century, but its role remained unknown, particularly due to the difficulty involved in cultivating it. In the 1980s, two Australian researchers, Barry MARSHALL (b. 1951) and Robin WARREN (b. 1937), managed to obtain pure cultures of this bacterium, and they suggested that it might have a role in ulcerous conditions. To demonstrate this, in true Pastorian fashion, MARSHALL went so far as to swallow a large bowlful of a culture of *Helicobacter pylori* **to see what would happen**. In the days following his action, he developed acute gastritis, which was authenticated by X-radiography. The discovery of the role played by *Helicobacter pylori* in ulcer pathology resulted in the 2005 NOBEL prize for Physiology and Medicine.

The experiment that is carried out to see what happens is often dictated by a sort of premonition that is called **intuition**, which, in the curious mind, leads to the elaboration of an idea after a long period of ripening in the subconscious.

2.2.2. The decisive experiment

"We know that there are good experiments and bad ones. They accumulate in vain. Whether one hundred have been done or one thousand, a single piece of work by a true master, a PASTEUR, for example, is enough to make us forget all the rest. BACON understood this well when he invented the term "experimentum crucis."

Henri POINCARÉ
Science and the Hypothesis - 1902

At the end of the 1850s, the polemic concerning the spontaneous generation of germs that had stirred up the scholarly world in the 18th century (Chapter II-3.4.1) came to the fore again, encouraged by the naturalist Felix Archimede POUCHET (1800 - 1872), author of the theory of heterogenesis. This theory, which was in vogue at the time, maintains that microorganisms are born spontaneously from organic material arising from the decomposition of living beings.

L. PASTEUR
(1822 - 1895)

Louis PASTEUR, who had already encountered microbial chemistry, and was convinced that life could only be born from life, decided to use experimentation to prove that the spontaneous generation of germs was a myth. A first experiment involved boiling "albuminous sugar water" in a glass flask, the slightly inclined neck of which was connected to a hollow platinum tube heated until it was red hot. During boiling, air was expelled from the flask, and it returned to the flask upon cooling. Under these conditions, no bacterial proliferation could be observed. It was argued that the heat given off by the metal tube destroyed the elements of the air that were indispensable to life. In order to answer this charge PASTEUR, advised by chemist Antoine Jérôme BALARD (1801 - 1879), carried out the famous **swan-necked flask experiment** (Figure III.2). The flasks that were used had a stretched neck that formed a long tube that was curved into a U, rather like the shape of a swan's neck. Meat bouillon placed in the flask was brought to the boil, with the neck of the flask remaining open to the air. The water vapor that was given off condensed in the U-shaped part of the neck. After cooling, air entered the flask, but the germs carried by the dust that it contained were held by the water droplets stuck to the tube. In this experiment, the air had not been subject to the effect of heat, yet, nevertheless, the contents of the flask were sterile, and remained so for months. However, if the liquid present in a flask chosen at random was agitated so that it came into contact with the U part of the pipe, or if the neck of the flask was broken, an abundant growth of microorganisms was visible after a few days. An obvious explanation was that the agitation or the contact of the medium with the ambient air had provoked an intentional contamination of the liquid in the flask by germs that were initially held in the tube. **The results of this swan-neck flask experiment were decisive in refuting the myth of spontaneous generation.**

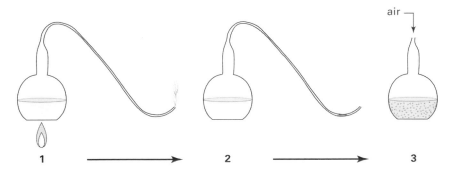

This illustration represents three steps in an experiment carried out by PASTEUR, the results of which led to the refutation of the idea of the spontaneous generation of microorganisms.
1 - Sterilization of the medium by heating to boiling point.
2 - In the flask left at rest after sterilization, the bacteria brought by the dust in the air are trapped in the lowest part of the U-shaped tube, which contains droplets of water vapor. The medium remains sterile.
3 - The neck of the flask tube is broken. The medium, which is in direct contact with the air, receives the dust and microbes that the air carries. Within a few days, the medium is polluted by an intense bacterial proliferation.

Figure III.2 - Swan-neck flask experiment carried out by PASTEUR

Caution is necessary when considering a given experiment as **decisive**, or as we commonly say, **crucial**. Thus, when PASTEUR commented on the results of his experiment that led to the theory of the spontaneous generation of microorganisms being refuted, he said "I do not claim that spontaneous generation never exists. In subjects of this kind, it is not possible to prove the negative. However, I do claim to show that in all those experiments in which it has been thought that the existence of spontaneous generation could be recognized […], the observation was the victim of illusions or of causes of errors." This relative caution in human understanding makes considerable sense in the life sciences.

2.2.3. Serendipity and the unexpected discovery

"Without doubt, the researcher must show considerable perspicacity and imagination during all work that is preparatory to a discovery. He must ignore nothing of what occurs during his experiments and his mind must always be alert. He must pay attention to everything unexpected."

André LWOFF
Games and Combats - 1981

It may happen that mechanisms are revealed by non-linear logic, and that the experimenter finds something he or she has not been looking for. English-speakers have named this "**serendipity** (Chapter III-2.2). It is possible for serendipity to lead to accepted dogma being disproved. This was the case for Claude BERNARD's discovery of the **glycogenic function of the liver**, which was a fundamental discovery in the domain of metabolism, and was the subject of BERNARD's scientific thesis, which he submitted in 1853. In his work *Lessons on the Phenomena of Life that are Common to Animals and Plants*, which is a summary of the classes that he gave at the Natural History Museum in 1876, Claude BERNARD tells how he originally wished to verify the concept of a functional dichotomy between the animal kingdom and the plant kingdom that had been postulated by two eminent French chemists, Jean-Baptiste DUMAS and Jean-Baptiste BOUSSINGAULT. According to this concept, plants synthesize and accumulate sugars, fats and proteins. These substances are consumed by animals, acting as nutrients and sources of energy. At this time the animal kingdom/plant kingdom functional dichotomy was recognized and taught as dogma. This theory held its own due to its Manichean nature: the plant kingdom transforms light energy from the sun into chemical energy, stored in organic molecules, and the animal kingdom lives at the expense of the plant kingdom by feeding on the organic molecules of the latter.

When Claude BERNARD tackled the problem of metabolic transformations in animals, he first looked at what happened to glucose, particularly with respect to its breakdown, in agreement with the ideas of his time. The experimental procedure headed directly for this goal. Blood was taken from dogs, with samples being taken from veins and arteries at different parts of the blood circuit. The glucose content was determined using the cupropotassic solution developed by chemist Charles Louis BARRESWIL (1817 - 1870). The content was determined by color changes, from blue to red, that were due to the reduction of copper. "From the first tests," wrote Claude BERNARD, "I was very surprised to find that the dogs' blood always contained sugar (glucose), no matter what they had eaten, and even if they had been fasting. This fact is easy to observe," he added, "and it is astonishing that it had not been seen before." He remembered that the German chemist Friedrich TIEDEMANN (1781 - 1861) had shown in 1832 that starch from food is transformed into glucose in the intestine, the glucose being found in the blood. Unfortunately TIEDEMANN had omitted to carry out a **control**, but Claude BERNARD did carry out a control.

This showed that animal blood does normally contain glucose, no matter what type of food is eaten, and of particular interest, when there is no intake of carbohydrates. When he fed a dog exclusively with boiled meat, which corresponds to a mostly protein diet, he observed that glucose, which was practically absent in the portal vein, was present in noticeable quantities in the blood of the subhepatic vein. Traveling through the liver, therefore, the blood had become loaded with glucose, thus giving rise to the idea that the liver might be an organ in which glucose is synthesized.

The **washed liver** experiment that was carried out in 1855 provided additional information. A dog's liver was removed and washed under a strong current of water to remove all the sugar that it contained. This "washed" liver was inadvertently left overnight on the lab bench. Wishing to continue his experiment to the end, Claude BERNARD again perfused the liver with water. To his surprise, he found that there was a noteworthy quantity of glucose in the perfusate, which appeared paradoxical, as the glucose had been removed the day before when the liver was washed. The material that was suspected of liberating glucose (glycogen) was isolated and purified in 1857 by precipitation with ethanol, in collaboration with the chemist Jules Théophile PELOUZE (1807 - 1867). One century later, it would be shown that glycogen is synthesized from glucose, which comes partly from glycogen-forming amino acids.

This fundamental discovery of glycogenesis gives rise to two remarks. The first concerns the content determination of glucose, a reducing sugar, using BARRESWIL's solution. This shows the efficacy of an **alliance of chemistry and physiology**. Generally speaking, in the life sciences, discoveries are dependent on methods that are most often borrowed from physics and chemistry and develop along with these disciplines. The second remark concerns the intellectual thought process adopted by Claude BERNARD. This was more-or-less dictated by the simple desire to check whether or not the dogma of a functional dichotomy between the animal kingdom and the plant kingdom was well-founded. As far as glucose is concerned, this dichotomy, which had previously been accepted as creed, was invalidated. The same was true, it was later found, for fats and proteins. Animals, like plants, are able to manufacture all these substances. To sum up, it was banal experiments concerning variations in the glucose content in dogs undergoing different diets that led to the existence of an animal/plant dichotomy being disproved and the unexpected but fundamental discovery of glycogenesis. These experiments were not part of some grandiose project, and there was no *a priori* theory involved. They were carried out mainly with the idea of a simple verification, adhering to a methodology that took care to note details and to look for an *a posteriori*, unbiased rationale in the experimental results. Serendipitous discoveries illustrate an opinion, widely held in the scientific community, that the pathways to discovery are not always visible from the outset, and that something unexpected that is encountered and investigated by the researcher is often more promising than a carefully planned project.

2.2.4. Advantages and traps involved in reasoning by analogy

"An analogy can deceive us as our instinct can lead us astray, and as we can stumble when walking: this is not a reason to renounce the use of our legs or to rebel systematically against our instinct and the senses."

Antoine Augustin COURNOT
Materialism, Vitalism, Rationalism - 1875

In the Middle Ages, analogical reasoning, which was honored in the Hippocratic tradition, was considered to be essential for building knowledge. The use of analogy in contemporary biological science has often had beneficial results and contributed to interesting discoveries. Sometimes, however, the analogy, an overly easy piece of mental gymnastics, can lead to the inappropriate interpretation of experimental results and give them an erroneous meaning. Constructive audacity in analogical reasoning and mistaken interpretation due to too hasty a comparison are illustrated here by memorable examples.

In the second half of the 19th century, while microbiology was establishing itself as a new discipline, certain bacteriologists convinced themselves that white blood cells, which were first called microphages and then neutrophils because of their tinctorial properties, were vectors of bacteria and spread infection (Figure III.3). In fact, **neutrophils** are in charge of the destruction of bacteria. They are **cellular** or **innate immunity** cells, in the same way as **macrophages**, cells that reside in the tissues.

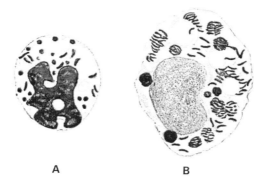

A   B

A - Guinea-pig neutrophil (microphage) filled with choleric vibrions, most of which are transformed into granules.

B - Guinea-pig macrophage filled with choleric vibrions not transformed into granules.

Figure III.3 - Internalization of choleric vibrions by the neutrophil or macrophage, two innate immunity cells
(from E. METCHNIKOFF - *Immunity in Infectious Diseases*, Masson, Paris, 1901)

This discovery was only accepted after a long controversy in which the naturalist of Russian origin, Elie METCHNIKOFF (1845 - 1916) was a pioneer. In this context, it is interesting to remember that in 1876, the German microbiologist Robert KOCH (1843 - 1910) had been the first to show a relationship between a well-defined pathology, in this case anthrax, and the invasion of the organism by a specific bacterial species, *Bacillus anthracis*. Microscopic examination of the blood of animals infected with this bacterium, and suffering from septicemia, revealed the presence of bacteria in neutrophils. This apparently sensible firsthand reasoning led to the conclusion that the neutrophils transported the bacteria and were responsible for the septicemia. Researchers of the time were far from suspecting that the neutrophils are particularly efficient agents in the antimicrobial defense system, and it was at the end of the 19th century that research carried out in Germany by the brilliant immunologists, Emil VON BEHRING (1854 - 1917), Shibasaburo KITASATO (1852 - 1931) and Paul EHRLICH (1854 - 1915), led to the discovery of **humoral** or **adaptive immunity** in which the antibodies that are secreted into the blood play a primordial role.

E. METCHNIKOFF
(1845 - 1916)

The idea of a **second type of immunity, cellular** or **innate immunity,,** came to the fore after METCHNIKOFF discovered that, in contrast to what was thought at the time, neutrophils and macrophages are protective cells, veritable shields against infection. This discovery, which was the fruit of audacious reasoning by analogy, followed some roundabout routes. METCHNIKOFF, who was a zoologist by training, was studying the digestion process in a small flatworm, a fresh water **planarium** (Figure III.4). The digestive tube of this flatworm does not have an anus, but has numerous closed diverticulae that are rich in digestive ferments. The planarium feeds off the blood of animals to which it attaches itself. METCHNIKOFF showed first that the red blood cells of goose blood (nucleated cells that are easy to find under the microscope because of their refringent nucleus) are sucked into the planarium's digestive tube and then internalized in the end cells of the digestive tube diverticulae. After a certain time, the blood cells disappear by a process of lysis. Reasoning by analogy, METCHNIKOFF's thinking moved from the digestion of goose red blood cells by the planarium to the digestion of bacteria by specialized cells, neutrophils in the blood and macrophages in the tissues. To back up this reasoning, experiments were extended to include invertebrates such as the May Bug and the snail, into which were injected goose red blood cells. Microscopic examination showed that after 24 hours the red blood cells had been digested by white cells present in the lymph of these invertebrates. The same phenomenon was shown in a lower vertebrate, the goldfish, and then in the guinea-pig. Intraperitoneal injection of goose red blood cells into the guinea-pig peritoneum triggers, within a few hours, an exsudation into the peritoneal cavity of neutrophils and macrophages that encompass and destroy the red blood cells. "The analogy," wrote METCHNIKOFF, "between the modifications undergone in the neutrophils and macrophages (internalization and lysis of the bacteria) and the phenomenon that takes place in the intestinal cells of the planarium, indicates that the resorption of the represented elements (by neutrophils and macrophages) must really be considered as a true intracellular digestion."

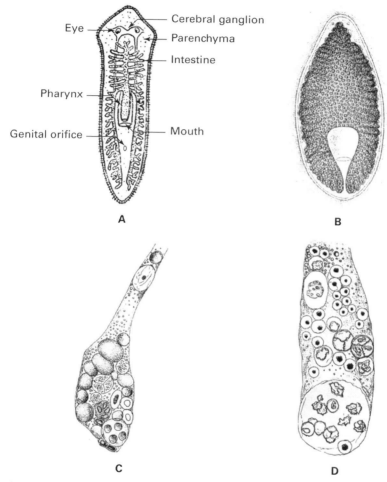

A - Fasting planarium. Note the morphology of the intestine with its numerous diverticulae.
B - Planarium a short time after sucking goose blood. The goose red blood cells can be seen because of the refrigent characteristics of their nuclei.
C - Planarium intestine cell filled with goose red blood cells being digested.
D - Planarium intestine cell in which some of the goose red blood cells have been digested.

Figure III.4 - Digestion of goose red blood cells by the Planarium
(from E. METCHNIKOFF - *Immunity in Infectious Diseases*, Masson, Paris, 1901)

Continuing his investigation, METCHNIKOFF demonstrated that an extract of guinea-pig macrophages is able to lyse goose red blood cells. The factor responsible for this lysis, contained in the extract, was named **cytase**. The following step was decisive, and concerned microbial infection. It answered the following question: what is the reaction of the neutrophils and macrophages with respect to microorganisms? The first experiment concerned the **infection of the daphnia, or freshwater flea, with a microfungus** of the blastomycetes species. The daphnia was chosen because, being transparent, its organism can be examined under the microscope. The daphnia was infected intestinally after fungus spores, equipped with sharp points, had pierced its intestines. It was then possible to observe a typical phagocytosis process carried out by the daphnia's white cells, which captured, internalized and digested the fungus spores. The daphnia's fight against parasitic infection is a typical example of cellular (innate) immunity. METCHNIKOFF widened his exploration to cover mammals. In particular, he proved that the rat's resistance to anthrax was dependent on the phagocytosis of anthrax bacteria by this animal's neutrophils and macrophages.

Another phenomenon enlightened METCHNIKOFF. While carrying out an experiment to see what would happen, he made a microscopic observation of the behavior of the **amoeboid cells** of a star-fish larva after a rose thorn had been implanted in its epidermis. He observed that the amoeboid cells migrated towards the location where the thorn had been implanted. This was a phenomenon that was reminiscent of the migration of blood neutrophils from the capillaries to a site where bacteria are proliferating, such as an infected wound, for example. METCHNIKOFF gathered these observations and drew conclusions from which would be formulated the **theory of phagocytosis**. This theory stipulates that neutrophils are able to perceive bacteria from a distance, to migrate towards these bacteria and to internalize them into vesicles: the internalized bacteria are destroyed by **digestion** in the same way that red blood cells are digested by the digestive tube cells of the planarium. Here we see both the association of two phenomena, the migration of star-fish larva amoeboid cells towards a zone of irritation and the digestion of red blood cells by the planarium, and the comparison of these two phenomena to the migration of neutrophils towards a zone infected by bacteria and the destruction of these bacteria by the neutrophils. Thus, by means of analogy-based reasoning, METCHNIKOFF was able to solve the puzzle of the steps of cellular (innate) immunity, the basic principle of which is **phagocytosis**.

In contrast to this happy example of the masterly discovery of phagocytosis, there are examples of analogical reasoning that has gone astray in a manner that has been temporarily prejudicial. It was by analogy with the studies of René Just HAÜY (1743 - 1822) on the growth of mineral crystals, studies which had been undertaken a few years earlier, that one of the promoters of cell theory, Matthias Jakob SCHLEIDEN, imagined that the nucleus of the cell, called the cytoblast, initiates crystallization within an unstructured fluid made up of mucus, the cytoblastema, the final result being the creation of a cell. This erroneous idea persisted for nearly twenty years, until Robert REMAK demonstrated that all cells arise, by division, from preexisting cells, a demonstration that was the true foundation stone of **cell theory**.

To sum up, although reasoning by analogy can be a source of errors, it should be recognized that it also has its virtues, provided that its conclusions are always subject to experimental approval. To use COURNOT's imagery, "we can stumble when walking: this is not a reason to renounce the use of our legs."

2.2.5. The part played by luck in the experimental method

"Although everything, in a sense, happens at random, that is, without premeditation, nothing happens by accident, that is, for no reason."

Georges CANGUILHEM
Studies on the History and Philosophy of the Life Sciences - 1994 (7th edition)

If there is one extensively advertised example in which luck came to the aid of the experimenter, it has to be that of the discovery of penicillin, the credit for which was shared between a bacteriologist, Alexander FLEMING (1880 - 1955), a physiologist, Howard FLOREY (1898 - 1968) and a biochemist, Ernst CHAIN (1906 - 1979). History has designated FLEMING as the main actor in this discovery. In fact, he was the innocent but wise discoverer of a mold, *Penicillium notatum*, which had the remarkable property of secreting an antibiotic that was very active with respect to a large number of bacteria, i.e., penicillin. In the 1920s, FLEMING's work in the bacteriology laboratory of St Mary's hospital in London involved the study of staphylococcus variants that differed from the wild type with respect to the shape and color of their colonies, as well as their virulence. On September 3rd, 1928, an event occurred that would lead to the start of the **penicillin adventure**. At the end of July of that same year, FLEMING had inoculated a suspension of staphylococci on some nutritive gel placed in circular, lidded, glass dishes that were called PETRI dishes (named for Richard PETRI (1852 - 1921), who invented them). When FLEMING returned from his vacation on September 3rd he examined the fifty or so PETRI dishes that he had negligently left out on his work bench. His attention was drawn to one of the dishes, in which a magnificent mold had grown, clad with long filaments. What struck FLEMING was the fact that no bacterial colony had developed around this mold (Figure III.5). Obviously, the mold had secreted an inhibiting substance, which FLEMING called **penicillin**.

FLEMING was not a chemist, and his attempts to purify the penicillin failed. He kept to experiments concerning the antibiotic effect or absence of antibiotic effect of the *Penicillium notatum* extract on a whole series of bacterial species. In particular, he noted the absence of a penicillin effect on the growth of a microbe, *Bacillus (Haemophilus) influenzae*, which at that time was considered to be the agent that was responsible for influenza, an infection that a few years later would be discovered to have a viral etiology. By destroying bacterial contaminants, the *Penicillium notatum* extract made it possible to obtain pure cultures of *B. influenzae*, giving rise to the title of the article published by FLEMING in the *British Journal of Experimental Pathology* (1929, vol. 10, pp. 226-236): "On the antibacterial action of cultures of a penicillium, with special reference to their use in the isolation of *B. influenzae*."

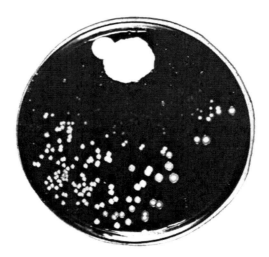

The figure shows a plate of nutritive gel in a PETRI dish that FLEMING inoculated with staphylococci. By chance, the gel was contaminated with the fungus *Penicillium notatum*. On the upper part of the plate, we can see the absence of bacterial colonies around the *Penicillium* colony. This is due to the fact that the *Penicillium* secretes an antibiotic that diffuses outward and prevents bacterial growth.

Figure III.5 - Photograph of the nutritive gel plate on which Alexander FLEMING observed the antibiotic capabilities of a mold, *Penicillium notatum*, with respect to colonies of staphylococci
(from R. TATON - *Causality and Accidents in Scientific Discovery*, Masson, Paris, 1955, all rights reserved)

The *Penicillium notatum* affair came to the fore again around ten years later, when FLOREY and CHAIN, at the Oxford Institute of Pathology, who were interested in bacterial antibiosis and were searching through the literature for documents relating to this subject, came upon FLEMING's article. Captivated by the enigma of the properties of penicillin, they set out to purify it. It was at Oxford that the first samples of penicillin were prepared and their antibiotic power, with respect to different types of serious infections, was demonstrated. After this, the United States took over, with mass production.

Although everyone knows the history of the discovery of penicillin, most people are unaware of the part played by **chance** in this discovery, which was later revealed in an unexpected manner. When the Nobel prize for Physiology and

Medicine was awarded to FLEMING, FLOREY and CHAIN in 1945, for their contribution to the discovery and isolation of penicillin, scientific historians took an interest in the FLEMING's first observation, and, for pedagogical purposes, tried to reproduce the contamination of a staphylococcus culture on a nutritive gel in a PETRI dish with the mold *Penicillium notatum*. The dishes, which had been inoculated with both staphylococci and fungal spores, were placed in an incubator at 37°C. No bacterial lysis occurred. Faced with this failure, which was repeated and unexplainable, everyone was puzzled. Then one of the experimental participants had the idea of looking at the weather reports for London in the summer of 1928, the year that FLEMING discovered penicillin. A sudden temperature change drew his attention. From July 27th to August 6th, the temperature was relatively low, oscillating between 16 and 20°C, but then, from August 6th until the end of the month, it rose to 30°C, with points that were even higher. A crosscheck was made of the optimum conditions for the development of *Penicillium notatum* (temperature from 15 to 20°C) and those for staphylococcus (temperature from 30 to 37°C). It then became possible to reconstruct the mode of development of the *Penicillium* spore that contaminated FLEMING's staphylococcus culture and to reproduce the scenario of the discovery. In 1928, FLEMING inoculated around fifty PETRI dishes with a suspension of staphylococcus and left them on his work bench at ambient temperature. During the inoculation, one of the dishes was inadvertently contaminated with a *Penicillium* spore coming from the mycology laboratory on the floor above, in the same building where FLEMING was working. Until August 6th, the temperature did not exceed 20°C, a condition that was favorable to the development of the *Penicillium*, but unfavorable to the development of the staphylococcus. After August 6th there was a heat wave that favored the multiplication of the staphylococcus. However, in the PETRI dish where the *Penicillium* had proliferated and secreted penicillin, the growth of the staphylococcus bacteria was blocked, which explains the ring around the mold that was empty of bacterial colonies. Applied *a posteriori*, using known meteorological data, the experimental method made it possible to understand the process by which the development of staphylococcus was stopped by *Penicillium notatum*. This somewhat caricatural example shows that the reproducibility of experimental results can depend upon **environmental circumstances** that are not necessarily obvious to the researcher. The idea of luck that is often associated with the discovery of the antibiotic effect of *Penicillium notatum* should not give rise to confusion. In fact, this discovery, and the lucky chance involved, arose from a fortuitous climatic phenomenon, a sudden variation in temperature during the London summer of 1928. This lucky chance spurred the perspicacity of FLEMING, who showed the right reaction in keeping the contaminated PETRI dish, instead of throwing it away, and who also had the truly scientific curiosity to take an interest in the nature of the antibiotic substance secreted by the *Penicillium*. It is often in this way that luck helps science.

3. THE IMPACT OF TECHNOLOGY ON THE LIFE SCIENCES IN THE 19TH CENTURY

"With Lavoisier and Laplace, physics and chemistry entered into the study of the phenomena of life, and experimenters had to make use of the instruments and apparatus of physics and chemistry. As science moves forward, we increasingly feel the need for specific installations that bring together the tools necessary for physical and chemical experiments and for vivisection, by means of which physiology penetrates the depths of the organism. The method needed to direct physiology is the same as that required for the physical sciences. It is the method that belongs to all experimental sciences, and is still today what it was in the time of Galileo. Finally, most questions in science are solved by the intervention of appropriate tools: the man who discovers a new procedure, a new instrument, often does more for experimental physiology than the most deep-thinking philosopher and the most powerful of generalizing minds."

Claude BERNARD
Lessons on the Phenomena of Life that are Common to Animals and Plants - 1878

During the first decades of the 19th century, physiology resolutely committed itself to methodical experimentation. If this experimentation was to become more incisive, it required specific, accurate analysis methods and appropriate instrumentation. The 19th century saw the rise of engineering which, although it was still artisanal in nature, showed superb creativity, with the invention of apparatus that was cleverly adapted to the study of the multiple functions of living organisms. The operational approach of physiology was thus able to base itself upon accurate, objective analyses and to benefit from the use of measurement instruments that were inherited from physics, and which replaced observation by means of the senses.

3.1. RATIONALIZATION OF OPERATIONAL PHYSIOLOGY

For François MAGENDIE, methodology in animal physiology was necessarily based on two conjoined processes, **vivisection** and the **chemical exploration of humors and tissues**. His pupil, Claude BERNARD, continued in this direction, aided by the chemical expertise of BARRESWIL and PELOUZE. In order to study the chemical mechanism of digestion, SPALLANZANI procured gastric juices from birds by making them swallow sponges that he later removed in order to squeeze out the juices (Chapter II-3.3). With MAGENDIE, Claude BERNARD and others, **physiology became operational**. A researcher wished to study digestion? First of all it was necessary to produce a stomach or pancreatic fistula, and then it would be possible to carry out chemical analyses of the secretions and finally to analyze the structure of the glands that secrete the gastric juices. The **pancreatic fistula** operation that Claude BERNARD carried out on a dog involved opening the pancreatic duct close to where it entered the duodenum, after having isolated it from the glandular tissue. A canula was put into it, and attached to the lips of the intestinal wound (Figure III.6). A bladder was fixed to the other end of the canula, and the secreted juices accumulated in this.

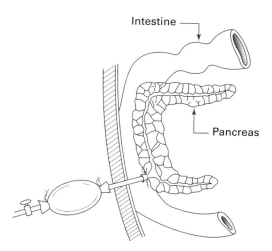

The practice of producing a pancreatic fistula in the 19th century, which had been tried out in the previous century, was codified and used systematically in different European physiology laboratories to collect pancreatic juices and study their digestive powers with respect to proteins, fats and sugars. The progress made in analytical chemistry in the 19th century made sense of the practice of producing a pancreatic fistula.

Figure III.6 - Pancreatic fistula produced in the dog
(from E. HÉDON - *A Precis of Physiology*, 13th edition, Doin, Paris, 1943)

Experiments carried out from 1850 - 1860 on the pancreatic juices that were collected in this way showed three types of ferments that are able to digest polysaccharides, fats and proteins, respectively. In 1877, the German physiologist Wilhelm KÜHNE (1837 - 1900) purified the substance that is responsible for the digestion of proteins from crushed beef pancreas, which led him to discover that there is an active proteolytic form of this substance and an inactive form. KÜHNE named the active factor trypsin. Later, it would be discovered that trypsin is formed by the cleavage of a non-active precursor, trypsinogen, acted upon by another factor secreted by the duodenum, enterokinase. At the end of the 19th century, two competing terms, diastase and enzyme, were used to designate the catalytic substances responsible for metabolic transformations in living tissues. The term **diastase** (διάστασις, separation) was proposed by PAYEN (1795 - 1871) and PERSOZ (1805 - 1868), in order to characterize the substance contained in a extract of germinated barley, the effect of which was the dissolution of the contents of grains of starch, leaving the envelopes separated from the unmodified grains. The term enzyme [2] (ἐν ζύμη) was proposed by KÜHNE to characterize the catalytic substance which, in yeast, , is responsible for alcoholic fermentation. The example of the discovery of trypsin, selected from among many, shows how efficacious chemical methodology can be when applied to living beings. Already, a new discipline was taking shape, **physiological chemistry**, which would later be called biological chemistry.

What about the study of the nervous system? Here again, operational physiology was taking over. This involved the sectioning of nerves and the ablation of regions of the brain that would be recognized as specific functional territories. These operations were carried out on different types of animal, particularly the cat, dog and monkey, the latter being used because of its high level of encephalization. The role of the cerebellum had been glimpsed by WILLIS in the 17th century, but it was necessary to await the experiments of the Italian anatomist Luigi ROLANDO (1773 - 1831), at the beginning of the 19th century, confirmed and broadened by those of Pierre FLOURENS, in France, ten years later, before it would be understood that the cerebellum controls the coordination of voluntary movements, its ablation leading to the loss of voluntary movements as well as to problems with equilibrium and posture. In the 1850s, an experimental method applied to the nervous system led Claude BERNARD to make some important discoveries, including the role of the sympathetic nerve in arterial vasodilation, the creation of artificial diabetes by lesion of the vagus (pneumogastric) nerve in the fourth ventricle of the brain, and the inhibition of motor nerves by curare. At the end of the 19th century, the English physiologist Charles Scott SHERRINGTON (1857 - 1952), one of the founders of modern neurology, invented the residual sensitivity method that made it possible to map cutaneous areas that were dependent on synapses in the spinal chord. This method, which was carried out on monkeys, involved sectioning three spinal sense nerve roots above and three others below a single root that was left intact. The cutaneous area that maintained its sensitivity was called the dermatoma, i.e., the region whose sensitivity was controlled by the nerve root that had been left intact. In this exploration of the nervous system, it was found that the surgical procedure could be rivaled by the localized application of an inhibitor poison, strychnine, on the nerve roots. At the turn of the 20th century, in the Saint Petersburg Institute of Experimental Medicine, Ivan Petrovitch PAVLOV (1849 - 1936) demonstrated the existence of conditioned reflexes. His experiments, which were carried out on dogs that had been psychologically conditioned, showed the beginning of a secretion of saliva as well as a secretion of gastric juices, the latter being shown by means of a gastric fistula.

---

2   In French, the gender of the words *"enzyme"* and *"coenzyme"* remained optional. They were considered masculine for more than a century, and then became feminine following a decision by the Académie des Sciences in January 1970. The original French edition of this work used the masculine gender for reasons of euphony.

The arrival of **anesthetics** (ether, nitrogen protoxide, chloroform), which were products of chemical synthesis, made operational physiology easier. The excision of organs could be carried out without any major difficulties. This showed previously unknown functions of internal secretion glands. After having carried out a pancreatectomy in an anesthetized dog, Oscar MINKOWSKI (1858 - 1931) and Joseph VON MERING (1849 - 1908) were surprised to observed that the dog drank and urinated continuously. Analysis of the dog's urine showed a huge amount of glucose. Here were all the typical signs of human diabetes, the causes of which were unknown at the time. The **pancreas** is a gland that is characterized by a double secretory function, an **exocrine secretion**, which releases the enzymes necessary for the digestion of proteins, lipids and carbohydrates into the intestine, a function that had been discovered in previous years, and an **endocrine secretion** that poured two vital hormones into the blood circulation, insulin and glucagon, the precise roles of which were unknown at that point. Insulin is secreted by special cells that are gathered together in islets, which had been noted from 1869 by a German cytologist, Paul LANGERHANS (1847 - 1888), these cells having remained forgotten due to the lack of an experimental means of understanding their function.

As with diabetes, which was found to be a consequence of the ablation of the pancreas, it was found that the metabolic problems caused by the ablation of other glands in the dog resembled certain human pathologies that had not yet been explained, and thus the experimental method was able to provide the key to understanding their etiology. Nevertheless, **experimental surgery**, which can be considered as the touchstone of functional physiology, still had some surprises, and even disappointments, in reserve. Thus, the thyroidectomy carried out for the first time in a dog by the French physiologist Emile GLEY (1857 - 1930) was a failure for reasons that remained obscure until the moment when careful examination of the thyroid showed that, hidden behind this organ, and barely visible, there were two little pea-sized glands. These are the parathyroid glands and the parathormone that they secrete controls vital processes. In 1835, in the *London Medical and Surgical Journal*, an Irish doctor, Robert GRAVES (1793 - 1853) reported the existence of hypertrophia of the thyroid in patients who also presented exophtalmia and paroxystic tachycardia. This illness, which was given the name exopthalmic goiter or GRAVES disease, the cause of which was presumed to be hyperactivity of the gland, was at this time treated by surgical exeresis. Two Swiss surgeons, Theodor KOCHER (1841 - 1917) and Jean-Louis REVERDIN (1842 - 1929), who carried out thyroidectomies in patients suffering from exophtalmic goiter, while maintaining the parathyroids, observed that after the operation a myxœdema syndrome developed, characterized by a swollen face, the slowing of growth, trophical problems with teguments and phanerae, hypothermia, and finally a slowing of respiration and blood circulation. A connection was made between this set of symptoms and those of spontaneous myxœdema in Man, which, therefore, could be linked to a deficit in the functioning of the thyroid. These first observations and experiments marked the beginnings of **endocrinology**, a science that studies the structure and functions of the endocrine glands as well as their secretion products, hormones, and that often associates experimentation in animals and research on clinical and biological symptoms in Man.

The exploration of an endocrine gland is typical of a deterministic process. First, the gland is removed from the animal. The physiological and biochemical effects that follow are compared to the clinical syndrome in Man that has been linked to the hypofunction of the same gland. Then the removed gland is replaced by administration of a glandular extract. If too much of this extract is administered, adverse effects are the result, and here again comparison is made with a clinical state of hyperfunction. This is followed by a step in which the active substance, i.e., the hormone, is isolated and characterized, and then synthesized. Finally, if a purified hormone is available, and possibly "labeled" with a radioactive isotope, it becomes possible to monitor the metabolism of the hormone, i.e., its modifications inside the organism. Nineteenth century experimental physiology added **intoxication** of animals with poisons such as curare or carbon monoxide to surgical exeresis, with the aim of creating lesions in certain organs and analyzing the resulting modifications in function.

At the time of Claude BERNARD, operational physiology in mammals mainly involved dogs and rabbits, and occasionally cats and monkeys. However, in embryology, the usefulness of other models, such as the frog or toad egg, was recognized, and new techniques were developed, such as micromanipulation and embryonic grafts. This was a considerable change from the thought processes of the previous century in which it had been considered that human medicine could only progress by experimenting on animals close to Man. "This is a mistake," writes Claude BERNARD in *Lessons in Operational Physiology* (1879), "the rabbit's ear is a thousand times better than Man's ear for studying the phenomena of vasomotor innervation. With respect to the general laws of physiology, the frog has rendered far greater service than Man himself could have done."

From the status of an art combined with empiricism that had prevailed in the past, 19th century medicine moved on to the status of a science, which was partly based on the achievements of the operational physiology carried out on animals, and partly based on establishing correlations between clinical observation of malfunctions of the organism and the existence of tissue lesions that are detected *post mortem*. In the human neurosciences, correlations were beginning to be made between motor or cognitive deficiencies and cerebral lesions that were observed during autopsies. Thus, the French surgeon Paul BROCA (1824 - 1880) located a zone of degenerescence in an area of the left hemisphere of the brain in an aphasic patient, after death. The German neurologist and psychiatrist Alois ALZHEIMER (1864 - 1917) identified zones of diffuse atrophia in the brains of deceased patients who had suffered from senile dementia.

To sum up, as an addition to the observations that arose from clinical settings and human pathological anatomy, animal operational physiology came into its own in the 19th century, establishing itself, within a few decades, as an outstanding methodological discipline in biology. Practiced correctly and interpreted rigorously, it revealed relationships between organs and their functions that had sometimes been suspected, but had never been demonstrated. These experimental results would be built upon by the physiology of the 20th century.

3.2. THE EMERGENCE OF INSTRUMENTAL ENGINEERING ADAPTED TO PHYSIOLOGICAL EXPERIMENTATION

In the 19th century, the instrumentation used in physiology and in associated sciences became **industrialized**. Measurement instruments that were remarkable for their accuracy became more rapidly available, and in greater numbers. Efforts were made to make them easier to handle. Physiology owes a great deal to this nascent engineering, which a few examples will demonstrate. In France, Henri Victor REGNAULT (1810 - 1878) developed apparatus for closed-circuit **measurement of gas exchanges** in animals (Figure III.7). A bell jar in which the experimental animal is placed communicates on one side with reservoirs containing potash to trap carbon dioxide and on the other side with a calibrated receptacle half-filled with water and topped with an oxygen atmosphere. Oxygen consumption is evaluated by observing the rise in the water level, and the amount of CO_2 given off is evaluated by weighing the amount of potassium carbonate that is formed. A similar principle was used when, in the 1920s, the German chemist Otto WARBURG (1883 - 1970) constructed a much more sensitive manometric instrument dedicated to the study of gas exchanges, i.e., O_2 consumption and the giving off of CO_2, in fine cross sections (< 0.5 mm) of living tissues (see Figure III.7). Towards the end of the 19th century, Max RUBNER (1854 - 1932) combined **measurements of heat production** and of **gas exchanges** in the same apparatus. The energy contribution of each type of food (proteins, lipids, carbohydrates) to the economy of the organism could therefore be determined (9.3 kilocalories/g of lipids, 5.3 kilocalories/g of proteins, 4.1 kilocalories/g of carbohydrates).

Etienne Jules MAREY (1830 - 1904) spread the use of **graphical records** that made it possible to ensure accurate monitoring of phenomena that an observer's eye could only note approximately. **Graphical recording** of experimental data can be considered as a major advance in the birth of biotechnology in the 19th century. Experimentation was no longer a slave to continuous visual observation of evolving phenomena. The graphical trace is there to give objective information about the progress of a phenomenon that is under study. In Germany, Carl LUDWIG, a pioneer in renal physiology, developed a mercury pump system to **extract gases from the blood** so that they could be analyzed. LUDWIG also invented the **kymograph**, a manometric instrument that makes it possible to monitor the pulsations of blood in an artery by recording them on a rotating drum (Figure III.8).

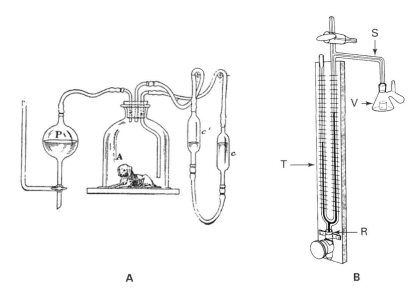

A   B

A - Diagram of REGNAULT and REISET'S macrorespirometer (middle of the 19th century). The animal being experimented on is placed under a bell jar (A) connected by tubes to a glass bulb, P, filled with water, and to a system of two bulbs, c and c', filled with a potash solution. The potash reacts with the carbon dioxide, CO_2, given off by the animal, forming potassium carbonate. Consumption of oxygen, O_2, by the animal creates a negative pressure and the column of water in P rises (from E. HÉDON - *A Precis of Physiology*, 13e édition, Doin, Paris, 1943).

B - Diagram of WARBURG's manometric microrespirometer (20th century). The U-shaped tube, T, calibrated in millimeters, is mounted on a plank. The lower part of it is attached to a rubber reservoir, R, which is filled with a colored liquid. Reservoir R can be compressed by means of a screw, making it possible to adjust the level of liquid in the tube T. The top of the left arm of the tube is open, while the right arm has a lateral branch, S, which ends in a ground opening onto which is attached the experimental flask (V). At the end of the right arm is a faucet that can be placed in the open or closed position. The flask, V, has a central compartment filled with potash, which is designed to absorb the carbon dioxide and a lateral diverticulum that can contain a particular metabolisable and respirable reagent. The cell preparation is placed in the main compartment of the flask, V, just before fitting the manometer onto a shaking device. The flask is bathed in a thermostat-controlled bath. At a given moment in the experiment, the manometer with its attached flask is removed from the shaking device, and by a slight tipping movement, the reagent is poured from the lateral diverticulum into the central compartment of the flask. The manometer with its flask is immediately replaced on the shaking device. In principle, it is possible to measure the consumption of oxygen within a range of a ten microliters. The apparatus generally includes around twelve manometers holding experimental flasks, thus allowing comparative analyses. Here we can see the miniaturization of the respirometer within the space of half a century from the time of REGNAULT and REISET, and how it was adapted to take series of measurements.

**Figure III.7 - Measurement of gas exchanges.
From macrorespirometry to microrespirometry**

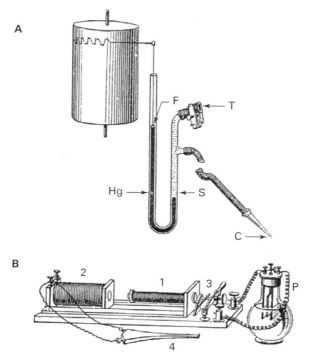

A - LUDWIG's kymograph: manometric apparatus designed to measure arterial blood pulse intensity and frequency. Using tube T, equipped with a mouthpiece, the branch of the manometer connected to the artery by means of a canula, C, is filled with an anticoagulant solution, S. There is a floater, F, floating on the mercury, Hg, in the left arm of the manometer. This is attached by a light stem to a stylet connected to a rotating drum covered with soot. The blood pulsations are shown by an undulating trace on the drum.

B - DU BOIS-REYMOND's induction sledge: apparatus used to produce electrical stimulation of the nerves. (1) inducing coil; (2) induced coil; (3) trembler; (4) spark drawer; P: battery. The battery P generates an electrical current that flows into coil (1) either by means of a hand-operated switch or by means of a trembler if rapid rythmn excitation is desired. The induced coil (2) slides along a groove in the wooden base, so that it can be closer or farther away from inducing coil (1), until it is contained in the cavity from which it is hollowed out. This movement makes it possible to freely regulate the current.

Figure III.8 - Types of instruments invented and used by 19th century physiologists
(from E. HÉDON - *A Precis of Physiology*, 13ᵉ édition, Doin, Paris, 1943)

At the beginning of the 20th century, the use of osmometry as a physical technique for determining the molecular weights of macromolecules was in fashion until more sophisticated and more accurate methods, such as analytical ultracentrifugation, appeared. In fact, the process of **osmosis** had been recognized a hundred years earlier. Its discovery was the result of a fortuitous observation made by Henri DUTROCHET (1776 - 1847) and described in his *Memoires for the Service of the Anatomical and Physiological History of Plants and Animals* (1837) and the reproduction of this observation by him using an ingenious instrument called an **osmometer**. Under the microscope, DUTROCHET observed curious movements of fluids in the transparent filaments of a mold that had developed on the flesh of a fish. "Small globules" were expelled at one end of the filaments, while at the other end water entered as if it had been pushed by the plunger of a syringe. "This observation," writes DUTROCHET, "made me think that I could achieve an analogous result with the intestines of small animals, into which, before I immersed them in water, I would introduce an organic liquid that was denser than the ambient fluid. Guided by this suspicion," he continues, "I took the cecums of young chickens, filled them with liquids that were denser than water, such as milk or a solution of gum arabic, and, after closing them with a ligature, I immersed them in water. These intestine did not take long to swell, becoming turgid because of the introduction of water to their insides." DUTROCHET decided to measure the pressure that resulted from this entry of water. In order to do this, he constructed an instrument that he called an osmometer (Figure III.9).

This instrument comprises a semi-permeable bladder acting as a reservoir, which is connected to a glass tube that is open to the outside. If the bladder is filled with a sugar solution or a gum arabic solution, and is then immersed in a receptacle filled with pure water, there is a rapid ascension of water in the tube. The level that is reached depends on the osmotic pressure, which itself depends on the concentration of the solution in the bladder. This osmotic phenomenon, which is so important in physiology, was reproduced with different types of animal and plant membranes. It provided a simple explanation for such phenomena as the rise of sap in the stems of plants, a phenomenon that had been shown a century earlier by the English naturalist Stephen HALES. Some time later, it was discovered that osmotic pressure depends on the number of molecules in solution, i.e., on the concentration or dilution, and that it is independent of the substance dissolved.

Emil DU BOIS-REYMOND and Hermann VON HELMHOLTZ, who were eminent neurophysiologists, built electrical stimulation apparatus to be used for **electrophysiology** experiments (see Figure III.8). VON HELMHOLTZ showed that an electrical current spreads through nerves at a measurable speed of around thirty meters per second. Half a century later the nerve impulse transmission mechanism would be discovered, expressed in terms of an **action current**, corresponding to spreading differences in the concentrations of sodium and potassium ions on each side of the neuron membrane. The action current thus definitively took the place of "animal spirits".

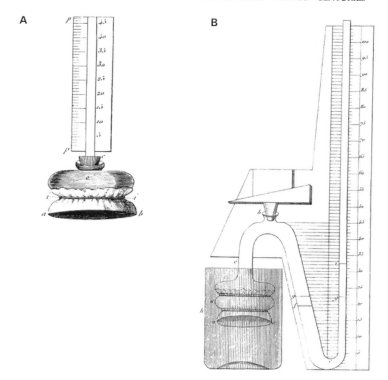

A - Simple osmometer. A semi permeable bladder, connected to an open graduated glass tube, is filled with a sugar solution. It is immersed into a water-filled receptacle. We then observe that water rises in the tube as water enters the bladder. The attraction of water into the sugar solution is due to the osmosis phenomenon, which tends to balance the concentrations of sugar on both sides of the semi permeable membrane.

B - Manometric osmometer. A semi permeable bladder (a), acting as a reservoir, is connected to a U-shaped manometer tube (e) the upper branch of which is fixed to a graduated plank. If the large branch of the manometer is opened, mercury can be poured into the lower curve (c). The levels reached by the mercury are represented as g, g. At the top of the small branch of the manometer, a sugar solution is introduced through opening (b) (which can be closed with a bung). This solution fills the reservoir comprising the semi permeable bladder, as well as branches e, b and b, g of the tube. The bladder is placed in a water-filled receptacle (h). The entry of water into the bladder by osmosis creates a backflow of liquid in the U-shaped tube, the level moving from position g, g to a position f, i.

Figure III.9 - Two types of osmometer constructed by DUTROCHET
(from H. DUTROCHET - *Memoires for the Service of the Anatomical and Physiological History of Plants and Animals*, J.B. Baillière, Paris, 1837)

Optical instrumentation was not without its own innovations. Using **polarimetry** (Figure III.10), the principle of which owes a great deal to the French physicist Jean-Baptiste BIOT, it became possible to analyze the deviation of polarized light by solutions of organic molecules that were known to be optically active.

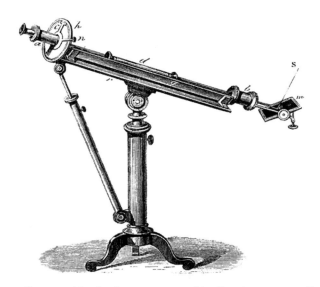

The polarimeter illustrated in the figure was used by BIOT to measure the optical rotation of optically active organic molecules in solution. In the copper trough, g, fixed to the support, r, there is a 20 cm long tube holding the solution to be experimented on. A black glass mirror, m, receives the light coming from a monochromatic source at a certain angle (polarization angle). The polarized light is reflected in the direction bda. In tube a, in the center of disk h, which holds divisions, there is a birefringent prism (Iceland spar) which is in a position perpendicular to the bda axis. This prism can be rotated around the axis of the apparatus using the button, n, which is attached to a tilting level, c, with a vernier. Before the experiment, a check is made that the extraordinary image given by the birefringent prism disappears when the tilting level corresponds to zero on the graduated scale (this is the case when the main cross section of the birefringent prism coincides with the plane of polarization of the light). Tube d now being filled with an optically active solution, and placed in the trough of the polarimeter, the extraordinary image provided by the polarized light reappears. In order to make it disappear again, the tilting level is rotated to the right or the left by a certain angle, depending on whether the solution is dextrorotatory or levorotatory. At the turn of the 20th century, polarimetry was used by Victor HENRI to analyze the procedure for enzymatic cleavage of saccharose into glucose and fructose in the presence of invertase. Victor HENRI's classical experiments marked the birth of the kinetic analysis of enzyme reactions.

**Figure III.10 - The polarimeter in the 19th century,
and its use in the analysis of optically active solutions** (from A. GANOT -
Treatise on Experimental and Applied Physics, 7th edition, A. Ganot, Paris, 1857)

Research relating to optical polarization was at the origin of **stereochemistry**. LE BEL and VAN'THOFF showed that all organic molecules that had rotatory powers included an asymmetric carbon, i.e., a carbon to which are attached four different elements (atoms or groups of atoms).

An instrumental assembly that was to lead to **spectroscopy** was developed, from 1814, by the German optician Joseph VON FRAUNHOFER (1787 - 1826). Sunlight, filtered by a narrow slit, was shone on a prism, an optical system that made it possible to observe the dispersion of the rays of this light. In the 1850s, Robert BUNSEN (1811 - 1899) and Gustav KIRCHOFF (1824 - 1887) constructed the first **prism optical spectroscope**. In the beginning, the spectroscope used light provided by the flame of a BUNSEN burner (Figure III.11).

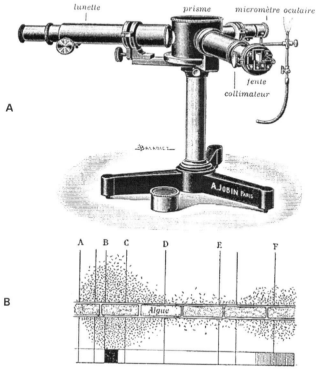

A - Prism spectroscope used in the 19th century. The instrument comprises a collimator, a prism housed inside the turret and an observation telescope. The eye observes, *via* the telescope eyepiece, in the focal plane of its objective, the spectral rays of light that is dispersed by the prism. The light is provided by the flame of a BUNSEN burner. A micrometer with a graduated scale makes it possible to observe the position of the different light rays (from G. SIMON and A. DOGNON - *Precis of Physics*, Masson, Paris, 1941, all rights reserved).

B - Microspectroscopic experiment. A green alga filament made up of an alignment of identical cells is deposited in a drop of water on the glass slide. The drop of water is sown with mobile bacteria that are sensitive to the action of oxygen. The glass slide is placed on the microscope stage. The alga filament is irradiated with a light spectrum of microscopic dimension. After a certain time, the bacteria accumulate along the alga filament in the locations that correspond to specific rays of the spectrum, particularly, as shown on the left of the figure, red radiation. It should be noted how the ingenuity of the device constrasts with its rusticity (experiment published in 1883 by the German cell physiologist Theodor W. ENGELMANN, in "Bacterium photometricum. Ein Beitrag zur vergleichenden Physiologie des Licht- und Farbensinnes", *Pflügers Archiv*, vol. 30, pp. 95-124).

Figure III.11 - The spectroscope in the 19th century, and its use in the analysis of photosynthesis

Both the spectroscope and a variation adapted for microscopic examination (**microspectroscope**) were rapidly brought into use as analysis tools by biochemists. This was the instrument used by Felix HOPPE-SEYLER when he discovered spectral modifications in the hemoglobin of blood from animals poisoned by carbon monoxide and also used by the British doctor and chemist Charles Alexander MCMUNN (1852 - 1911) to show the presence of pigments that are characterized by their absorption of certain rays of white light, in extracts of animal tissues. These pigments would be re-discovered forty years later, in Cambridge (England), by David KEILIN (1887 - 1963), who would name them cytochromes and show their function in cell respiration.

Another brilliant application of spectroscopy, in the domain of photosynthesis, was made by the German Theodor Wilhelm ENGELMANN (1843 - 1909), who showed the efficacy of certain radiations arising from the decomposition of white light in the production of oxygen by a photosynthetic green alga. This filamentous alga, which was made up of an alignment of cells, was placed in a drop of water on a glass slide. Bacteria were added that are able to move by chemotactism towards an oxygen source. The glass slide was placed on the stage of a microscope and illuminated with a light spectrum of microscopic dimension. Within a few minutes, the bacteria moved towards the region of the alga that was receiving the red radiation, and which was giving off oxygen by photosynthesis. This phenomenon was explained by the selective absorption of red radiation by the chlorophyll (see Figure III.11).

At the end of the 19th century, the **exploration of cells** took over from general physiology, with the use of **optical microscopy**, which had come into common use. Nevertheless, some scholars were reticent about accepting the microscope. Certain faults involving optics led to the work of cytologists being discredited, to the extent that Auguste COMTE, in his *Course in Positive Philosophy*, denounced "the abuse of microscopic research and the exaggerated credit that is too often given to a means of exploration that is so dubious." However, ingenious modifications were going to make the optical microscope an instrument that was completely reliable, outwitting COMTE's somber prognoses.

With the **immersion objective** invented by Giovanni AMICI (1786 - 1863) in the 1820s, both the lens of the objective and the object to be examined were immersed in cedar oil with the same refraction index as the glass of the objective, the whole forming a single spherical diopter lens, which eliminated some of the artifacts of refraction. A few decades later, using the combination of silicates of differing diffraction index, according to the principle laid down by DOLLOND (Chapter II-3.2), the German engineer Ernst ABBE (1840 - 1905) managed to produce **apochromatic lenses** that had practically the same focal distance for blue, green and red light. Such lenses eliminated the colored fringes that are due to chromatic aberration. Hired by the small optics company that would become the prestigious Carl Zeiss company of Iena, ABBE improved the performance of the microscope and successfully dealt with resolution problems involved in distinguishing two very close points in an object. He was responsible for developing the **condenser** (condensing lens) that allows light to converge on the object.

In addition to the improvements made in the physical performance of microscopes came the use of **selective dyes**. In the 1860s, the German cytologist Joseph GERLACH (1820 - 1896) discovered that carmine, in an ammoniated solution, fixes specifically to the nucleus of a cell. Another German cytologist, Martin HEIDENHAIN (1864 - 1949) developed a rival technique for coloring the **cell nucleus** with ferrous hematoxylin. From the end of the 1850s, the English chemist, William PERKIN (1838 - 1907) synthesized a series of dyes based on oil tar. These dyes were manufactured on an industrial scale in Germany from 1870 onwards. The German chemist Paul EHRLICH observed that certain dyes, which were said to be basic, color the nucleus of a cell (safranin, basic fuschin, thionein blue, toluidine blue, methylene blue, methyl green, gentian violet), and others that were said to be acidic (eosin, acid fuschin) color the organelles of the **cytoplasm**. EHRLICH differentiated the white blood cells that take part in immune defenses into neutrophils, basophils and eosinophils, according to their affinity for precipitated dyes. One ingenious experiment based on the use of dyes suggested to him that living tissues consume oxygen. This experiment involved injecting animals with **non-lethal dyes that are sensitive to oxidoreduction conditions**: alizarin blue, indophenol blue (colored in the oxidized state, colorless in the reduced state). After the animal had been killed, different tissues (liver, muscle,...) were removed and fine cross-sections made. At first, no coloration was detected, but gradually, because of the exposure to air, the surface of the cross section turned blue, a sign of oxidation. This was corroborated by the fact that the blue coloration was enhanced in an oxygen-enriched atmosphere. Prophetically, EHRLICH supposed that when dyes became fixed to certain cell structures, the dyes could influence operation of the cell structures, a notion which, applied to microorganisms, caused him to discover that Trypan red is active on the trypanosome responsible for sleeping sickness, and that an arsenic derivative, Salvarsan, was able to stop the evolution of syphilis, one of the scourges of this period.

Given solid support by the flourishing synthetic chemistry industry and by clever optical microscopy technicians, the German school of cytology became particularly innovative. Thanks to a mastery of microscopy and related techniques (use of **specific dyes**, practice of making fine cross sections of tissues using **microtomes**), several of the essential components of the cell were identified: the nucleus with its chromosomes and its nucleolus, mitochondria which together make up the "chondrioma", the endoplasmic reticulum, called the ergastoplasm, a network of sacculae that would be called the GOLGI apparatus after the person who discovered them, Camillo GOLGI (1844 - 1925). Walther FLEMMING (1843 - 1905) gave the name **chromatin** to the colorable material of the nucleus present in thread-like structures called **chromosomes** by Wilhelm WALDEYER (1836 - 1926). FLEMMING showed that the chromosomes separated longitudinally during cell division. The broad lines of the phenomena of cell division, **mitosis** for somatic cells, **meiosis** or reductional division for germ cells, were understood. Thus, mitosis is shown under the microscope by a series of complex phenomena located in the nucleus and the cytoplasm, with the result that the substance of the parent nucleus is divided to give two daughter nuclei (Figure III.12).

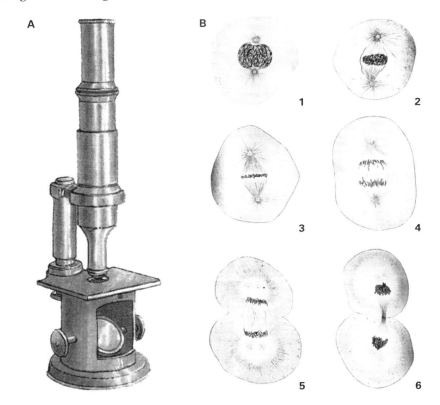

A - The drawing represents a 19th century optical microscope.
B - Cell division (mitosis) in a fertilized trout egg (*Trutta fario*), three days after fertilization. The different steps of mitosis, as represented in the drawing of observations made using an optical microscope, are as follows: **(1)** prophase; **(2)** beginning of the metaphase; **(3)** metaphase where we see the chromosomes come together along an equatorial plate; the achromatic spindle is finally put together; **(4)** start of the anaphase; **(5)** end of the anaphase; **(6)** telophase and reconstruction of the nuclei inside the daughter cells.

Figure III.12 - The optical microscope in the 19th century, and its use in the study of cell division (from A. PRENANT, P. BOUIN and L. MAILLARD - *Treatise on Histology*, Librairie C. Reinwald, Schleicher Frères & Cie, Paris, 1904, all rights reserved)

The reaching of an understanding of cell division, which is the basis of cell theory, illustrates the role played by the engineering of optical microscopes and dyes at the end of the 19th century. Nevertheless, the limits imposed by the resolution of optical microscopy (0.25-0.30 m), were reached. For further insights into the fine structure of endocellular organelles, it would be necessary to await the development of the electron microscope in the 1930s. In order to understand the functions of these organelles, it would be necessary to open the black box of the cell and find the appropriate methods for separating and isolating them. This decisive step would be taken in the middle of the 20th century (Chapter III-6.2.3).

3.3. APPLICATION OF ANALYTICAL CHEMISTRY TO PHYSIOLOGICAL EXPLORATION

The progress made in analytical chemistry in the 19th century ran almost parallel to that made in physiology. The **elementary composition** of many biomolecules was determined with precision, after they were broken down in a combustion chamber, and the resulting gases (O_2, H_2, N_2, CO_2) analyzed. In Giessen, Justus VON LIEBIG's laboratory, which was very well equipped with analysis apparatus, became a reference center for training in the teaching of and research into organic chemistry. Having been trained at a young age in GAY-LUSSAC's school in Paris, LIEBIG, once he had returned to Germany, was offered the chair of chemistry at the university of Giessen, at the age of 21 years. The fruitfulness of his ideas, the richness of his teaching and the practical sense that he brought to it, as illustrated by the perfecting of organic chemical analyses, attracted a multitude of students to his side. Using chemical analysis and content analysis as a basis, the exploration of **cell metabolism** was begun.

Experiments carried out by Claude BERNARD showed that, in animals, the constituents of proteins (later identified as glycogen-forming amino acids) can be converted to glucose, and experiments carried out by Jean-Baptiste BOUSSINGAULT revealed that food sugars are transformed into fats. In the 1890s, the Russian physiologist Sergei WINOGRADSKY (1856 - 1953) discovered that anaerobic soil bacteria that are responsible for nitrification take their oxidation energy from ammonia, and that they use this energy for the synthesis of organic molecules from carbon dioxide. Other bacteria take their energy from the oxidation of hydrogen sulfide, H_2S, in mineral sulfur. These were the first examples of **bacterial chemosynthesis**.

During the last decades of the 19th century, cytologists and biochemists became interested in the fermentation of sugars by yeasts. The impetus had been given by Emil FISCHER, whose work had shown the specificity of the recognition of sugars by different species of yeasts. Nevertheless, the breaking down of a sugar into ethanol and carbon dioxide by fermentation, upon contact with yeast cells, remained a mysterious phenomenon that the experimental method had not yet tackled, and that caused well-known scholars to lose themselves in conjectures. In 1837, three cytologists, the Frenchman Charles CAGNIARD-LATOUR (1777 - 1859), and the Germans Friedrich KÜTZING (1807 - 1897) and Theodor SCHWANN came to the conclusion that beer yeast comprised a mass of thermosensitive globular corpuscles that were able to divide. Experimentation had shown that there are different species that can be differentiated according to specific modes of fermentation. This conclusion, which was thought to favor vitalist theses, was the basis of an impassioned debate let loose by organic chemists including Justus VON LIEBIG and Friedrich WÖHLER (1800 - 1882). The argument used by LIEBIG was that fermentation is a putrefaction process that results from the action of air on an unstable ferment present in the yeast. In response to this argument, Louis PASTEUR remarked that at the same time as glucose is transformed into ethanol, the yeast mass increases due to cell growth, cell division being an undeniable criterion of the living state.

The yeast quarrel finally came to an end in 1897, an important date in the history of **cell enzymology**. That year, Eduard BUCHNER (1860 - 1917) discovered unexpectedly that glucose is fermented to ethanol by an extract of beer yeast with the cells removed. This **acellular extract** was prepared from beer yeast mixed with fine sand and infusory earth, and then crushed in a mortar with a pestle. The crushed result was subjected to extraction in a hydraulic press with a pressure of approximately 500 atmospheres. The juice that was obtained was clarified by being passed through infusory earth. BUCHNER observed that sugar (saccharose), deliberately added to preserve this extract and to use it for therapeutic purposes, was rapidly fermented, with the production of ethanol and the giving off of CO_2. The substance responsible for this fermentation was called **zymase**. It was different from the catalytic factors that were already known, such as invertase, which transforms saccharose into fructose and glucose. Emerging onto **experimental reductionism** (Chapter III-7.2), BUCHNER's discovery left a hope that, once the black box of the cell was opened, the organelles that the optical microscope had hinted that it contained could be separated and analyzed in terms of structure and function.

4. NEW DISCIPLINES IN THE LIFE SCIENCES IN THE 19TH CENTURY AND THEIR METHODOLOGICAL SUPPORT

"Modern science is categorized by its ever-increasing specialization, necessitated by the enormous amount of data, and the complexity of techniques and of theoretical structure within every field. Thus, science is split into innumerable disciplines that continually generate new subdisciplines. As a consequence, the physicist, the biologist, the psychologist and the social scientist are, so to speak, encapsulated in their private universes, and it is difficult to get word from one cocoon to the other."

Ludwig VON BERTALANFFY
General System Theory - 1968

In the second half of the 19th century, the different, related aspects of biology began to branch off from one another, with the scientific community rapidly coming to regard them as disciplines in their own right, with specific research methods and teachings. Thus, **cytology** and **histology, biochemistry, embryology, genetics, enzymology, microbiology, immunology, endocrinology, vitaminology** and **pharmacology** came into being. Within this branching out, **physiology** remained a science that analyzed integrated functions, both in animals and in plants, necessarily helped along by the contributions of biochemistry and biophysics.

At the turn of the 20th century, in reaction to a stifling isolation, interactions began to be formed between the new disciplines. Cytology, associated with cell physiology, became **cell biology**. In the middle of the 20th century, the merging of biochemistry and genetics led to a new discipline, **molecular biology**, from which were born **structural biology**, which is based on the deciphering of the anatomy of biomolecules, and **biocomputing**, which is devoted to the comparative analysis of identity data arising from genetic material and macromolecular structures. **Developmental biology**, which was developing rapidly, with its roots in experimental embryology, was helped along by techniques from cell biology and molecular biology.

J. LISTER
(1827 - 1912)

At the end of the 19th century, microbiology had made remarkable progress. Pathology was able to explain the causes of previously mysterious infectious diseases, from the moment at which it was possible to master the technique of **pure bacterial cell cultures**. The English surgeon Joseph LISTER (1827 - 1912), who was a pioneer in spreading the use of asepsis and antisepsis in human surgery, had, in 1878, described a method for obtaining pure bacterial cultures. He used a microsyringe to put a very small volume of around one microliter of a dilute bacterial suspension onto a glass slide. The bacterial density in the microdrop was evaluated by examination under a microscope. LISTER made a dilution according to this density. An aliquot fraction of the diluted suspension that was considered statistically to contain one or zero bacteria was injected into a culture medium.

This technique, based on the **limit dilution** principle, was adopted by bacteriologists until, in 1881, the German physician and bacteriologist Robert KOCH developed a more efficient technique using a **solid medium**. In fact, in 1872, a military physician Joseph SCHROETER (1835 - 1894) had already cultivated chromogenic bacteria, colonies of which were easy to identify because of their color, on starch paste. The first solid medium used by KOCH comprised slices of potato kept away from bacterial contamination by the dust in the air. Later, gelatin, poured into dishes with covers (PETRI dishes) was used (Chapter III-2.2.5). However, at 37°C, the temperature often used for bacterial cultures, gelatin is still soft, which makes it difficult to spread bacterial suspensions. So gelatin was replaced with agar-agar,

R. KOCH
(1843 - 1910)

which is an extract of a marine alga. In current practice, the agar is sterilized and fluidified beforehand by heating, and the nutrients necessary for bacterial growth are added to it. Cooled to a temperature below 40°C, the agar sets to form a solid gel. A droplet of a very dilute bacterial suspension is deposited on the gel and spread uniformly across it by means of a thin, curved spreader, which acts rather like a rake. The dish is covered with its lid and is placed in the oven and brought to 30° or 37°C. After 24 hours, it is found that, scattered across the gel, little mounds have formed. These correspond to bacterial colonies, each colony (or clone) arising from the iterative division of a single bacterial cell into several million daughter cells. Nowadays, the technique of producing bacterial isolates on PETRI dishes seems so obvious that it is difficult to imagine its beginnings. The first microphotographs of bacterial cells coming from pure cultures of the anthrax bacillus were taken by Robert KOCH from 1867 onwards (Figure III.13). Thus, making use of a technique that was astonishingly simple and very inexpensive, bacteriology established itself as an experimental discipline.

A 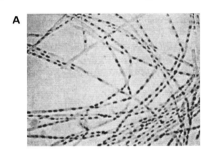 B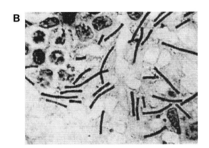

At the beginning of the microbiological era (end of the 19th century), the anthrax bacillus was a model microorganism for the study of infectious pathologies (easily identified illness in the animal, typical rod shape of bacilli).

A - Cells of anthrax bacillus (not-colored) showing the existence of spores that are refringent in light.

B - Smear of cells from the spleen of an animal infected with the anthrax bacillus, after coloration, showing the characteristic rods of the anthrax bacillus, in the middle of the spleen cells.

Figure III.13 - First photomicrographs of bacteria taken by Robert KOCH in 1877

A mastery of the production of pure cultures made it possible to make definite identifications of many previously-unknown bacterial species, to establish relationships between a particular bacterial species and the infection it causes in animals, and by extension, in Man, and also to protect against infection and contagion by applying the principles of asepsis. Experimental surgery in animals and human surgery benefited directly. It was during this period that the criteria regarded as

necessary to associate a well-identified germ with a specific disease were formulated, these criteria being labeled KOCH's postulates:
- the germ must be present in each diagnosed case of the disease;
- the germ must be isolated and characterized from a diseased host animal being used for experimentation;
- a specific disease must be reproduced by inoculation of a pure culture of this germ into a healthy host animal;
- the germ must be retrieved from the infected host animal.

These specificity criteria arose straight out of Bernardian determinism. The rapidity with which bacteria proliferate and the ease with which cultures of bacteria can be handled led to bacteria becoming the preferred biological **models** in the 20th century for the exploration of metabolism reactions and the study of the mechanisms of protein synthesis.

While KOCH was developing his technique for producing pure bacterial cultures on solid media, the Dane Emil HANSEN (1842 - 1909), at the Carlsberg laboratory in Copenhagen, was perfecting a method for culturing yeast, also on a solid medium. Starting with a complex mixture, different species of yeasts were isolated and differentiated on the basis of their morphology and the nature of their fermentation products. In the last two decades of the 19th century, more than a hundred species of yeasts were identified, listed and stored, and were thus available for requesting laboratories. One of these yeasts, *Saccharomyces cerevisiae*, would later be chosen as a **unicellular eukaryotic microorganism model** for the study of physiological processes that are specific to eukaryotic cells, such as cell division, the secretion de neosynthesized proteins or the traffic of macromolecules between the nucleus and the cytoplasm. It is interesting to note that it was with *S. cerevisiae* that enzymatic adaptation was described for the first time, at the turn of the 20th century. It was discovered by a French microbiologist Frédéric DIENERT (1874 - 1948), who noticed that glucose added to a suspension of yeast is fermented immediately, whereas the fermentation of galactose only begins after a latency period of several minutes. The significance of this period of latency with respect to the repression of gene expression was revealed at the beginning of the 1950s by Jacques MONOD and François JACOB, based on studies of the bacterium *Escherichia coli*. In the first decades of the 20th century, the results of experiments carried out on yeasts, bacteria and also on animal tissues, provided evidence that **similar chemical and enzymatic mechanisms** were implemented at all levels in Nature, in accordance with DARWIN's theory of evolution.

In 1879, Louis PASTEUR made a fortuitous observation of the way in which the virulence of the bacteria responsible for cholera in chickens disappeared by simple aging of the bacterial culture. Reproducing this bizarre result experimentally led to the concept of **vaccination**, which was confirmed in a resounding manner by the Pouilly le Fort experiment carried out by PASTEUR in 1881. Sheep, which had previously been inoculated with anthrax bacteria killed by heat, were resistant to an injection of live anthrax bacteria, the same living bacteria triggering fatal septicemia in control sheep. While discoveries were accumulating in microbiology, the concepts of **cellular** or **innate immunity** and of **humoral** or **acquired immunity** were born. Here again, the experimental method, which was so often a case of "provoked observation", showed its power. Elie METCHNIKOFF discovered the **phagocytosis** of bacteria by certain white blood cells, the neutrophils, and by tissue macrophages (Chapter III-2.2.3), while the observations of Emil VON BEHRING and Shibasaburo KITASATO concerning the immunization of guinea-pigs with denatured diphtheria toxin led to the idea of **antibodies** neutralizing the fatal active toxin. Today, antibodies have become tools that are commonly used in cell biology to recognize specific proteins, whether the proteins are present in a cell extract or are located in a compartment of the cell.

The **descriptive embryology** that was demonstrated at the beginning of the 19th century by the work of Christian PANDER (1794 - 1865) on the first divisions of fertilized eggs in different animal species (batrachians, fish, birds, mammals) gave way, a few decades later, to **experimental embryology**, the pioneers of which were Wilhelm ROUX (1850 - 1924) and Hans DRIESCH (1867 - 1941). Analysis of the first stage in the development of the fertilized egg led to a recognition of iterative divisions that lead to an accumulation of cells, the **blastomeres**, their arrangement in a single layer, the **blastoderm**, which surrounds a central cavity, the **blastocele**. The small organism that corresponds to this stage is called the **blastula**. This is followed by a stage of **gastrulation**, during which the blastoderm becomes invaginated, to form a primitive intestinal cavity. DRIESCH showed that each isolated blastomere of the sea urchin egg blastula is able to develop into an adult organism, and is said to be **totipotent**. Once the blastula stage has passed, this totipotentiality disappears.

For reasons that were mainly technical, the **batrachians** egg became the material of choice in embryology. In fact, it is after the eggs of these animals are laid that they are fertilized by sperm. The fact that the fertilized egg develops in fresh water, outside the female organism, was an obvious advantage for observation of the first stages of fertilization. Between 1925 and 1929, using the fertilized eggs of amphibians at the blastula stage, Walter VOGT developed a technique for **labeling** small groups of cells **with vital dyes**, neutral red or Nile blue, in order to be able to monitor what happened to them during gastrulation. This very simple but ingenious technique was a great help in understanding the mechanism of the formation of **germ layers** and their evolution towards a defined organ in an adult organism. For the first time, it became possible to map the "presumed" territories of the blastula. In the last decades of the 20th century, with the convergence of the methods and principles of classical genetics and molecular genetics, experimental embryology took the label **developmental biology**. Among the findings of this rapidly-expanding discipline, let us mention the discovery of **homeogenes** in drosophila, these being genes that control the specificity of the location of organs in the different segments of an insect body.

In the nutritional life sciences, the use of the experimental method led to the opening of an unexpected domain, **vitaminology**. It had been noted that, despite being fed with enough protein, lipids, carbohydrates and mineral salts to maintain a caloric equilibrium, an individual could not be kept in good health. Scurvy, which is characterized by ulcerations of the mucosa of the mouth and digestive hemorrhages, and was later identified as being avitaminosis C, had been known for a long time. This condition was found, in particular, in sailors who went on long sea voyages. In order to prevent scurvy, a Royal decree at the beginning of the 19th century in Britain made it mandatory to add citrus fruit to the sailors' daily rations, a practice that arose from an observation that was mainly empirical. The first vitamin that was identified according to experimental criteria was vitamin B1. In the 19th century, a lack of this vitamin was found to cause Beriberi, an illness that was endemic in Far Eastern countries, and was characterized by peripheral polynevritis, paralysis, cardiac insufficiency and edema of the extremities. The incidence of this illness had gradually increased as unpolished rice, which was the basic food of eastern countries, was replaced by white or polished rice that the food industry produced by removing the outer husk of the rice grain. We now know that the rice grain husk contains a hydrosoluble vitamin, thiamine, or vitamin B1, which forms the carbon framework of thiamine pyrophosphate, a coenzyme involved in the decarboxylation of α-cetonic acids inside cells. **Experimental proof** of a relationship between neurological problems and a specific nutritional deficiency was provided in the 1890s by the Dutch physician Christiaan EIJKMAN (1858 - 1930). When he fed chicks with polished rice, he observed the appearance of nerve problems similar to those seen in human Beriberi. These symptoms regressed rapidly when the polished rice was replaced with unpolished rice or even when rice husks were added to the polished rice. In 1912, a biochemist of Polish origin, Casimir FUNK (1884 - 1967), gave the name **vitamin** to the purified anti-Beriberi factor, when it was recognized that this factor had the properties of an amine and that it was indispensable to life. The term "vitamin" continued to be used to designate factors of an organic nature, which did not necessarily carry amine groups, the discoveries of which followed one another over the years, each factor being shown to be indispensable, in small quantities, for keeping the individual in good health. The originality of vitamins is that they are the constituents of coenzymes or prosthetic groups of enzymes.

While biology, which up until then had been confined to experimental physiology, burst forth into many different disciplines, medicine was undergoing a revolution of ideas that led to a revision of its **nosological classification**, which had been somewhat hesitant, if not chaotic, up until then. The idea of deleterious miasmas gave way to that of infections, with the germs that caused these infections being identified, one by one. The antisepsis that SEMMELWEIS had recommended for the prevention of the spread of infectious diseases, a recommendation based purely on empirical experiments (Chapter III-2.2.1) was shown to make sense and be based on an unarguable rationale. Symptoms that had previously seemed scattered and without correlation with one another were grouped together into **syndromes**, a disease being defined as a particular, well-characterized and easily-identifiable syndrome. The nosological framework of medicine was becoming consistent (Chapter IV-3).

5. THE IDEA OF QUANTIFICATION IN THE LIFE SCIENCES

Although the **idea of number** in the evaluation of experimental results was beginning to make its way to the fore in the 19th century, how far this was necessary was not yet unanimously agreed upon. The first comparative tests in medicine were the subject of ironical comments by Auguste COMTE: "To compare two curative methods according only to statistical tables of their effects, disregarding any healthy medical theory, this is absolute empiricism based upon cold mathematical considerations," he wrote, or "no notion of fixed numbers, or even more so, of numerical laws and finally, above all, of mathematical investigation, can be regarded as compatible with the fundamental character of biological research." Claude BERNARD, despite his inspired approach to physiology, was no more enamored of statistics. Nevertheless, the experiments carried out at this time by MENDEL, on the hybridization of peas, showed the usefulness of working with large numbers, in this case several thousand plants.

G. MENDEL
(1822 - 1884)

In contrast to other naturalists, MENDEL was concerned about statistical rigor. This sense of rigor no doubt arose from the teaching he had received from the physicist Christian DÖPPLER (1801 - 1853) at the University of Vienna. A few years before, the German physiologist Rudolph WAGNER (1805 - 1864), had suggested applying the principles of **statistics** and the calculation of **probabilities** to the analysis of traits inherited from parents. When MENDEL began his experiments in the 1850s, a German botanist, Carl GÄRTNER (1772 - 1850), had already published a monograph on plant hybridization, dealing especially with peas, about ten years beforehand. In fact, the accepted view at this time was that the characteristics of the parents were mixed, and appeared combined in the children. Taking the pea, *Pisum sativum*, as his model, MENDEL selected seven pairs of characteristics that were particular in that each characteristic in a pair is present in one variety of pea, but not in another, for example, the shape of the seed, smooth or wrinkled, the color of the seed, yellow or green, the color of the covering of unripe seeds, white or brown, the shape of the pod, smooth or notched, the color of the unripe pod, green or yellow, the position of the flowers, axial or terminal, and the length of the stem, more or less than one meter. It is probable that the choices were dictated by the results of many preliminary tests. MENDEL's experiments were spread out over a period between 1854 and 1865, i.e., around ten years, during which more than 12 000 plants were analyzed. In 1865, the results were reported in a first communication that was presented to the Brno Society of Natural Sciences, and then an article was published in 1866 in the reports of the same Society. For example, cross-breeding pure race plants, some of which had smooth seeds and others wrinkled seeds, gave rise to hybrids (F_1) that all had smooth grains, no matter which type of parent the ovule or the pollen had been taken from. The smooth seeds of the hybrid plants were then sown. They produced second generation, F_2, plants that, by self-fertilization, provided 5 474 smooth seeds and 1 850 wrinkled seeds, i.e., three times as many smooth seeds as wrinkled ones. Obviously, in the F_1 generation, the wrinkled characteristic was masked and only the smooth characteristic was shown. In the F_2 generation, the wrinkled characteristic reappeared. MENDEL gave the name **dominant characteristic** to the smooth type and **recessive characteristic** to the wrinkled type. He observed the same phenomenon for the other pairs of characteristics that were studied. For example, the "yellow" type is dominant, while the "green" type is recessive. MENDEL continued his experimentation and demonstrated that dominance and recessivity are also observed when cross-bred plants present two different characteristics (shape and color, for example). Based on these observations, one conclusion was obvious: the characteristics under study (smooth, wrinkled, yellow, green...) **segregate independently**, thus excluding the theory of heredity by mixing that was widely accepted at this time. Thus, **the use of a large number of samples**, by providing interpretable results that were not the result of chance, led to the formulation of the first laws of inheritance, and to thinking in terms of **individual hereditary traits**.

MENDEL's laws, which were rediscovered at the beginning of the 20th century by Hugo DE VRIES (1848 - 1935), Carl CORRENS (1864 - 1933) and Erich VON TSCHERMAK (1871 - 1962), acted as the basis for the genetics of populations illustrated a little later by the work of Thomas Hunt MORGAN (1866 - 1945) and his co-workers on the drosophila (fruit fly). Many cross-breeding experiments involving thousands of normal drosophila and drosophila affected by mutations led to the **chromosome theory of heredity**. In 1909, the Danish botanist Wilhelm JOHANNSEN (1857 - 1927) published the results of experiments that he had carried out on the reproduction by self-fertilization of two pure lines of the bean *Phaesolus*, and the analysis of their seed size. He showed that over several generations these lines were maintained with the same characteristics. He concluded from this that the members of a line were genetically identical, and to convey this idea, he coined the term **genotype**. Nevertheless, he did note that within the same line, seed size was distributed according to a **Gaussian curve** with the maximum point corresponding to the average size.

Nowadays, it is no longer necessary to demonstrate the need to use a **large number of samples** or the importance of **statistical data** in certain areas such as pharmacology. Before it can be distributed on a large scale, any new drug must be subjected to a statistical study of its effects on patients, in association with a parallel study using a placebo. The results are then subjected to mathematical analysis, which allows a rigorous and objective analysis of the characteristics of the drug.

6. *A NEW EXPERIMENTAL ORDER FOR THE LIFE SCIENCES IN THE 20TH CENTURY*

When we examine the evolution in protocols and the reports of experiments carried out on living organisms over the last two centuries, we see that the same logic, marked by determinism, governs the development of ideas and the analysis of results. However, **differences** appear over time, involving the **technical facilities** that are implemented. In the 19th century, in physiology, these facilities were limited to the scalpel, apparatus for unrefined volumetric measurements of oxygen and carbon dioxide, perfected calorimeters and rudimentary electrophysiology instruments. Despite significant progress, the optical microscope only gave an approximate vision of cell contents. Biochemists had a few pieces of apparatus inherited from physics, such as the polarimeter (see Figure III.10) and the spectroscope (see Figure III.11). The acidity and alkalinity of media were estimated approximately using colored indicators (sunflower dye, for example). The only animal models used for experimentation were the dog, the rabbit, the frog and the planarium.

The first decades of the 20th century saw the appearance of measurement instruments that have since come into everyday use, such as the **pH meter** constructed by the Dane SØRENSEN (1868 - 1939), the **filter photometer**, and then the **spectrophotometer** made popular by many optical companies, or even WARBURG's **manometric respirometer** (see Figure III.7B). These pieces of equipment helped the chemistry of living beings to make spectacular progress. The work of the German biochemist Leonor MICHAËLIS (1875 - 1949) demonstrated the unexpected effect of the pH of an incubation medium on enzyme activity. Using colored reagents, photometry made it possible to monitor the evolution of chemical reactions in cellular homogenates. WARBURG's manometric apparatus was immediately successful because it was able to measure the consumption of oxygen and the emission of carbon dioxide by suspensions of living cells, on a micromolar level.

The makeshift nature of the facilities available for experimentation in the 19th century, which might be considered as surprising given the lightning progress of knowledge at that time, is in considerable contrast to the profusion of facilities that are available to modern biology more than a century later, with a large variety of experimental models, and instruments capable of probing into the details of cell organization and pushing back the limits of the structural exploration of macromolecules to the dimensions of the atom and of analyzing kinetics on a scale of a picosecond. The manufacture of scientific instruments has moved from an artisanal to an industrial scale. The exeresis of organs in order to determine their role in the economy of the individual has given way to the isolation of intracellular organelles in order to discover their specific functions. Mendelian hereditary traits have been given a concrete form in terms of sequences de nucleotides in DNA. Labeling of organic molecules with radioactive atoms or fluorescent ligands makes it possible to follow their metabolic pathways. Three-dimensional structures, on a scale of one angström (Å), and thousands of protein species are known and listed. Each year, several dozen new structures are added to this list. Instrumentation that is more and more efficient, miniaturized and even robotic, makes it possible to manipulate genes, to cut them up and stick them back together. Where is it possible to situate the transition between two periods that are so different? While the beginnings can be seen between the two world wars of the 20th century, the situation only really took off in a noticeable way in the 1950s and 1960s, and expanded even more in the following decades.

6.1. A REASONED CHOICE OF MODEL ORGANISMS

"The great book of Nature resembles the fabled book of the Sybil of Cumae, the separated and scattered leaves of which asked to be looked for, to be put back together by the sagacity of a laborious investigation. The physiologist must therefore seek to read into the organism of all living beings, without exception; he must question these beings separately; each of them will tell him its word, each of them will raise before his eyes a particular portion of the veil with which Nature covers her mysteries. It is from the universality of these searches, from the gathering together of these scattered documents, that an understanding of life will arise."

Henri DUTROCHET
*Memoires for the Service of the Anatomical
and Physiological History of Plants and Animals* - 1837

In the 19th century, before the idea of **animal or plant models** for experimental research was clearly understood, reasoned choices were made, making the experimental approach easier and the interpretation of results more direct. If MENDEL chose the pea as a model, this was probably following a series of preliminary tests and a programmed selection of characteristics (plant color, shape, and size...) that would later be known to be controlled by genes located on different chromosomes. The discovery of mitosis in somatic cells and meiosis in germ cells, made by Edouard VAN BENEDEN (1846 - 1910) and Theodor BOVERI (1862 - 1924) definitely benefited from the choice of a worm from the horse intestine, *Ascaris megalocephala*, the somatic cells of which only had two pairs of chromosomes. At the beginnings of experimental embryology, the frog's egg was immediately preferred both because of its large size (1 to 2 mm in diameter) and because the eggs laid by the female are fertilized outside the organism.

In the 20th century, the range of experimental models expanded, including not only the animal and plant kingdoms (Figure III.14) but also microorganisms. Biochemists who were interested in the metabolism and explored its reactions and catalytic enzymes first preferred two models, **yeast** and the **rat**.

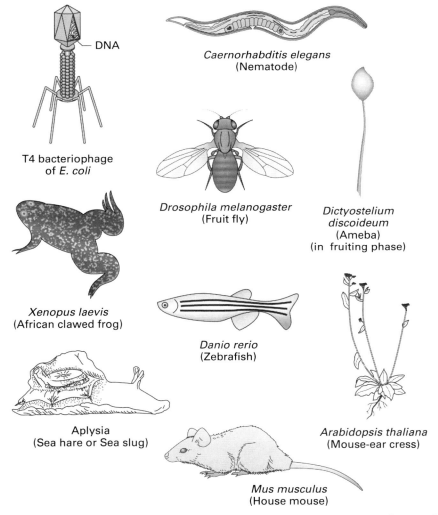

The experimental models represented in this figure were chosen according to the questions concerning specific functions (genetic, metabolic...).

Figure III.14 - Different experimental models used in biological research

Depending on the progress in genetics, questions were asked about what role was played in the metabolism by genes. To answer these questions, it was necessary to find an experimental model that could be explored by both chemical and genetic analyses. BEADLE (1903 - 1989) and TATUM (1909 - 1975) chose the **mold** *Neurospora crassa*, a **heterothallic ascomycota**. Its cells only had a single set of chromosomes and genes. **It was therefore a relatively simple species, from the genetic point of view, with neither dominance nor recessivity.** As there are no homologous chromosomes in a haploid species, mutation of a gene leads to a loss of function that is not compensated for by the corresponding allele, as could be the case for a diploid species. The sequence of the *N. crassa* genome, published in April 2003, includes 40 megabases and 10 000 genes coding for proteins. *N. crassa* has a sexual reproduction cycle, but can also reproduce vegetatively, giving rise to thousands of spores. This asexual spore reproduction corresponds to a type of grafting. At the time when BEADLE and TATUM were carrying out their experiments, it was known that irradiation with X-rays greatly increased the percentage of natural mutations. Certain of these mutations could affect the genes involved in the synthesis of enzymes catalyzing the reactions of the metabolism and create disturbances in vital functions, thus blocking cell growth. In fact, if a certain reaction is essential to a microorganism in order for it to develop, a mutant that is incapable of carrying out this reaction cannot survive.

BEADLE and TATUM used a solid synthetic **medium**, known as a **minimal medium**, made up of agar to which was added glucose as a carbon and energy source, an ammonium salt as a source of nitrogen, mineral salts containing phosphorus, sulfur, potassium, calcium and a few oligoelements (iron, magnesium…), and finally biotin, a B group vitamin. From the elementary components of this minimal medium, a wild type strain of *N. crassa* is, in fact, capable of synthesizing all the chemical species (carbohydrates, lipids, proteins, nucleic acids) involved in the composition of its molecular structures, with the exception of biotin. After irradiation of a wild type strain of *N. crassa*, **mutants** were isolated that were **unable to proliferate on the minimal medium** (Figure III.15). Proliferation started up again after the addition of a specific vitamin or amino acid. Based on these results, BEADLE and TATUM postulated that irradiation had led to the mutation of a gene coding for an enzyme that is indispensable to the synthesis of the vitamin or the amino acid concerned. The famous aphorism "One gene, One enzyme" arose from this occasion. By demonstrating the efficacy of an interaction between **genetics and biochemistry**, this type of experiment paved the way for **molecular biology**, a hybrid discipline that the physicist Warren WEAVER (1894 - 1978), at the Rockefeller Institute of New York, a few years earlier, had prophesized would come to prominence. In order to measure the fantastic progress that was made in knowledge concerning this domain, over a period of one century, it is enough to remember that at the end of the 1930s, the problem of the control of genetic determinism was still under debate.

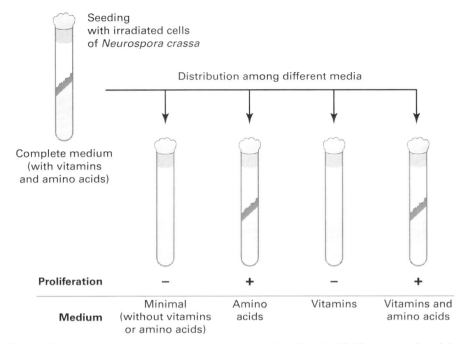

Cells of the mold *Neurospora crassa*, previously irradiated with X-rays or ultraviolet rays in order to increase the rate of mutations, are deposited on a complete culture medium (solid agar). After proliferation, the cells are distributed among different tubes, one of which contains a minimal culture medium (synthetic medium without vitamins or amino acids), and the others of which contain the minimal medium to which has been added a mixture of vitamins or of amino acids, or a mixture of vitamins and amino acids.

The absence of proliferation signifies a defect in the synthesis either of the vitamin(s), or of the amino acid(s). Here there is a defect in amino acid synthesis because the addition of the mixture of amino acids starts the proliferation up again. The operation is continued in order to identify precisely for which amino acid the synthesis has been blocked because of a mutation in one of the enzymes involved in the pathway of this synthesis. The synthesis of any amino acid involves a chain of precursors, which are generally well identified. The addition of these precursors, one by one, to the minimal culture medium makes it possible to locate the site of the mutation, by identification of the precursor for which the synthesis is blocked.

Figure III.15 - Principle of production and detection of mutants in the microscopic mold *Neurospora crassa*
(adapted from G.W. BEADLE (1946) *American Scientist*, vol. 34, p. 31)

In the 1940s, **microbial chemistry** was expanding rapidly. Bacteria became a preferred terrain to which the principles of molecular biology could readily be applied. A non-pathogenic strain of the enterobacterium *Escherichia coli* (*E. coli*) became the reference experimental model in numerous laboratories around the world. It is surprising to note that the **genetic code** (which was later recognized to be universal, or nearly so) was first deciphered based on experiments on *E. coli*, and that this same microorganism would provide the information that was going to lead to the **central dogma of molecular biology** being postulated, involving a transcription of genes into messenger ribonucleic acids and the translation of the latter into proteins. What confusion there would have been if a eukaryotic model had been chosen, with its chromosomes carrying coding sequences (exons) interrupted by noncoding sequences (introns)!

The structures of viruses are the simplest to be found in the living world, at the frontier between this world and the inanimate one. **Bacteriophages**, viruses that infect and kill bacteria, and which have a genome that is twenty times smaller than that of *E. coli*, are models that can be used in addition of the bacterial model for the study of gene expression. Bacteriophage T7, which is commonly used, comprises a icosohedral protein envelope inside which is housed the genetic material, comprising a long strand of DNA. The envelope is extended by hooked appendages that allow the bacteriophage to catch on to the surface of a bacterium and a hollow tube through which the virus DNA passes, to be injected into the bacterium (see Figure III.14). The injected DNA is replicated and diverts the bacterium's protein synthesis machinery in such a way that dozens of neoformed bacteriophages accumulate inside this bacterium. When there are about a hundred of them, the bacterium bursts, liberating bacteriophages that continue to spread the infection. The genome of bacteriophage T7 contains 39 937 base pairs distributed among 56 genes that are classified into early, delayed early and late genes. Among the early genes is the one coding for the RNA polymerase that transcribes phagic DNA into messenger RNAs. The replication genes of phagic DNA are among the semi-early genes. Finally, the late genes code for the proteins of the envelope of the bacteriophage. The different steps involved in the infection of *E. coli* by bacteriophage T7 have recently been modeled (Chapter IV-4.2).

Nowadays, **prokaryotic organisms** remain the preferred tools for the exploration of mechanisms in molecular genetics. Their rapid reproduction in identical form is a definite advantage. At 37°C, *E. coli* divides every 20 to 40 minutes, depending on the culture medium. When cultivated in a PETRI dish on a solid medium comprising agar to which has been added nutritive substances, a single isolated cell of *E. coli* can, in 24 hours, provide, by iterative division, a **colony or clone** with a population that can be as high as one hundred million cells. Most of our knowledge of **molecular biology** – replication of DNA, genetic code, gene expression, protein synthesis – comes from research carried out on *E. coli*). This research has been helped by the small size of the genome (around 4 million base pairs) mainly made up of exons coding for proteins as well as ribosomal and transfer ribonucleic acids. While *E. coli* still remains a prototype that is widely used, many other prokaryotes are the subject of investigation. It should not be forgotten that the first transformation experiments were carried out with pneumococcus. Each year, species with unexpected properties are added to the list. Each year new findings are there to surprise us, making microbiology a science that is forever young, continu ously asking new questions. For example, a curious discovery in the 1980s was that of the programmed death of certain species of bacteria, in keeping with the apoptosis of eukaryotic cells. It was found that the decision to commit suicide made by a bacterium depended on the presence in its chromosome or in the plasmid that it hosts of genes coding for either a stable, lethal toxin (T), in one, or a relatively unstable, antidote protein (A), in the other. The association of TA controls the survival of the bacterium, while its dissociation leads to its death, the stable toxin carrying out its lethal action while the unstable antidote has already disappeared. This became something of a puzzle when, after methodical analysis, it was shown unexpectedly that expression of the TA pair occurs in the majority of cases of bacteria that are able to develop outside eukaryotic cells, while bacteria that have a development that must occur in eukaryotic cells are lacking in the TA system. Thus, KOCH's bacillus (*Mycobacterium tuberculosis*), which proliferates in extracellular media, contains 36 pairs of genes coding for TA proteins, while the leprosy mycobacterium (*Mycobacterium leprae*), which only develops in the presence of living tissue, does not have any. Does the presence of the TA system in certain bacterial species have any relation to the process of programmed death, or to a survival differential between cells that carry the TA system and those that do not, or even to a state of stress, bacteria hosted in a eukaryotic host benefiting from stable food and temperature conditions? The solution of the TA puzzle is doubtless still some distance away, but it could have implications in human pathology.

The **yeast** *Saccharomyces cerevisiae*, a **unicellular eukaryotic microorganism**, was chosen early on as an experimental model for experiments dealing with cell metabolism. *S. cerevisiae* mostly exists in the form of diploid cells that reproduce by budding. When the medium is lacking in carbon nutrients, the cells enter meiosis (reductional division). The **haploid cells** and pseudosexual cells that are derived are distributed into type a and α cells that can propagate in haploid form or conjugate to give diploid cells again. After the second world war, there was renewed interest when genetic analysis of *S. cerevisiae* provided information to be added to the knowledge arising from biochemical data. **Many mutants** of this yeast were and are useful tools for the study of the mechanisms of signaling, the control of cell division and the molecular sorting of neosynthesized proteins. It has recently been observed that derivatives of oxygen, such as hydrogen peroxide, significantly modify the protein equipment of *S. cerevisiae*, by acting on the expression of certain genes. For this raison, *S. cerevisiae* is a good study model for looking at **oxidative stress**. Other yeast strains are currently under investigation, such as *Saccharomyces pombe*, which is an excellent model for studying the cell cycle, and *Pichia pastoris* and *Pichia augusta*, which, after transfection by human genes, are able to produce human proteins.

Another eukaryotic organism, the **ameba** *Dictyostelium discoideum*, provides interesting information for the study of **cell differentiation**, **chemotactism** and **intercellular interactions**. This ameba lives in the soils of forests in the form of independent mobile cells that feed on bacteria or yeasts that they detect by chemotactism. In the absence of nourishment, the amebas no longer divide, but join together to form 1 to 2 mm long multicellular microstructures that take the form of a "slug" The differentiation process continues and produces a structure comprising a stem ending in a ball, the fruiting apparatus, which contains a multitude of heat and drought resistant spores. Under favorable conditions, these spores germinate and produce mobile amebas, initiating a new cycle.

The **fruit fly**, *Drosophila melanogaster*, which was used by MORGAN at the beginning of the 20th century, remains a preferred model in developmental biology, for several reasons. **It has a short life cycle** of around two weeks. **It is highly fertile**, which allows **statistical analysis** of characteristic traits in descendants. There are **only four pairs** of **chromosomes**, which allows easy study of gene recombinations between chromosomes. It was in the drosophila that **development genes** or **homeogenes** that control the positioning of organs in the embryo were discovered. Located on one of the four chromosomes, the homeogenes form a complex, and the order in which they are arranged on the chromosome corresponds curiously to that of the segments that they control along the insect's body. Any modification in a homeogene leads to the formation of "monsters" that have, for example, feet on the head or eyes on the thorax. Homologues of the development genes in drosophila exist in numerous species that have been listed, including mammals. The science of **infectiology** has recently recognized the drosophila as an interesting model for the study of the pathogenesis associated with a certain number of microbial species, including the fearsome *Listeria monocytogenes*, and for the identification of proteins defending against infection. In fact, the drosophila has an innate **immune system** capable of producing antimicrobial peptides.

In the last few decades, the **nematode** *Caenorhabditis elegans*, the **aplysia** (sea hare or **sea slug**) , the **planarium** and the **zebrafish** have become experimental models) that have been adopted for their specific characteristics. *C. elegans* was introduced by Sidney BRENNER (b. 1927) at the end of the 1960s. It is a small, 1 mm long worm that lives in the ground, feeds on bacteria and reproduces every 3 to 4 days. It is a hermaphrodite. A first surprise occurred when it was observed that out of the 1 090 cells of the larva, only 959 cells remained in the adult. Thus, 131 cells had disappeared, a strange phenomenon caused by a process of **programmed cell death** or **apoptosis**, the study of which acted as a model for other organisms. *C. elegans* is also a good model for the study of the functioning of the nervous system. Compared with the hundreds of billions of neurons in the human brain, and a thousand times as many synapses, the **nervous system** of *C. elegans* has only 302 neurons and approximately 7 000 synapses. The fact that the body of *C. elegans* is **transparent** makes certain experimental approaches easier. Thus, it is possible to use a laser beam to selectively destroy a neuron, in order to evaluate its function using electrophysiological techniques. In addition, the early embryo of *C. elegans* has recently been chosen as a model to study certain aspects of cell division, such as the duplication of the centrosome. When compared with *C. elegans*, another nematode, *Caenorhabditis briggsae*, provides useful information that helps in the understanding of speciation phenomena that have played a role in evolution.

The **aplysia**, or sea slug, has a rudimentary nervous system, but its **neurons are exceptionally thick** (200 to 500 micrometers in diameter), which makes it possible to implant microelectrodes in them and use electrophysiological techniques to study the response to tactile stimulation. The aplysia becomes accustomed to sensitivity with respect to stimuli, which means that it is possible that it could be used for elementary analysis of the mechanisms of the **memory**.

For a long time, anatomists have been attracted to the **planarium**, a small fresh water flatworm, because of its anatomy (see Chapter III-2.2.3) and its ability to **regenerate**. When a planarium is sectioned along a bilateral symmetrical plane, each of the halves, left or right, regenerates that half that is missing. The same thing happens when the planarium is sectioned transversally. Regeneration is due to **stem cells** that are not yet differentiated, and are thus capable of reprogramming. It is now known that such cells, particularly embryonic stem cells, can evolve to form an entire individual. What is interesting about the planarium is that even an **adult** individual is capable of **regeneration**.

With the **zebrafish**, *Danio rerio*, which is a 3 cm long tropical fish with pretty blue and beige stripes, we move into the domain of vertebrates. As with the planarium, the zebrafish is able to regenerate any amputated part of its body, hence the interest of this animal with respect to the study of **stem cells**. The zebrafish has other, non negligible, advantages. It is **easy to rear**, it **reproduces rapidly** (a female can lay 200 eggs in one week), and it has a **transparent embryo**, so that any phenotypical anomalies that may occur during its development may be seen. The embryo, which is from 1 to 2 mm long, can easily develop in the wells of a microplate and survive on its food reserves. The main organs are formed in around thirty hours. The beating of the heart can be observed with a magnifying glass. After injection of a fluorescent derivative into the heart, it is possible to see the blood vessels by microangiographic techniques. In comparison, analyzing the normal or abnormal development of a mouse embryo necessitates sacrificing the mother and dissecting the embryo. Another interesting point is that a not inconsiderable number of **human pathologies** show very similar phenotypes in the zebrafish. For example, a zebrafish mutant carries a condition that corresponds to human hereditary telangiectasia, which is characterized by arteriovenous malformations. In addition, human infectious diseases can be reproduced in the zebrafish. When it is infected with *Mycobacterium marinum*, which is very similar to the tuberculosis bacillus, *Mycobacterium tuberculosis*, the zebrafish forms granulomas that mimic those of pulmonary tuberculosis. The zebrafish can also be infected by bacteria such as salmonella and listeria, and by viruses. It defends itself against these infections by activation of immune defense systems close to those of mammals. In fact, it has **innate (cellular) immunity** that includes phagocytotic cells (neutrophils and macrophages) and **adaptive (humoral) immunity** based on antibody-producing lymphocytes.

In the 16th and 17th centuries, the transformation of the fertilized **hen** egg into an embryo was the subject of in-depth analyses by prestigious anatomists such as Ulisse ALDROVANDI, FABRICIUS D'ACQUAPENDENTE and William HARVEY. Embryos of the chicken and other birds remain frequently-used models in developmental biology.

The **mouse** and the **rat** are mammals that are used with increasing frequency in experiments dealing with certain aspects of human physiology. From the beginning of the 20th century, the transmission of easily identified phenotypical characteristics, particularly fur coloring, in the **mouse** (*Mus musculus*), was the subject of genetic research carried out by the French biologist Lucien CUENOT (1866 - 1951), this research leading to the rediscovery of laws that had been formulated by MENDEL, working on peas a few decades earlier. After being neglected for a time, the mouse has once again become an experimental model in genetics, now that researchers have viable homogeneous strains that have been obtained by in-breeding over several generations, and now that transgenesis techniques, i.e., the introduction of foreign genes into the fertilized egg at the first stages of its development, have been mastered. **Certain mutations in the mouse reproduce human pathologies** that might be treatable by gene therapy, thus offering a good experimental model. This is the case for the SCID (Severe Combined ImmunoDeficiency) mutation, which is responsible for a particularly severe immune deficiency that leaves the organism no longer able to defend itself against infections.

From the end of the 19th century, the Norwegian brown rat (*Rattus norvegicus*) has been used for cardiovascular exploration. Today, we have pure lines, some of which carry metabolic anomalies such as diabetes and arthritis, and are useful models for human pathologies. The **brown rat** is the animal of choice in studies concerning behavior and acquired reflexes. In 1906, a line of **white rats** was created at the Wistar Institute of Anatomy and Biology in Philadelphia. In the 1940s, the hepatic tissue of Wistar rats was used to prepare, for the first time, after grinding and differential centrifugation, different types of endocellular organelles; nuclei, mitochondria, ribosomes and endoplasmic reticulum.

In the **plant kingdom**, mouse-ear cress (*Arabidopsis thaliana*) has been adopted widely as a model in studies of molecular biology and developmental biology. This is a plant in the Brassicaceae (formerly Cruciferae) family, which has hermaphroditic characteristics and is highly prolific, which is a **significant advantage in genetics**. Certain plant species are particularly well-adapted to the study of hydric stress or saline stress on gene expression. Another advantage of the plant kingdom is that the cells of adult plants are **totipotent**, which means that a single somatic cell can give rise to a whole plant, a phenomenon that makes it possible to explore the mechanism of cell regeneration. In addition, plant **transgenesis** has become an everyday technique, adapted to practical ends. Genetically Modified plants (GM plants) are employed not only in agriculture (resistance to herbicides and to insects) but also in pharmacology (production of medications) (Chapter IV-2). *A. thaliana* was recently used, unexpectedly, in a study of virulence factors for pathogenic bacteria such as *Pseudomonas aeruginosa, Enterococcus faecalis* and *Staphylococcus aureus*. These bacteria infect the plant and proliferate in biofilms on the surface of the roots. The plant defends itself by a deviation of its metabolism, manufacturing secondary metabolites that have a bacteriolytic activity. These metabolites might prove to be new antibiotic agents for the treatment of microbial infections in human medicine.

In the catalogue of model living organisms, a space must be kept for bizarre microorganisms such as the **paramecium**, which is characterized by nuclear duality, i.e., the presence of a micronucleus with primarily germ-line functions and a macronucleus carrying the genome responsible for the expression of proteins.

The interest given to a number of eukaryotic experimental models has increased with the amount of **genome sequencing data**, and with the availability of comparative information added to the list of genes having the ability to code for proteins. **The nuclear genome sequence of *S. cerevisiae*,** distributed over 16 pairs of chromosomes and with 6 000 genes counted, was published in *Science* (vol. 274, pp. 5463-5567), in 1996. **This was the first known sequence of a eukaryotic genome.** It was also the fruit of a mammoth operation coordinated by the biochemist André GOFFEAU (b. 1935) at the University of Louvain-la-Neuve in Belgium, an operation involving the concerted efforts of 641 yeast geneticists belonging to 92 small European laboratories and large DNA sequencing units located in Canada, the United States and Japan.

The genome sequence of the nematode *C. elegans*, which was published in 1998, was the first genome sequence in the animal kingdom to be decoded. This performance gave a decisive impetus to the work being carried out on the human genome sequence, with the concrete result in 2001 of the publication of a first draft of this sequence, followed, in 2003, by a more elaborate form. If we add that **nearly half of the genes of *C. elegans* have a homologue in Man**, we can see the advantages of studying them, particularly as certain of these genes encode proteins, which, when mutated in Man, lead to serious pathologies. This is the case for the protein Dys-1, related to human dystrophine, a molecule that is present in the membrane of the muscle fiber, and for which a mutation leads to severe myopathy. It was in *C. elegans* that interfering microRNAs were isolated for the first time. These are a new type of RNA that were discovered recently, and their function is to repress the translation of messenger RNA into protein (Chapter IV-1.2.2). The publication in May 2005 of the genome sequence for the ameba *Dictyostelium discoideum* excited a considerable amount of interest, for the same reasons as the publication of the genome sequence of *C. elegans*. In fact, many homologies with the human genome have been highlighted, particularly with respect to genes involved in human pathologies. Recent publications (2004 and 2005) have shown strong analogies between the genomes of the mouse (2.6 billion base pairs), the rat (2.75 billion base pairs), the chimpanzee (3 billion base pairs) and the genome of Man (3 billion base pairs). There are only 35 million single differences (i.e., 1%) between the genome of the chimpanzee and that of Man. However, there are significant differences between the latter genomes and that of the chicken, which is three times smaller (1 billion base pairs), although the coding potential is practically the same (because of a smaller proportion of non-coding sequences). In December 2005, the genome sequence of the domestic dog (*Canis familieris*) appeared in *Nature*. This sequence includes 2.4 billion base pairs and a high percentage of homology with the human genome. Pathologies that are found in Man, some of which are of genetic origin, can be found in dogs as well; cancers, cardiopathies, hearing problems, cataracts... Gene therapies that have been tried with success in dogs might, in the future, be usable in human medicine. At present, other genome sequences are being explored, including the cow, the pig and the rabbit, for reasons that have as much to do with economics as they have to do with academic curiosity (evolution of the genome of vertebrates, consequences in the medical sphere, particularly with the xenotransplantation of pig organs [Chapter IV-2.4]).

As far as the plant kingdom is concerned, 25 000 genes coding for proteins have been identified in the genome of *A. thaliana*, according to the sequence published in 2000. Hundreds of interfering microRNAs are present. The genetic instability due to transposable elements discovered in corn by Barbara MCCLINTOCK (1902 - 1992), has been found in many other plants, including the snapdragon and the petunia, and has been analyzed in detail, based on variations in flower color and shape.

Genetic and molecular exploration of a large variety of species with modern methods has led to the accumulation of data for a considerable number of molecules of DNA and proteins. Based on this inventory, it is possible to recognize analogies or even identities that help to understand the evolution of living beings, in molecular terms. This is the principle upon which the study of **comparative genomics** is based. One of its successes has been the highlighting of the existence of **homeogenes**, that is, regulatory genes that are involved in the organizational plan of the body.

It is not possible to over-emphasize the role that has been played by, and continues to be played by, model organisms in the conceptual and pragmatic advances made in biology. Today, the biologist has a vast range of animal and plant models available to choose from, and has only to select one that is ideally suited to his or her well-formulated question or appropriate study. The importance of the choice of a good experimental model was underlined by the work of the Danish physiologist, August KROGH, who became illustrious for his use of isotopes in metabolic studies (Chapter III-6.2.4), from which we get the term **KROGH's principle**, which is sometimes given to such a reasoned choice.

6.2. A BREAKTHROUGH IN TECHNIQUES FOR EXPLORING THE FUNCTIONS OF LIVING BEINGS

"Current experimentation excels by the sophistication of its apparatus, the statistical analysis of quantified results and the new, sometimes hypercritical, logical, psychological and sociological considerations of the methodology of scientific discovery."

Mirko D. GRMEK
Médéa's cauldron - 1997

The term "molecular" was adopted in the 1950s to characterize a biology that was trying to meet the challenge of understanding living beings beyond the cellular level. The term physicochemical biology could just as well have been used, insofar as we measure the progress made in our understanding of living beings according to the yardstick of technical advances. Newly-invented, remarkably high-performance instruments allow biologists to isolate endocellular organelles and biological macromolecules [3] as well as decode metabolic networks.

6.2.1. Imagery of the infinitely small

The first **electron microscope** was constructed in 1932 by the German physicist Ernst RUSKA (1906 - 1988). Transmission electron microscopy, which was developed in the 1940s, breathed new life into cytology. Electron microscopy presents analogies with optical microscopy, with respect to its elementary principles (Figure III.16). The specimen, which is quite thin, is lit by a beam of electrons that is focused by means of a magnetic lens acting as a condenser, as in optical microscopy. A second magnetic lens, which is analogous to the objective of the optical microscope, gives an enlarged image of the specimen. A third lens, the so-called projection lens, which is equivalent to the eyepiece of the optical microscope, makes it possible to project the image of the specimen either onto a fluorescent screen, where it can be observed directly, or onto a photographic plate, where it can be fixed. While the power of resolution of the optical microscope does not go below 0.3 micrometers, that of the transmission electron microscope is around one nanometer. Successive improvements have made electron microscopy a technique of choice for the exploration of the structural details of endocellular organelles, either inside the cell, or when they have been isolated. George PALADE (b. 1912), Keith PORTER (1912 - 1997) and Fritiof SJÖSTRAND (b. 1912) were the pioneers of this new technique. Various tricks and stratagems that have arisen from the imagination and sense of observation of different experimenters have led to improvements in performance. The inclusion of fixed, dehydrated samples in relatively hard

---

[3] It should be noted that the concept of the macromolecule was introduced around 1925 by the German chemist Hermann STAUDINGER (1881 - 1965). His studies on the structure of rubber led him to postulate that carbon atoms can bind covalently to form huge, perfectly-stable chains, and that such chains should be found in biomolecules.

epoxy-type resins has facilitated the production of ultrafine cross sections (0.1-0.2 micrometer) by ultramicrotomes. Derivatives of heavy metals, uranyl acetate or osmic acid, combining with certain molecular motifs, thanks to their electron diffraction power, make it possible to obtain images in which the motifs are shown up in black on a light background. It has also been possible to locate a protein of interest in a cell by means of a treatment with specific antibodies into which gold microbeads that are dense to electrons have been incorporated.

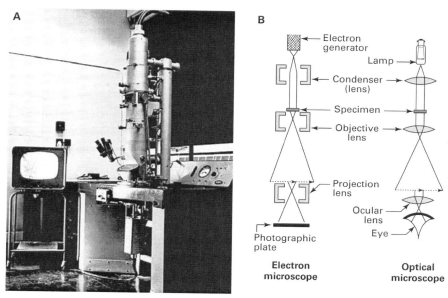

A - EM6 Transmission electron microscope (reproduced from A.W. AGAR and R.W. HORNE - *Techniques for Electron Microscopy*, Desmond H. KAY Ed., 2nd edition, Blackwell Scientific Publications, Oxford, 1965, with permission of Blackwell Publishing Ltd).

B - The diagram shows the analogies between the principle of optical microscopy and that of electron microscopy. In optical microscopy, the light from a lamp is focused on the specimen by means of a condensing lens or condenser, made of glass. The electron microscope uses a beam of electrons delivered by a red-hot tungsten filament. These electrons, propelled through a high potential field, are focused on the specimen by means of a magnetic lens. The magnetic lenses of the electron microscope, known as objective and projection lenses, are analogous to the glass lenses in optical microscopy.

Figure III.16 - Comparison of the principle of the Transmission Electron Microscope and that of the Optical Microscope

In the 1970s, the method of **cryofracturing biological membranes** was adopted for the study of transmembrane proteins. A mechanical shock applied to a membrane that has been frozen in liquid nitrogen separates the two sheets of the lipid bilayer, allowing the proteins that were buried in it to appear. After depositing a fine layer of carbon and then shadowing at a certain angle of the fracture zone with platinum, followed by dissolution of the biological material with an acid, the platinum imprint is examined by electron microscopy. On this imprint, the membrane proteins appear in relief (Figure III.17).

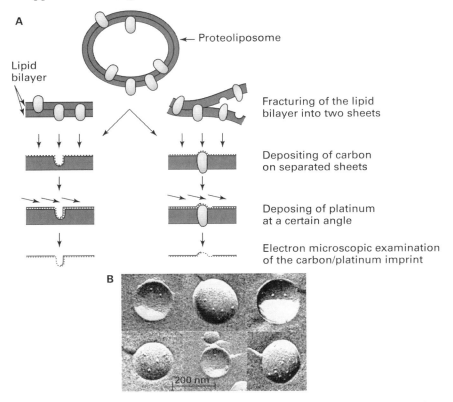

A - The different steps involved in cryofracturing biological membranes are shown diagrammatically in this figure. Cryofracturing of a biological membrane (lipid bilayer "scattered" with proteins) makes each of the two sheets of the bilayer appear isolated, with either the proteins that stick out, or with cavities left empty by the proteins after deinsertion. The sheets are treated successively by pulverization with carbon, and then platinum at a certain angle. The carbon-platinum imprint is examined by transmission electron microscopy.

B - The picture represents proteoliposomes after cryofracturing. The proteoliposomes contain the adenine nucleotide mitochondrial transporter, in a functional form, this membrane protein catalyzing the transport of ATP/ADP through the internal membrane of mitochondria. Protein particles of an average diameter of 7.5 nm, corresponding to the ADP/ATP transporter, stand out clearly against the background of membrane lipids (reprinted from G. BRANDOLIN, J. DOUSSIÈRE, A. GULIK, T. GULIK-KRZYWICKI, G.J.M. LAUQUIN and P.V. VIGNAIS © (1980) "Kinetic, binding and ultrastructural properties of the beef heart adenine nucleotide carrier protein after incorporation into phospholipid vesicles", *Biochim. Biophys. Acta*, vol. 592, pp. 592-614, with permission from Elsevier).

Figure III.17 - Principle involved in cryofracturing biological membranes, used in electron microscopy and application to the highlighting of protein particles incorporated into liposomes

Cryofracturing can be followed by **cryoblasting**, during which the frozen water is sublimed, which makes it possible to uncover deep structures. Cryofacturing is used not only on cell and endocellular organelle membranes, but also on proteoliposomes, i.e., vesicles limited by a lipid bilayer into which membrane proteins that have been purified from natural membranes beforehand have been incorporated. In the author's laboratory, this technique was carried out on proteoliposomes in which the incorporated protein species was an adenine nucleotide transport protein that catalyzes the exchange between ADP and ATP through the internal mitochondrial membrane. On the electromicrograph, it is possible to see protein particles of a diameter of 7 to 8 nanometers standing out against a background of membrane lipids (Figure III.17). These proteins are functional: in proteoliposomes preloaded with ATP, they are able to exchange intravesicular ATP for the [^{14}C]ADP added to the incubation medium.

Scanning electron microsocopy, which is complementary to transmission electron microscopy, provides three dimensional type images of the surface of biological samples under examination, and shows the fine anatomy of particularities such as phagocytotic cell filopods or pseudopods.

The use of the **serial cross-section** technique for tissues or unicellular organisms such as yeast has made it possible to reconstruct the three-dimensional organization of the interior of a cell, using photographs of different cross-sections. Nevertheless, this technique is not free from various artifacts, particularly faults in the juxtaposition of cross-section images. **High resolution electronic tomography** of frozen samples helps to eliminate this difficulty. After cryofixation, samples are photographed from different angles of incidence that are obtained by means of step-by-step rotation, each step being limited to a few degrees. Computer processing of the images provides a three dimensional reconstruction of the intracellular structures, with a resolution of less than 10 nanometers.

In brief, the implementation of a technology for examining the ultrastructure of the living cell by means of electron microscopy, associated with physicochemical approaches, has resulted in a new chapter in the study of cell biology, which describes the progress of cell events on a macromolecular scale. The move from optical microscopy to electron microscopy is shown here by what could and can be observed of the mitochondrion, a cell organelle whose function is to oxidize metabolites and retrieve the resulting energy at the end of an ATP synthesis. Before electron microscopy became available, mitochondria could be visualized by optical microscopy as minuscule rods that could be colored with stains such as Janus green (Figure III.18A). Electron microscopy showed the existence of a double membrane and, above all, the extent of the surface area of the internal membrane that is accordion-pleated inside the mitochondrial matrix (Figure III.18B). This membrane contains the enzyme that catalyzes the synthesis of ATP, ATP synthase, which is also called ATPase because its action is reversible. This enzyme has a membrane sector Fo, and an extramembrane sector, F1, which sticks out of the membrane (Figure III.18E). By means of a staining trick, it is possible to highlight sector F1, which is represented by small spheres of a diameter of around 10 nanometers (Figure III.18C). Examining these spheres at a very high resolution shows an assembly of seven sub-units, six of which are organized in a crown around a central sub-unit (Figure III.18D). Thus, the mitochondrion that optical microscopy had shown as a cytoplasmic organelle, but had been unable to characterize, has been shown at the macromolecular level by means of electron microscopy.

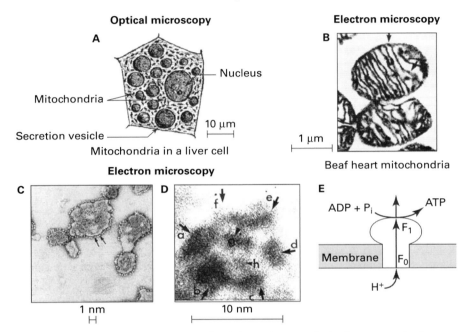

(A) shows a guinea-pig liver cell examined through an optical microscope. In the cytoplasm, we can distinctly see small rods that can be stained with Janus green: these are mitochondria. (B) shows mitochondria isolated from a beef heart, examined by electron microscopy. There is a marked contrast between the richness of detail of the image obtained with an electron microscope, particularly the accordion-pleated structure of the internal membrane (arrow), and the absence of any detail when optical microscopy is used. In (C), negative staining of mitochondrial internal membrane fragments obtained by treating mitochondria with ultrasound reveals small spheres (arrows) that stick out of the membrane that is turned over: this is the F1 sector (catalytic sector) of mitochondrial ATPase (or ATP synthase). Very high resolution examination of these small spheres (D) reveals the presence of seven sub-units, six of which are organized into a crown, with the seventh in the center. (E) shows a diagram of ATPase (ATP synthase) with its Fo sector (proton channel) inserted in the membrane and the F1 sector that sticks out of the membrane. Figures (B), (C) and (D) refer to beef heart mitochondria prepared in the author's laboratory. This work on beef heart mitochondria was carried out in collaboration with Jean-Jacques CURGY, Jean ANDRÉ and Christian COLLIEX at the University of Orsay in France.

Figure III.18 - The change from optical microscopy to electron microscopy in the middle of the 20th century. The Mitochondrion.

Although electron microscopy may seem to have supplanted optical microscopy, in the second half of the 20th century there was a renewed interest in the latter form of microscopy, including ingenious variations involving its association with video cameras and image processing. Phase contrast microscopy, invented in 1932 by Fritz ZERNICKE (1888 - 1966), and adapted to the morphological study of living cells, made use of the fact that regions of different composition in a cell have different refraction indices. In 1952, the contrast was improved by using differential interference optics (NOMARSKI system). Above all, during the 1980s, **confocal microscopy** made the tomography of cells possible, meaning that their contents could be examined without any destructive maneuvers. In the confocal technique, a thin laser beam sweeps the interior of a cell at different depth levels, thus carrying out a series of "optical cross-sections". If an intracellular protein has been "labeled" beforehand, for example by fixation of a **specific antibody to which a fluorescent ligand is attached**, such as fluorescein or rhodamine, the light emitted by the optical cross section will give a colored image of the protein on a video screen. Recently, the GFP (*Green Fluorescent Protein*) present in the *Aequora victoria* jellyfish has come to occupy a privileged position in fluorescence microscopy. Its chromophore, illuminated with a blue light, emits a green fluorescence. This protein can be expressed in a cell after insertion of its gene. What is more, genetic engineering makes it possible to produce chimeric proteins comprising the GFP associated with proteins that we wish to study the destiny of.

When it was noticed in the 1970s that laser light was able to trap and move latex micro-beads, thought was given to using lasers to carry out operations on isolated cells or on organelles inside cells. This is how **optical tweezers** came into being. These are able to remove cells and move endocellular organelles, for example mitochondria, along microtubules. These nanomanipulation techniques use low power (0.1-1 watt) lasers with a wavelength of approximately 1 000 nanometers. The force developed is within a range of a few piconewtons.

The latest development in the exploration of nanostructures by imagery is near-field microscopy. In this original technique, an **atomic force microscope** evaluates the subnanometric three-dimensional morphology of biological samples. Although quite small in size (around thirty centimeters high) (Figure III.19), this microscope accomplishes remarkable things. A metal probe comprising a tapered point with a diameter of a few nanometers, connected to a very flexible spring, scans the sample surface laterally and vertically, at a distance of 5 to 10 Å, without touching the surface. During this scanning operation, variations in the curvature of the spring are recorded using a laser beam and a photodiode. The operating principle depends on the tunnel effect. For a distance of less than 10 Å, an electrical current passes between the point of the probe and the surface it is scanning, and the nearer the point is to the surface, the greater this current is. For a distance greater than 10 Å, the electrical current disappears. The data are recorded, processed by computer and translated into a relief map of the surface of a membrane, with a resolution that is greater than one nanometer.

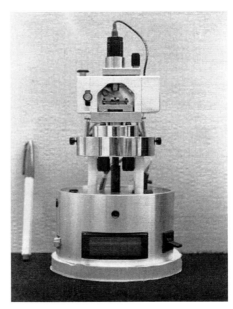

Height of the apparatus: approximately thirty centimeters (a ballpoint pen is photographed next to it). This apparatus is able to analyze the relief of a macromolecule with atomic resolution.

Figure III.19 - Atomic Force Microscope, Nanoscope II
(reproduced from S. KASAS (1992) "Atomic Force Microscopy in Biological Research", *Médecine/Sciences*, vol. 8, n° 2, pp. 140-148, with permission)

Beyond the exploration of macromolecules on a nanometric scale, **X-ray crystallography** allows access to the anatomy of macromolecules on a scale of one angström (Å). In 1953, the X-ray diffraction diagrams on hydrated DNA crystals obtained by Rosalind FRANKLIN (1920 - 1958) in the laboratory of Maurice WILKINS (1916 - 2004) led James WATSON (b. 1928) and Francis CRICK (1916 - 2004) to propose the famous double helix structure of DNA. Even more laborious was the ultrastructural approach taken with proteins. John KENDREW (1917 - 1997) obtained a first success with the myoglobin of the sperm whale, a hemoprotein that is present in red muscles (Figure III.20). Myoglobin was chosen because of its relatively small mass (17.8 kilodaltons) and the ease with which it crystallizes. In 1957, John KENDREW published a first model of the structure of myoglobin at a resolution of 6 Å, followed in 1959 by a more refined model with a resolution of 2 Å. In 1962, the resolution reached 1.4 Å. In 1968, the structure of horse hemoglobin, a tetrameric molecule that is four times bigger than myoglobin, was resolved down to 2.8 Å by Max PERUTZ (1914 - 2002). This exploit rang in a new era in structural biophysics, devoted to the study of the three-dimensional architecture of biological macromolecules and the details of their interaction with specific ligands.

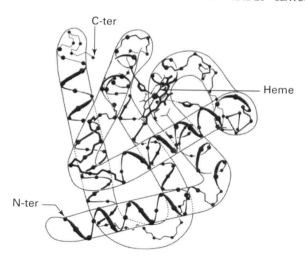

Model of sperm whale myoglobin reproduced in a stylized form, in which the locations of the amino acids are represented by dots. The model was constructed on the basis of X-ray diffraction data of a myoglobin crystal, and also an understanding of the amino acid sequence of this protein (J.C. KENDREW, H.C. WATSON, B.E. STRANDBERG, R.E. DICKERSON, D.C. PHILLIPS and V.C. SHORE (1961) "A partial determination by X-ray methods and its correlation with chemical data", *Nature*, vol. 190, pp. 666-670).

Figure III.20 - Three-dimensional structure of sperm whale myoglobin resolved by X-ray crystallography
(reproduced and adapted from R.E. DICKERSON, "X-Ray analysis and protein structure", in H. NEURATH Ed. © (1964) *The Proteins*, p. 634, Academic Press, with permission from Elsevier)

With the arrival of **synchrotron radiation**, which covers a range of wavelengths from X-rays to infrared, biology entered the arena of "big science" (Figure III.21). As synchrotron X-radiation is much brighter than classical sources of X-rays, diffraction diagrams of macromolecules give much more information and exposure times are greatly reduced. Most importantly, protein microcrystals that barely diffract classic X-rays become analyzable.

In addition to the X-ray diffraction technique, **Nuclear Magnetic Resonance (NMR)** can be used for proteins in solution. Considerable progress has been made in NMR techniques in the last fifty years, rising from a power of 50 MHz to 500 MHz and more. From the exploration of the three-dimensional structure of small proteins (around 10 kilodaltons) we have moved on to molecules with a mass greater than 50 kilodaltons. NMR can be applied to biological molecules, certain atoms of which contain a nucleus with a spin of 1/2, such as ^1H, ^{31}P, ^{13}C and ^{15}N. As natural proteins have a low ^{13}C and ^{15}N content, they must be enriched with these isotopes before they can be analyzed by NMR. To this end, these proteins are overproduced by bacteria that have been genetically modified and cultivated in the presence of bicarbonate labeled with ^{13}C and/or a nitrogen substrate labeled with ^{15}N.

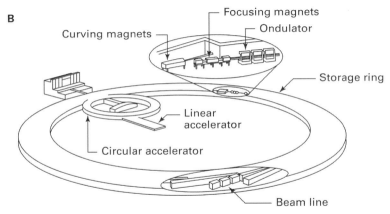

A - Panoramic view of the outside (© Studio de la Revirée - Gryn.); **B** - Schematic representation of the production of synchrotron radiation (© ESRF).

The electrons emitted by an electron canon are accelerated first in a linear accelerator and then in a circular accelerator, in which they reach an energy level of around six billion electronvolts. These very-high-energy electrons are injected into a large-capacity storage ring (844 meters in circumference, see photo A), where they circulate in a vacuum at a constant energy level for hours. The trajectory of the electrons in the storage ring is defined by the magnetic fields that they travel through, which are generated by different types of magnets; curving magnets, undulator and focusing magnets. This image illustrates the enormous size of modern instruments devoted to experimental research.

Figure III.21 - The era of enormous instruments used in biology.
The Synchrotron in Grenoble, France (European Synchrotron Radiation Facility - ESRF)

At the time of writing, the three-dimensional structures of more than 30 000 protein species are accessible in databanks, and this number grows each month. We understood the function of most of these proteins, but we did not know their structure. An understanding of the structure of a protein sheds light on the mechanism of its operation, in the same way as in the 17th century, the highlighting of anatomical relationships between the skeleton, the muscles and the tendons led to explanations of movement functions given in mechanical terms such as the operation of winches, pulleys and levers (Chapter II-3.3). In both cases, the thought processes were similar, the difference being that with the progress of technology at the end of the 20th century, it became possible to determine the ultrastructure of living beings on an atomic scale.

6.2.2. Enumerating and isolating macromolecular structures

Until the 1930s, the isolation and purification of proteins from tissue extracts and humoral liquids (blood serum, for example) involved fractional precipitation with mineral salts, particularly ammonium sulfate. This technique, brought in at the turn of the 20th century, was very popular, and remains appropriate for proteins that are present in such quantities that their purification does not lead to excessive losses. Nevertheless, the stumbling block continued to be the isolation of proteins of interest that were present in very small quantities in certain biological media. Between 1930 and 1950, while looking for a solution to this problem, three new approaches to fractioning, **electrophoresis**, **chromatography** and **ultracentrifugation**, completely revolutionized analytical proteinology and, in addition, contributed to an understanding of certain enzyme mechanisms involved in cell operation. Modern biology has built itself up on the basis of the systematic use of these three techniques, frequently associated with the use of radiolabeled molecules.

Electrophoresis involves the differential migration of molecules carrying positive or negative charges in an electrical field. The speed of migration of the proteins depends on the ratio of their net charge to their mass. The first liquid vein electrophoresis apparatus was constructed by the Swede Arne TISELIUS (1902 - 1971) in the 1930s. After this, electrophoretic migration used solid supports that were easy to handle and adapted to everyday use (paper, agarose or polyacrylamide gels). The first application of electrophoresis to **human pathology** took place at the end of the 1940s, when Linus PAULING (1901 - 1994) recognized that there is a difference in migration in an electrical field between normal human hemoglobin (A) and abnormal hemoglobin (S). The hemoglobin S came from subjects who were suffering from a particular type of anemia that is known as sickle-cell anemia or drepanocytosis (from δρέπανον = sickle) because of the sickle shape of the red blood cells, which is exacerbated in a low-oxygen medium. In 1956, the German-born British biochemist Vernon INGRAM (1924 - 2006) discovered that the electrophoretic anomaly of hemoglobin S was due to replacement of a glutamic acid (negatively charged amino acid) in hemoglobin A with a valine (uncharged) in hemoglobin S. This was the first demonstration of a **relationship between a pathology and a molecular lesion of a hereditary nature**.

Electrophoretic migration in a gel (polyacrylamide gel) makes it possible to separate proteins according to their size, migration of a protein through the gel becoming slower as size increases. The protein mixture is first treated with sodium dodecyl sulfate (SDS), a detergent that denatures the proteins when fixing to them, at an average rate of one molecule of SDS per amino acid residue. After migration, the proteins are displayed in the gel by staining with a compound such as Coomassie blue or a silver salt. Protein amounts of around 0.1 microgram are detected using Coomassie blue, five to ten times fewer with a silver salt. Two-dimensional electrophoresis, developed in the United States in the 1970s by Patrick O'FARRELL (b. 1949), aims to separate the hundreds or even thousands of proteins present in a cell extract, depending on their charges and their masses. The first step (first dimension) involves isofocusing, which separates the proteins according to their charge, or more specifically, to their isoelectric point. The second step (second dimension), which is perpendicular to the first, involves migration in the presence of SDS, which differentiates the proteins according to their molecular mass (Figure III.22) [4]. Gel zones displayed after staining are excised. Once a protein has been extracted from an excised gel fragment, the following step involves determining the sequence of the amino acids that it is made up of. The principle is based on the iterative cleavage de peptide bonds from one end of the protein. The amino acids that are freed one by one are identified and their sequence is deduced based on the order of their appearance. This classical technique has tended to be replaced by the analysis of amino acid sequences by **mass spectrometry**, a technique that breaks up the macromolecules, ionizes them and separates the ions according to the ratio of their mass to their charge. Two-dimensional electrophoresis has become an essential tool for the analysis of all of the proteins in a cell (**cell proteome**), and any modifications that occur to them, depending on genetic mutations or environmental changes.

Over the same time period, chromatography has had as profound an effect as electrophoresis on the analytical approach to biomolecules. Introduced at the beginning of the 20th century by the Italian-born Russian biochemist Mikhaïl TSVET (1872 - 1919), chromatography gets its name from the fact that it was first applied to the separation of plant pigments of different colors. Nowadays, chromatography is used to separate all sorts of small molecules (amino acids, oses, nucleotides, lipids). The first type of chromatography, developed in the 1940s by the British researchers Archer MARTIN (1910 - 2002) and Richard SYNGE (1914 - 1994), was based upon partition between two phases, a stationary phase that remains fixed to the support and a migrating phase. For this reason, it was known as **partition chromatography**. In the classical method, the chromatographic support material is a sheet of filter paper. The cellulose that makes up this paper is bound with water.

---

4   P.H. O'FARRELL (1975) "High resolution two-dimensional electrophoresis of proteins", *Journal of Biological Chemistry*, vol. 250, pp. 4007-4021.

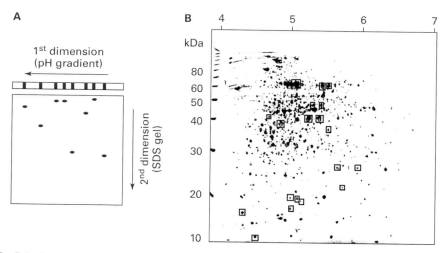

A - Principle of two-dimensional electrophoresis. The proteins of a biological extract are separated in a first dimension by isoelectric focusing in a gel column to which ampholines that differ in their isoelectric points have been added, in the absence of a denaturing agent. The proteins in this "native state" are separated according to their charges. After focusing, the gel strip containing the proteins is cut out and applied to the side of the surface of a flat gel containing sodium dodecyl sulfate (SDS), an agent that denatures the proteins. The latter are separated according to their molecular masses in a second dimension.

B - Separation of the proteins contained in an extract of the bacterium *Mycobacterium* 6PY1, which is able to degrade pyrene. The first migration was carried out in a pI gradient of between 4 and 7, the second in a 12.5% polyacrylamide gel. The proteins are shown up by staining with silver. The size of the molecular markers, expressed in kilodaltons, is shown on the left. Comparison of the composition of extracts obtained from bacteria that have grown on pyrene and those obtained from bacteria that have grown on other carbon sources makes it possible to identify the proteins that are specifically expressed for pyrene degradation (framed spots) (carried out and photographed by Yves JOUANNEAU and Christine MEYER, UMR 5092 Research Unit of the CNRS-CEA-UJF, Grenoble; from S. KRIVOBOK, S. KUONY, C. MEYER, M. LOUWAGY, J.C. WILLISON and Y. JOUANNEAU (2003) "Identification of pyrene-induced proteins in *Mycobacterium sp. strain* 6PY1: evidence for two ring-hydroxylating dioxygenases", *J. Bacteriol.*, vol. 185, pp. 3828-3841).

Figure III.22 - Separation of the proteins of a biological extract by two-dimensional electrophoresis

The migrating phase is generally a mixture of an organic solvent and water. In ascending chromatography, the lower part of the sheet of paper on which a mixture of different molecular species has been deposited is standing in a basin containing the migrating phase (Figure III.23). During migration of this phase up the filter paper, by capillary action, each molecular species has the choice between the hydrated cellulose of the paper and the organic solvent, depending on its polarity.

Another example, involving the separation of cyclical DNA bases, shows the immediate impact that chromatographic techniques had on analytical biochemistry. Until the beginning of the 1940s, the puric (adenine and guanine) and pyrimidic (cytosine and thymine) base composition of a given type of DNA was deduced, after complete hydrolysis, from the quantities of the crystallized derivatives of these bases obtained in the form of picrates, sulfates or brucine salts, estimated by weighing. This was a tricky and rather imprecise operation because of the inevitable losses that occurred during repeated crystallizations, and required working with samples of several tens of milligrams of DNA. With chromatographic separation on paper, a few micrograms of a DNA hydrolysate are enough to give precise information concerning the stœchiometry of puric and pyramidic bases. DNA bases absorb ultraviolet (UV) light. Thus, if chromatograms are looked at under UV light, it is possible to find the zones on the paper to which the DNA bases have migrated. These zones are cut out, eluted and the respective quantities of the bases that are present are measured by UV spectrophotometry. It was using this type of chromatographic analysis from 1949 onwards that Erwin CHARGAFF (1905 - 2002), observed that, no matter what the origin of the DNA (animal, plant or microbe), adenine and thymine always have a ratio of 1. The same is true for cytosine and guanine. This information probably helped WATSON and CRICK in drawing up their model of the double helix of DNA in 1953, a model in which adenine and thymine, cytosine and guanine are paired in a complementary fashion.

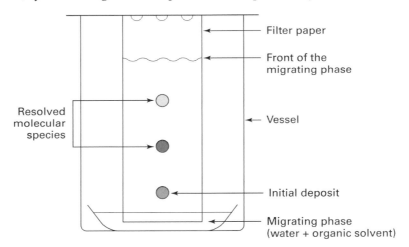

The chromatography operation is carried out within an enclosure. The sheet of chromatography paper (special filter paper) is suspended from the support at the top of the sheet. The bottom part of the paper is bathed in the migrating phase (a mixture of organic solvent and water) placed in a basin. Beforehand, a mixture of molecular species has been spotted onto the sheet of paper. As the migrating phase ascends by capillary action up the filter paper this carries along the deposited molecular species differentially, depending on their polarities (affinity for water, fixing preferentially onto the cellulose of the paper or affinity for the organic solvent that is migrating). The diagram shows the position of two molecular species after chromatography.

Figure III.23 - Principle of partition chromatography on paper (ascending chromatography)

At the beginning of the 1950s, research moved on to chromatography on columns and to the adaptation of its different variants to the separation of macromolecules such as proteins. Among these variants is migration as a function of charge (**ion-exchange chromatography**), which was developed by the Americans Stanford MOORE (1913 - 1982) and William STEIN (1911 - 1980). Filtration through a synthetic porous gel (molecular screen) so that small molecules that travel through the pores are more easily held than large ones (**gel filtration chromatography**), as well as the interaction of migrating molecules with a matrix having a specific affinity, for example specific antibodies of the protein to be isolated (**affinity chromatography**) were the subject of original studies carried out by the group of Jerker PORATH (b. 1921) at the University of Uppsala (Sweden). The use of automated systems, associated with the detection of protein material by absorption of ultraviolet light, made it possible to collect fractions that could be subjected to later detailed analysis.

The third fractioning technique that emerged in the 1930s was ultracentrifugation, which was initiated by the Swede Theodor SVEDBERG (1884 - 1971). Analytical ultracentrifuges that were equipped with an optical detection system were first used for the separation of protein species of different molecular masses. Ultracentrifugation was then adapted to the separation of endocellular organelles according to their size and their density (nucleus, mitochondria, smooth and rough endoplasmic reticulum, GOLGI apparatus, endosomes, lysosomes and peroxysomes) from crushed tissue (also called a homogenate) in a solution of isotonic saccharose (sucrose) (0.25 M). Depending on the size and density of the organelles, separation is carried out by centrifugation at increasing centrifugal forces which can reach or even exceed 100 000 g, either in a medium of homogenous density (differential centrifugation) or in a density gradient (isopycnic centrifugation) (Figure III.24). When the endocellular organelles of a well-defined species have been isolated, their morphology is examined by electron microscopy and their enzyme specificity is determined in order to evaluate function. Ultracentrifugation was a major technical breakthrough for cell biology, making it possible to collect and characterize endocellular organelles that the optical microscope had shown to the eye, but for which the functional identity was unknown. Ultracentrifugation in a saccharose (sucrose) gradient can be adapted to allow separation of protein macromolecules or protein complexes.

The counting, separation, isolation and identification of macromolecules or endocellular organelles from a tissue extract are now operations that are carried out routinely in biology laboratories, with the trend being increasingly towards robotization and miniaturization. These operations form part of the methodological arsenal of the **reductionist approach**, which involves the Cartesian principle of isolating parts from a whole, in order to be better able to characterize its properties, provided scattered information can be gathered together afterwards in order to provide a synthesis (Chapter III-7).

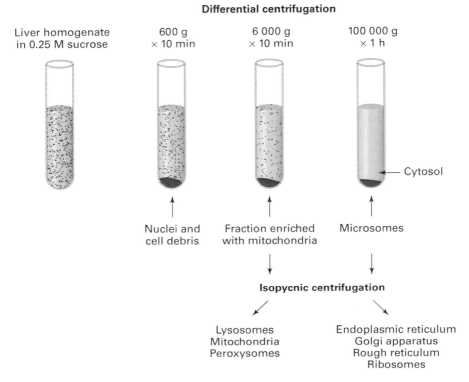

Differential centrifugation: a hepatic tissue homogenate in a solution of isotonic saccharose (0.25 M) is subjected to a series of centrifugations at increasing centrifugal forces, in order to sediment out, in succession, the nuclei and cell debris, then a fraction enriched with mitochondria but contaminated with lysosomes and peroxysomes, and finally a very heterogeneous fraction of particles called microsomes, which contain fragments of smooth reticulum (without fixed ribosomes) and rough reticulum (with fixed ribosomes), free ribosomes and fragments of the GOLGI apparatus. In practice, as shown by the figure, centrifugation of a rat liver homogenate at 600 g for 10 minutes at 4°C results in sedimentation of the nuclei and cell debris in the form of a pellet. The supernatant is collected and poured into another tube to be centrifuged. Centrifugation at 6 000 g for 10 minutes sediments a fraction rich in mitochondria. The operation is continued with centrifugation at 100 000 g for one hour, which sediments the microsomes. The supernatant is called the "cytosol".

Isopycnic centrifugation: both the microsomes and the fraction enriched with mitochondria are re-suspended in 20% saccharose solution. The specific suspensions are placed in centrifuge tubes containing a saccharose solution, the concentration of which varies along a gradient from 20% (top of the tube) to 50% (bottom of the tube). High-speed centrifugation separates the organelles according to their density, in equilibrium with the density of the gradient.

Figure III.24 - Separation of rat liver endocellular organelles by differential centrifugation and by isopycnic centrifugation

6.2.3. Isotopic labeling

The idea of analyzing the metabolic modifications of an organic molecule inside a living organism by incorporating an easily-traceable label goes back to the beginning of the 20th century. In 1905, the German biochemist Franz KNOOP (1875 - 1946), who was interested in the breakdown of long-chain fatty acids in animal organisms, had the idea of fixing a benzoyl residue that acted as a "**label**" to the methyl end of the fatty acids, in order to monitor their breakdown products labeled by the presence of the benzoyl residue. Experiments were carried out on dogs to whose food KNOOP had added "labeled" fatty acids that had either an even or an odd number of carbon atoms. The results led to the conclusion that the breakdown of fatty acids takes place by iterative removal of two-carbon units. The arrival of **isotopes** from the 1930s simplified labeling techniques. In 1932, the American chemist Harold UREY (1893 - 1981) developed a method for separating deuterium (^{2}H or D), a stable isotope of hydrogen. Other **stable isotopes**, ^{15}Nitrogen, ^{13}Carbon and ^{18}Oxygen, were isolated soon afterwards. The first **radioactive isotope**, ^{32}Phosphorous, was obtained by Irene JOLIOT-CURIE and Frederic JOLIOT in 1934. The first biological applications of synthetic molecules enriched with stable or radioactive isotopes soon came onto the scene. Within a decade, results accumulated that demonstrated the validity and power of this new experimental approach. With the exception of a few minor effects, which were called "isotope" effects, molecules labeled with isotopes behave like natural molecules.

In the 1930s, in Copenhagen, Georg HEVESY (1885 - 1966) and August KROGH (1874 - 1949) analyzed the passage of water through biological membranes using heavy water, D_2O. Experiments carried out by the Dane Hans USSING (1911 - 2000) on rats fed with deuterated amino acids showed that in less than three days these amino acids are incorporated into tissue proteins in non-negligible concentrations (10% in the liver, 2.5% in the muscle tissue). Similar studies carried out in New York by Rudolf SCHOENHEIMER (1898 - 1941) and David RITTENBERG (1906 - 1970) on rats fed with ^{15}N-labeled leucine and deuterized stearic acid gave rise to the concept of a dynamic state for biomolecules in living organisms. In 1941, KAMEN and RUBEN, using water labeled with ^{18}Oxygen ($H_2^{18}O$), provided proof that during photosynthesis oxygen is liberated by cleavage of the water, which was a logical continuation of the experiments on the photoproduction of oxygen by green plants carried out by PRIESTLEY and INGENHOUSZ in the 18th century (Chapter II-5.2).

At the beginning of the 1950s, mineral phosphate and nucleotides labeled with the radioisotope ^{32}P began to be used on a large scale in bioenergetics and in molecular biology, as did organic molecules labeled with the radioisotope ^{14}C, with the aim of exploring metabolic networks. As measurement of radioactive isotopes using a GEIGER counter, named after the inventor Hans GEIGER (1882 - 1945) is much faster than measurement of stable isotopes with a mass spectrometer, the use of radiolabeled biomolecules became very fashionable. Radiolabeled metabolites, separated by chromatography or electrophoresis on paper, are found by **autoradiography**, a technique in which the sheet of paper containing the radioactive metabolites is applied to a photographic film. Wherever a radioactive metabolite is located, the film shows a black spot after a time of exposure.

To sum up, the design and manufacture of ingenious, easy-to-use instruments that were adapted to specific purposes according to new ideas and hypotheses, both in the domain of imaging and in that of fractioning, no doubt had a great deal to do with the remarkable growth of the life sciences in the middle of the 20th century. Isotopic radiolabeling also contributed to this growth. It is likely that without these technical achievements, most of which arose from instrumental physics, many major discoveries would have had to wait.

6.2.4. The instrument and the method
The analysis of reality via the instrument

As we have seen, increasingly numerous and effective laboratory instruments were developed during the 20th century. The use of these instruments required that an optimal methodological approach be adopted, or, in other words, it was necessary for there to be a **good fit** between the **instrument** and the **method** applied to the biological object for which properties were to be determined. Thus, when, in the 1940s, ultracentrifuges that were capable of generating centrifugal forces of 100 000 g became available, biologists decided to use them to separate the endocellular organelles liberated in the grinding liquid of a crushed organ such as a rat liver, for example, in order to study their structures and functions. The idea was that the largest or densest organelles, such as the nuclei, would sediment out at a low centrigual force, while smaller or less dense organelles would sediment out at a higher centrifugal force. In order to simulate the intracellular medium, which is rich in potassium ions, researchers used an isotonic potassium chloride solution as a grinding medium. This led to a large-scale aggregation of organelles that prevented any separation by differential centrifugation. Because of the unfortunate results of this attempt, which had nevertheless been based on justifiable ideas, the use of a saline medium was abandoned and replaced with the use of a solution of saccharose (sucrose), an uncharged molecule. This procedure, which was valid from a physicochemical point of view but doubtful from a physiological point of view, was nonetheless the key to success in the isolation of fractions enriched with nuclei, mitochondria, lysosomes, reticular membranes, ribosomes, etc. Thus, by a set of trial and error operations, which were mainly carried out "to see what would happen", and led to potassium chloride being replaced by saccharose, cell biology benefited from a decisive experimental advance that resulted in the novel possibility of isolating each species of endocellular organelle, and of testing its structural and functional properties.

In a similar way, confocal microscopy would not have excited the degree of interest that it now arouses if it had not benefited from the development of the fluorescent labeling of proteins in order to determine their endocellular location. In addition, it was necessary to have antibodies that were able to recognize these proteins and an appropriate methodology for fixing fluorescent ligands onto these antibodies without altering their reactivity, and it was also necessary to perfect a cell permeabilization technique in order to allow an antibody that had been made fluorescent to reach its recognition site. There are many examples that show that the more complex an instrument is, the more complicated and **ruse-like** the procedure that follows on from the preparation of the biological sample becomes. Instrumentation and methodology are clearly two indissociable facets of the experimental method.

While nowadays the performance of laboratory instruments and the precision of chemical analysis methods make it possible to determine details of the operation of living beings that were previously inaccessible, the importance of the breakthroughs that were made by scholars at the end of the 19th century should not be underestimated. Despite their use of methods and tools that are now considered to have been antediluvian, these scholars, with the primitive methods available to them, were able to solve some of the enigmas of cell life. METCHNIKOFF, who was interested in the phagocytosis of bacteria by amebas, observed not only the encompassment of bacteria in vacuoles but also the acidification of the contents of these vacuoles. He managed to do this in an ingenious and very simple way, causing the amebas to ingest small blue sunflower seeds. After a certain time, the blue color changed to red, signifying acidification. A few years later, the sunflower was replaced with neutral red, a more sensitive indicator that turns a deep pink color in an acid medium. It was this type of approach that led, in the 1880s, to METCHNIKOFF's discovery of the acid characteristics of the phagocytosis vacuoles in neutrophils and macrophages, cells that are the main agents of innate immunity (Chapter III-2.2.4). One century later, the acidification of phagocytosis vacuoles in neutrophils and macrophages was looked at again, and quantified by means of analytical techniques using fluorescent reagents that are able to detect variations in pH or microscopic probes and electrophysiological techniques. Nevertheless, the primary discovery was made by METCHNIKOFF.

While an appropriate method is required in order to make judicious use of an instrument, the instrument itself, because of the sophistication of its design, must be recognized as the guarantor of the reality of the object being explored. When reflecting upon scientific realism, the Canadian philosopher Ian HACKING (b. 1936), taking the example of the optical microscope, recounts his own feeling about the necessary fit between the representation of an object seen by the eye *via* a microscope and the form of the object in its reality, i.e., between the observed and the real [5].

---

[5] "Do we see through a microscope?" (1981) *Pacific Philosophical Quarterly*, vol. 62, pp. 305-322.

This questioning is all the more pertinent in that many laboratory instruments, inherited from advances in contemporary physics, appear to the layman like so many mysterious machines that are capable of delivering a coded message, which the investigator is in charge of interpreting. This is the case for giant installations such as the ESRF (European Synchrotron Radiation Facility) in Grenoble, France, which harbors powerful sources of X-radiation to be used for the study of the structures of macromolecules (see Figure III.21). It is also the case for instruments that are used for tissue and cell imaging (Chapter III-6.2.1), which are types of black boxes that are interposed between the experimental fact, what is understood by the senses, and the interpretation of the researcher.

Nevertheless, confidence in the reliability of an instrument is sometimes counterbalanced by a sentiment of frustration that results from a lack of knowledge of how the instrument works, which is sometimes complex. This leads to a legitimate curiosity. Staying with the modest example of the use of the optical microscope, HACKING wrote that in order to understand how to "see" through a microscope, it is necessary to "do" so, that is, it is necessary to experiment with this microscope. Thus, in order to convince oneself that a certain part of a cell does in fact exist as it is represented as existing, it is a good idea to carry out the micro-injection of a fluid into exactly this part of the cell: "We see the tiny glass needle – a tool that we have ourselves hand crafted under the microscope – jerk through the cell wall." Nevertheless, HACKING adds that this does not mean that we are freed from philosophical perplexities. It is by doing and not by looking that we learn to differentiate between what is visibly an artifact due to the preparation of the instrument and the real structure that is seen with the microscope. What is true for the optical microscope is obviously also true for any other imaging instrument, from the different variants of optical imaging to sophisticated forms of cerebral imaging by magnetic resonance or positon emission (Chapter IV-3.2).

While the instrument is a required support for experimentation, and while the performance of an instrument provides the researcher with a justified sense of satisfaction, it is not free of artifacts. Only the experimental method, which is constantly questioning, is able to discern these artifacts and to correct them, or at least obviate them. Recently, attention was drawn [6] to the deleterious effects of extremely powerful synchrotron radiation on proteins having an oxidoreduction function, such as catalase or superoxide dismutase. This effect, due to the large-scale production of free radicals, is reduced at low temperature. Thus, certain structures that have been sent to databanks in the last few years may one day be revised.

---

6   O. CARUGO and K.D. CARUGO (2005) "When X-rays modify the protein structure: radiation damage at work". *Trends Biochem. Sci.*, vol. 30, pp. 213-219.

7. OPENING UP BIOLOGICAL EXPERIMENTATION TO REDUCTIONISM

"The biologist has long studied living organisms as wholes and will continue to do so with ever increasing interest. But these studies can tell us nothing of the nature of the physical basis of life, which no form of philosophy can ignore. It is for chemistry and physics to replace the vague concept "protoplasm" – a pure abstraction – by something more real and descriptive."

Frederick G. HOPKINS
Problems of Specificity in Biochemical Analysis - 1931

Experimental reductionism in biology refers to any process that reduces the complex functions of an organ, a cell or an endocellular compartment to simplified functions that depend on elementary components that can be made accessible to analysis by physicochemical methods. The full value of data obtained using a reductionist approach is seen when they are followed up by a **reconstitution experiment** involving the re-association of the isolated elementary components. Experimental reductionism proceeds from a Cartesian approach to the exploration of tissular, cellular or molecular mechanisms, i.e., "divide each of the problems I am examining into as many parts as I can, as many as shall be necessary to best resolve them" (Chapter II-6.3). In contrast to reductionism, **holism** postulates that the phenomena of living beings cannot be reduced to the sum of their elementary functions, and that the whole contains more information than the sum of its parts. While conceding that the functioning of an organ involves complex networks of elementary reactions, subject to regulations the least modification of which can have deleterious effects, it is nevertheless true that the biologist confronted with "black boxes" for which he or she does not see the overall function has no other alternative, as a first step, than to open these boxes and explore their contents. From timid beginnings at the turn of the 20th century, the reductionist approach to understanding the functioning of the structures of living beings became very popular within a few decades. This growth in popularity was obviously facilitated by technical advances that made it possible, starting with cells, to use fractioning to isolate the organelles that were contained in cells, or which made it possible, using methods from physics (electron microscopy, X-ray diffraction) to monitor and control the breakdown and then the reconstruction of macromolecular complexes.

7.1. THE FIRSTS STEPS IN EXPERIMENTAL REDUCTIONISM: FROM THE ORGAN TO THE CELL

At the turn of the 20th century, cell theory had finally been accepted. The cell was considered to be the structural and functional unit of both animal and plant living tissues. Optical microscopy showed that several cell species could cohabit in the same organ. Thus, in the brain, neurons cohabit with glial cells. Also, at this time, a certain number of unicellular organisms, bacteria, yeasts and protozoans, were identified and listed, using their morphological characteristics and certain of their functions that the nascent science of biochemistry had managed to analyze, for example their ability to ferment certain sugars and not others. In keeping with that which had already been accomplished with unicellular organisms, the biologists' dream was to **isolate cells** of the same type from organs, to **make pure cultures** of them, and, using these cultures, to **determine their functions**. At the beginning of the 20th century, the American Ross HARRISSON (1870 - 1959) managed to make cells from fragments of frog embryo tissue deposited on coagulated lymph obtained from adult frogs grow and divide. As well as providing a solid support medium, the coagulated lymph provided the nutrients necessary for cell life. This was the way in which, using fragments of spinal chord, HARRISSON was able to show the formation and lengthening of axons. In 1913, at the Rockefeller Institute in New York, Alexis CARREL (1873 - 1944) observed that the addition of crushed chick embryos to fibroblasts made it possible to cultivate the latter over several weeks. Following systematic testing, a **semi-synthetic** culture medium was developed in 1955 by the American Harry EAGLE (1905 - 1992). This medium contained amino acids, vitamins, mineral salts and oligoelements as well as a serum that provided growth factors that had not been identified at that point. It was discovered that in such a medium tumor cells could divide and proliferate indefinitely. This was the case for the HeLa cells obtained in 1952 from a cervical carcinoma. This was also the case for fibroblasts, epithelial cells and many other cell species that would be used around the world for purposes as varied as the exploration of the metabolism, the control of cell division, the operation of signaling sequences, or even the measurement of the activity of ion channels. In the 1970s there was a temporary trend involving the **impromptu use for metabolic purposes of isolated cells** from different tissues. Hepatocytes isolated from the rat liver were used to produce **metabolic balance data** concerning ureogenesis, glycogenesis and other reaction chains.

In certain experiments, the use of pure cell species required the use of **cell sorting** apparatus. These instruments are called **cytofluorimeters**. They are also known under the name FACS (Fluorescence-Activated Cell Sorter) because they make it possible to separate cells of the same type from a heterogeneous cell mixture, the former having been found because of the binding of specific marker with a fluorescent surface, for example an antibody associated with a fluorescent probe and directed against a protein of the plasma membrane. With an appropriate device that uses an electrical field, the cells that have been made fluorescent and are flowing through a channel receive an electrical charge that allows them to be differentiated from the non-fluorescent cells and to be separated from them.

Having isolated, homogeneous living cells available opens up the field of exploration to the possibilities of microinjections of **inhibitors** or **activators** of enzymes, or of signaling sequences, by micromanipulation under a microscope. As each cell is considered to be a laboratory where, every second, thousands of chemical reactions are taking place, the term "Lab In Cell" (LIC) is used. In order to analyze the functioning of isolated and immobilized living cells, **nanotechnology** tools are essential, for example, nanoneedles adapted to high-precision micropumps that are able to inject or remove volumes of a few picoliters.

Even though the cell is a complex entity, the **move from organ level to cell level** is nevertheless a **reductionist process**. As with any reductionist process, the information that it provides is restrictive. Thus, the very fact that the cells are isolated means that we lose any information concerning adhesion proteins, selectine, cadherin and others, which keep the cells associated in an organ, as well as any information about lacunary and communicating intercellular junctions that act as channels through which ion flow occurs. A typical case is the overall contraction of the heart, which results from an electrical impulse that is produced in a highly-localized region, the sinoauricular node, and spreads to all the cardiac tissue *via* lacunary junctions. With such limitations in mind, the biologist is free to undertake a more global approach to phenomena, and, at the same time, to develop a way of thinking and protocols that are a better reflection of reality.

7.2. A CELLULAR GLYCOLYSIS: A PROTOTYPE FOR A REDUCTIONIST APPROACH TO EXPLORING THE METABOLISM

In 1897, Eduard BUCHNER discovered that glucose is fermented and transformed into alcohol by a yeast extract that is without living cells (Chapter III-3.3). This was the beginning of a reductionist approach to the exploration of cell metabolism. Up until that point, whole cells had been used to study the fermenting power of yeasts with respect to certain sugars. At the end of the 19th century, even though most people who thought about it accepted that intracellular processes of a chemical nature take part in the transformation of glucose into alcohol, the most prevalent idea was that this is indissociable from the living state. This was why evidence for the **acellular fermentation of glucose** was greeted with amazement. This quite simply meant that once it was opened, the black box of the cell could give access to a chemical analysis of its contents. BUCHNER named the active substance that is present in the soluble yeast extract **zymase**. It would take around forty years to reach an understanding that this substance corresponded to a set of a dozen different enzyme species, each of which catalyzes the transformation of one metabolite into another in a chain of reactions that is called **glycolysis**. We will restrict ourselves here to looking at the very first reductionist-style experiment that was carried out by two British biochemists, Arthur HARDEN (1865 - 1940) and William YOUNG (1878 - 1942), an experiment to demonstrate that mineral phosphate (orthophosphate or inorganic phosphate) is involved as a reactant in the mechanism of glycolysis.

HARDEN and YOUNG had observed that the fermenting power of the yeast extract, which was measured by the amount of carbon dioxide given off, increased noticeably over the first half hour, and then suddenly slowed down. They had the idea of adding an aliquote fraction of the yeast extract, treated with heat, to the raw yeast extract. This addition had an observably stimulatory effect on the speed of production of CO_2. As the heat had denatured the protein macromolecules, HARDEN and YOUNG came to the conclusion that a **small thermostable factor**, present in the boiled extract, is necessary for the fermentation of glucose. Their next step was a dialysis of the raw yeast extract through a filter that was able to trap macromolecules. This filter comprised a porous porcelain bulb, the inside wall of which had been covered with gelatin. The bulb was filled with the yeast extract, and a pressure of 50 atmospheres was applied. After **dialysis**, the gelatinous residue sticking to the walls was dissolved in water. Its fermentation activity with respect to glucose had disappeared. The same was true for the dialysate. Nevertheless, when the mixture of dialysate and gelatinous residue was added to a solution of glucose, a considerable amount of CO_2 was given off. After carrying out numerous tests, proceeding by a process of elimination, HARDEN and YOUNG discovered that the active molecule was **mineral phosphate**. The idea that mineral phosphate can be used for the phosphorylation either of glucose or of derivatives of glucose was tested. At this time, mineral phosphate was differentiated from the phosphate incorporated into organic molecules by its ability to form a precipitate in the presence of magnesium citrate. Although this analysis was somewhat rough and ready, it did show that a large proportion of the mineral phosphate that was added disappeared during fermentation, no doubt captured by an organic molecule arising from the metabolism of glucose. These unexpected results, which were published in 1905, provided one of the keys to understanding the mechanism of glycolysis. It was at the end of the 1930s that German biochemists discovered that mineral phosphate is incorporated into glyceraldehyde 3-phosphate, a product of the catabolism of glucose, during an oxidation reaction that leads to the formation of 1,3-bisphosphoglycerate, a primordial step in glycolysis.

In the middle of the 20th century, a reductionist approach applied to cell metabolism contributed to the isolation and characterization of different species of endocellular organelles (nuclei, mitochondria, reticulum, GOLGI apparatus, ribosomes, endosomes, lysosomes, peroxysomes...), each of them with specific functions, from cell homogenates (Chapter III-6.2.3). This knowledge was a necessary prerequisite for the untangling of the web of interactions between organelles that control cell metabolism.

7.3. DECONSTRUCTION AND RECONSTRUCTION OF MACROMOLECULAR COMPLEXES

The **spontaneous assembling (autoassembling)** of components isolated from a macromolecular complex, which is a phenomenon that is encountered in certain viruses, has also been shown in the case of bacterial ribosomes. Autoassembling is made possible by the fact that the components of the resulting complex carry within themselves the information necessary for their mutual recognition, a necessary condition for their **interaction**. Applied to these systems, the reductionist method provides basic information that, by extension, makes it possible to understand the strategic steps in the structuring of a macromolecular complex from its components. A striking example is that of the **tobacco mosaic virus**, an enormous molecular complex comprising ribonucleic acid (RNA) and protein, with a mass reaching 40 000 kilodaltons. Examined under an electron microscope, this virus has a cylindrical shape that is 3 000 Å long and 180 Å in diameter (Figure III.25).

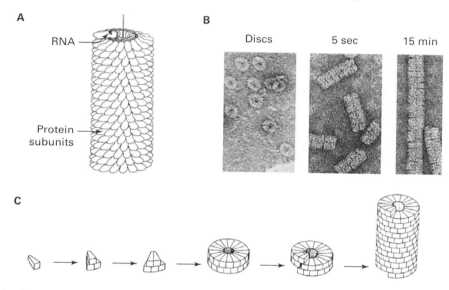

A - Diagram of the interpretation of the structure of the tobacco mosaic virus from X-ray diffraction spectra. The protein sub-units form a tight helicoidal network, with 16 and 1/3 units per turn of the helix. The RNA, which is represented in the form of a wire, is buried between the spires of the helix and interacts with the protein sub-units at a rate of 3 nucleotides per protein sub-unit.

B - Artificial structuring of the tobacco mosaic virus protein sub-units (without RNA) by pH transition (pH 7 → pH 5) into a cylindrical macromolecule that mimics the original virus.

C - Diagram representing steps in the artificial structuring of the virus protein sub-units (in the absence of RNA) into a cylindrical macromolecule, in a slightly acid medium (pH 5), using elementary sub-units, with transition from a disc form to a lopsided form.

Figure III.25 - Deconstruction and reconstruction of tobacco mosaic virus particles
(reproduced and adapted from A. KLUG © (1980) "The assembly of tobacco mosaic virus", *The Harvey Lectures*, Series 74, pp. 141-172, with permission from Elsevier)

The protein component is made up of an assembly of 2 130 identical sub-units, each of them with 158 amino acids. These sub-units wind in a helix around an RNA core comprising 6 400 nucleotides. They can be isolated from virus particles solubilized in an alkaline medium, by precipitation with ammonium sulfate. The RNA can be liberated from its protein coat by the action of a detergent, sodium dodecyl sulfate. After removal of the protein, the RNA is precipitated by the addition of ethanol in a slightly acid medium.

In 1955, at the University of California, Berkeley, USA, Heinz FRAENKEL-CONRAT (1910 - 1999) and Robley WILLIAMS (1908 - 1995) discovered that the isolated components of the tobacco mosaic virus, protein sub-units and RNA, after a few hours of contact, are able to associate to give infectious viral particles that are morphologically identical to the native virus (see Figure III.25). Prior treatment of the RNA with ribonuclease prevents the formation of viral particles. Based on these preliminary but revealing data, different investigators, including Aaron KLUG (b. 1926) in Cambridge (UK) and Rosalind FRANKLIN in London verified the effect of various parameters on the **virus reconstitution process** and elucidated methods for the first steps in the reconstruction. The first step is the autoassembling of around twenty protein sub-units into a two-storey flat disc. A slight acidification of the medium causes the disc structure to become lopsided, the beginning of a transformation into a helix. Under physiological conditions, with a neutral pH, transition to a helicoidal structure is caused by interaction of the RNA with the disc structure (see Figure III.25A). Artificial disc-helix transition in an acid medium therefore only mimics the effect of the RNA on the nucleation process under live conditions (see Figure III.25B). This result was nevertheless of fundamental importance in coming to the conclusion that tobacco mosaic virus protein sub-units possess the information necessary for the assembly of the whole viral particle.

In the 1960s, at the University of Wisconsin, Masayasu NOMURA (b. 1923) described the functional reconstruction of **ribosomes** of the bacterium *E. coli* from the protein and ribonucleic components that make them up. Ribosomes comprise two sub-units, each sub-unit containing one molecule of ribonucleic acid around which are organized a certain number of different protein species (32 for the large sub-unit and 21 for the small sub-unit). Reconstruction *in vitro* from isolated components makes it possible to characterize the intermediary steps and to determine those that play a strategic role. It should be noted that although autoassembly *in vitro* does produce a functional ribosome, this operation requires several hours, while in a proliferating bacterium the production of a ribosome only requires a few minutes. Thus, while the *in vitro* process mimics the *in vivo* one, and the overall process involved in autoassembly to produce a ribosome from components in a test-tube is similar to that implemented by the bacterium, there are differences, involving the speed of mutual recognition of components, the rapid elimination of errors in the choice of interactions and the methods of finishing the construction. The ease and rigor involved in the organization of autoassembly *in vivo* are mainly due to the subsidiary role played by molecules known as **chaperones**[7], which facilitate mutual recognition and decrease error rates by preventing illegitimate interactions. Generally speaking, chaperone molecules control the correct folding of the proteins in the final phase of their biosynthesis, which determines their three-dimensional architecture.

7.4. THE BIRTH OF VIRTUAL BIOLOGY - MODELING CELL DYNAMICS

At the turn of the 20th century, biologists who espoused a reductionist philosophy carried out morphogenesis experiments, in order to try to understand whether the organization of the structures of living beings depends on **known physicochemical laws that could even be mathematized.** In 1910, Stephane LEDUC (1853 - 1939), a professor at the school of medicine in Nantes, France, published *The Physicochemical Theory of Life*. This included magnificent photographs representing the diffusion of China ink droplets that are supposed to mimic the filaments of the typical achromatic spindle of cell karyokinesis. We also see artificial objects resembling living cells, shrubs and fungi. These structures were produced by mixing solutions of gelatin and mineral salts (Figures III.26 and III.27) and were the result of the osmotic growth process. Compared with typical structures of living beings, they were discussed on the basis of a **mechanistic explanation of morphogenesis.** We can understand the background for such an unbridled imagination when we look back over the fantastic explosion of knowledge in physics and in physico-chemistry at this time. "Biology," wrote LEDUC, "is part of the physico-chemistry of liquids. The physico-chemistry of life thus includes the study of non-electrolytic and electrolytic solutions and colloidal solutions, of the molecular forces involved in these solutions, osmotic pressure, cohesion, crystallization, and of the phenomena produced by these forces, diffusion and osmosis." Elsewhere, he added, "Any phenomenon that is manifested by a living being is a vital phenomenon, and even when it has been possible to reproduce it by physical forces, it does not cease to be a vital force." In the antivitalist climate of the beginning of the 20th century, the experimental performances of Stephane LEDUC did not fail to excite the imagination.

After this, mathematics, followed by information technology, would be brought into play in order to try to rationalize the **processes of morphogenesis** in terms of laws. The British naturalist D'ARCY THOMPSON (1860 - 1948) was one of the first to try to explain and translate into mathematical form the undulations of the corollas of flowers, the veins of leaves, the ornamental motifs of insects or of the coats of certain mammals and, above all, that which he considered to be the most illustrative, the spiral form of certain shells. From this point of view, the Nautilus, a

---

[7] The word "chaperone" (from the French *"chaperon"*) means an adult person who accompanies a young girl to prevent her from making any unfortunate encounters. This word first appeared in English-language publications to designate molecules that help in protein folding and prevent proteins from making prejudicial associations.

marine mollusk, was D'ARCY THOMPSON's favorite subject of study. In *On Growth and Form*, published in 1917, he wrote that the shell and the bone, the leaf and the flower, are portions of the living machine, and that it is in obeying the laws of physics that living matter has organized itself. He said that his sole aim was to bring together phenomena involving growth, structure and form, using mathematical considerations and certain laws of physics. Twenty-five centuries later, we find, once again, the philosophical preoccupations of PLATO.

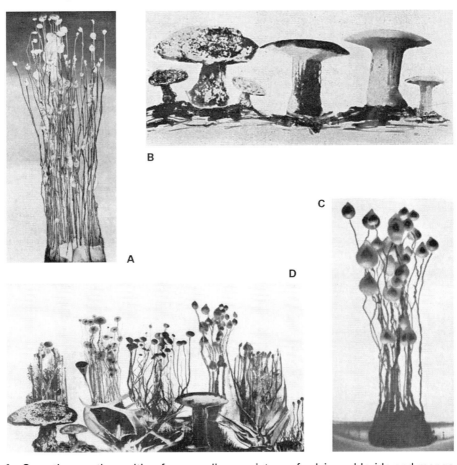

A - Osmotic growth resulting from seeding a mixture of calcium chloride and manganese chloride in a solution of potassium silicate, sodium carbonate and sodium phosphate. **B** - Fungus-shaped osmotic formations. **C** - Osmotic growth with terminal organs having pseudonuclei. **D** - Figure showing shrubs and fungi, obtained using a mixture of gelatin and potassium ferricyanide.

Figure III.26 - Osmotic formations imitating forms in Nature (plant material)
(from S. LEDUC - *Physicochemical Theory of Life and Spontaneous Generation*,
A. Poinat, Paris, 1910)

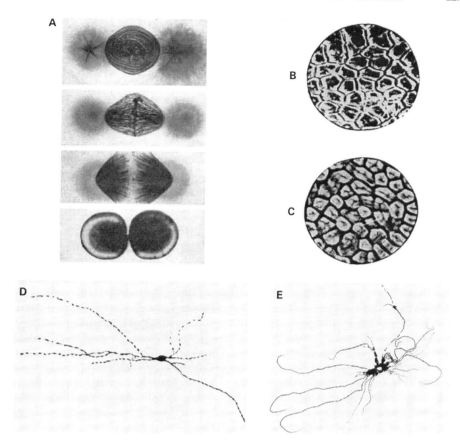

A - Reproduction of the successive stages of karyokinesis by diffusion of China ink droplets. **B** - Frog epithelial cells. **C** - Cells resulting from segmentation of the contents of an artificial cell. **D** - Nerve cell prepared by the GOLGI method (stained black using silver). **E** - Artificial cell mimicking a neuron.

Figure III.27 - Osmotic formations imitating forms in Nature (cells)
(from S. LEDUC - *Synthetic Biology*, A. Poinat, Paris, 1910)

Acting on the same inspiration as D'ARCY THOMPSON, the British mathematician and engineer, Alan TURING (1912 - 1954) tried to give a **physicochemical explanation, reduced to equations**, of the process of morphogenesis. At the end of the 1930s he constructed one of the first computers (TURING machine). During Word War II, he helped to decipher the German Enigma code, making it possible to intercept radio messages sent from Berlin to the German submarines that were attacking allied ships. The deciphering of the Enigma code changed the course of the war and ensured victory for the allies in the Battle of the Atlantic. After the war, using the computing methods that he had mastered, TURING decided to try to "model life". In 1952, in an article entitled "The chemical basis of morphogenesis" (*Philosophical Transactions of the Royal Society, London*, B-237, pp. 32-72), he postulated that in a homogeneous medium, chemical systems are able to **self-organize** into patterns by means of the coupling of chemical reactions with diffusion processes. A few imperatives are necessary, i.e., the existence of activating molecules (activator system) and inhibiting molecules (inhibitor system), the regeneration of the activator by self-catalysis and the braking of this regeneration by the inhibitor, and finally a more rapid diffusion of the inhibitor compared to the activator. These conditions are sufficient to trigger a self-organization shown by concentration peaks for the two species, activating and inhibiting. A short time afterwards, the feasibility of the self-organization of a simple chemical system, in the form of concentric and periodic oscillating waves, was demonstrated by the Russian Boris BELOUSOV (1893 - 1970) during the oxidation of citric acid by potassium bromate in the presence of cerium ions, and then confirmed by his compatriot, Anatol ZHABOTINSKY (b. 1938).

During the same period, the American Norbert WIENER (1894 - 1964), creator of **cybernetics**, likened the signaling messages in the cell to telegraph messages, thus making them capable of **modeling**. WIENER taught mathematics and engineering at the Massachusetts Institute of Technology (MIT). Working on the frontier between information technology and biology, he invented cybernetics, i.e., a control science based partly on the principle of **feedback**. This is a mechanism that makes it possible to stabilize a certain dynamic state, which, in living beings, is at the center of physiological processes such as the maintaining of body temperature in homeothermic organisms or molecular processes such as the control of the flow of metabolites inside cells. The feedback principle would play a large part in the information processing applied to **artificial intelligence**. In the 1960s, Ilya PRIGOGINE (1917 - 2003), founder of the School of Thermodynamics in Brussels, brought forward the idea that biosystems operate **outside the thermodynamic equilibrium**, that feeding them constantly with energy prevents them from evolving towards a state of maximum entropy and that close to molecular chaos an ordered organization that is characteristic of living structures is set up. These structures are said to be **dissipative** because they are only maintained if the system is constantly fed with reagents. During the same period, in the United States, mainly at Stanford University and the Massachusetts Institute of Technology (MIT), the first artificial intelligence programs appeared. The mathematics of artificial intelligence would very quickly be applied to the mysteries of human thought *via* the work of the Swiss psychologist Jean PIAGET (1896 - 1980) on the acquisition of language and logic in children. PIAGET showed that thought does not arise solely from formalized sequences, as in the case of the digital processing of information, but often proceeds in a back-and-forth manner, trial-error, trial-success. At the end of the 20th century, robots that were kinds of artificial animals, able to simulate the complex functions of living beings (the visual system of the fly, for example) made their appearance. These were followed by hybrid robots, which were a kind of chimera comprising a robot body equipped with high-performance electronics associated with a living tissue, such as a neuron network. A new, original way of exploring living beings had arrived (Chapter IV-4.3).

8. THE EXPERIMENTAL METHOD FACED WITH CONTEMPORARY TRENDS IN PHILOSOPHY AND IN SOCIAL LIFE

In the 17th century, the rational practice of experimental sciences was only taking its first faltering steps. It was therefore appropriate to impose standards. These standards, contained in the writings of BACON, DESCARTES and BOYLE, defined the principles of objectivity, openness to criticism and reasoning by induction. These made up the basis of the experimental method, which, within four centuries, spurred the explosive progress made in knowledge in the fields of physics, chemistry and biology. Nowadays, large, computerized, robotic laboratory instruments that require considerable technical skill and experience to handle, increasingly sophisticated analysis methods, the overwhelming accumulation of publications in innumerable scientific reviews and the often unpredictable consequences of the results of research in the varied areas of human activity are all challenges for the philosophers of science who try to analyze the scientific process of the present and to discover a new rationale for it. Before looking at a few of their thoughts, it would be useful to take a brief detour and look at the vitalist and mechanistic theses that were the sources of conflict in the life sciences in the 19th century.

8.1. CONFRONTATION BETWEEN VITALISTS AND MECHANISTS. THE EMERGENCE OF ORGANICISM

In the first decades of the 19th century, **vitalism** was still very much alive. Faced with vitalism, the **mechanistic doctrine** also had its sycophants. DESCARTES' animal-machine and DE LA METTRIE's man-machine are long-ago symbols of a mechanistic conception of Nature that held that the living being functions like a machine, under the direct control of the laws of physics. These ideas were taken up again in the 19th century. The **positive philosophy** of Auguste COMTE played a pivotal role. "Physiology," declared COMTE in his *Course in Positive Philosophy* (40th lesson, 1838), "only began to acquire a truly scientific character with the trend towards its irrevocable differentiation from any ideological or metaphysical supremacy, since the time, not very long ago, when vital phenomena were finally relegated to the status of being subject to general (physicochemical) laws of which they present only simple modifications." He went on to say that "during the century or so since biology began to try and finds its place in the hierarchy of fundamental sciences, it has constantly been bounced back and forth between metaphysics, which has tried to hold it back, and physics, which has tended to absorb it, between the spirit of STAHL and the spirit of BOERHAAVE." One of the postulates of the Comtian way of thinking is that human thought, over time, has passed through three stages, starting with a theological stage, followed by a metaphysical stage in which the mystery of Nature is substituted for God, and ending with a final scientific stage, known as the positivist stage, in which only the concrete part of knowledge is dealt with.

A. COMTE
(1798 - 1857)

Thus, the positivist philosophy of Auguste COMTE opposes the necessity of understanding that which is real and searching for laws that govern the dynamics of Nature with vain metaphysical discussions about "why", i.e., about the question of initial and final causes. COMTE asserted the subordination of the living state to chemical-type mechanisms. He wrote that "without the slightest exaggeration, we can really look at all chemical sciences as comprising, by their very nature, a spontaneous transition from inorganic philosophy (inanimate world) to organic philosophy (living world), despite the profound differences that must separate them." During the same period, syntheses carried out by chemists suggested that the barrier between the inanimate world and the living world is not insurmountable. In 1828, the German chemist, Friedrich WÖHLER, synthesized urea from ammonium cyanate, urea having previously been considered as a production of and a signature of the living world. In 1845, another German chemist, Hermann KOLBE (1818 - 1884), synthesized acetic acid, a product of microbial fermentation. At the end of the 20th century, Eduard BUCHNER demonstrated that glucose is fermented into ethanol by an acellular extract of yeast, a demonstration that finally eliminated any belief in a fermentation principle of a vital nature that could not be dissociated from the intact state of the yeast cell.

Some of the positions taken up by Auguste COMTE, which were rather abrupt, did not fail to cause surprise, and deserve comment in the light of methodological and conceptual advances that have been made in contemporary biology. Thus COMTE denied the validity of numerical laws in biological research (Chapter III.5). He rejected the application of statistics to medicine. In fact, he showed a marked disdain concerning the "medical arts", which he distinguished from "biological science". "Those who would consider it absurd to contemplate allowing navigators to cultivate astronomy will come to consider it strange to abandon, in an analogous manner, biological studies to the leisure of physicians, as the one is not, in itself, more rational than the other." Facts would give the lie to these pessimistic views. At the end of the 20th century, "systems biology" was blithely mathematizing the functioning of the cell and of the organs. In medicine, Claude BERNARD was already prophesizing the arrival of scientific medicine in his *Introduction to the Study of Experimental Medicine*. Contemporary molecular medicine, which is expanding rapidly, has shown him to be right (Chapter IV-3).

In the middle of the 19th century, the **positivist movement** was carried along by the remarkable performance and intensive development of machinistics. Theoreticians made equations for the conversion of heat into mechanical work. The interconversion of the different known forms of energy, heat, electricity, mechanical energy and magnetic energy, was demonstrated. The German physicists and physicians Julius Robert MAYER and Hermann HELMHOLTZ showed that the oxidation of nutrients in tissues is a source of energy given out in the form of heat or retrieved in the form of mechanical work. The analogy was made with the steam engine, which burns coal and activates a movement. At this time, scholars were far from suspecting the complexity of cell metabolism and the sophistication of the genetic control of the enzymes that catalyze the chemical reactions of this metabolism. In 1848, the positivist upsurge was given a concrete form in the foundation of the French Society of Biology under the aegis of Auguste COMTE and Charles ROBIN (1821 - 1885), holder of the chair of the histology of the Paris Faculty of Medicine. This would be a sounding board for the ideas of Auguste COMTE. At the same time, in Germany, the moderate vitalism of Johannes MÜLLER, founder of a famous school of physiology, was challenged by several of his pupils, including Emil DU BOIS-REYMOND and Ernst VON BRÜCKE and also by Matthias SCHLEIDEN, who was a supporter of Comtian positivism. One of the most ardent German converts to positivism was the famous physiologist Carl LUDWIG. Holding to an attitude that was essentially determinist in nature, VON HELMHOLTZ in Germany and Claude BERNARD distanced themselves both from speculative vitalism and from a mechanistic materialism that reduces the living being to an overly simplistic physicochemical system. In this strange concert of ideas in which the positivist score was dominant, the tempo was given by Charles DARWIN, with the publication of his major work, *The Origin of Species by Means of Natural Selection*. The idea of contingency that arose from this showed an unavowed contradiction with a teleonomic conception of Nature.

At the turn of the 20th century, a German physiologist who had emigrated to the United States, Jacques LŒB (1859 - 1924) discovered that if the eggs of a sea urchin are placed in a slightly acid solution, and then in a hypertonic salt solution, development begins without any fertilization having taken place, leading to the larval stage and sometimes to the adult stage. The Frenchman Eugène BATAILLON (1864 - 1953) showed that pricking the egg of a frog with a needle causes it to divide and develop to the larval stage. These various facts were taken up by mechanistic philosophers in order to spread their doctrine.

One of the postulates of positivism is that an understanding of the phenomena of Nature by means of laws should make it possible to predict their occurrence and even modify their course. The hope that science would one day be able not only to explain the functioning of the living being but also to control it, or even direct it, became crystallized in a radical form, **scientism**. Scientism was an amalgam of Comtian positivism and a materialistic physicalism that was violently opposed to vitalism. Set up as a philosophy of science, its intention was to explain the wholeness of the world, including social phenomena, according to the same fundamental laws. In France, political opportunity provided scientism with an unexpected prestige. Marcellin BERTHELOT (1827 - 1907), an organic chemist, senator and minister of public instruction, was its flag-bearer. It should be remembered that while vitalists and mechanists were arguing, a wave of political conflict was sweeping through the nations of Europe, initiated in France by the 1848 revolution, and fed by the economic difficulties arising from the competing development of the mining and mechanical industries. The increasing impact of State economies on political decisions was at the origin of a restructuring of society and of a new way of looking at science, with an increasing awareness of the emerging role of fundamental research in industrial development.

After World War I, in Austria, a neopositivist movement was born, spurred on by the German philosopher Moritz SCHLICK (1882 - 1936), and called **logical positivism**. Known as the **Vienna Circle**, this new positivist school brought together philosophers who had been won over to the values of mathematics and physics. The doctrine of the Vienna Circle, laid out in the *Manifesto on the Scientific Conception of the World* (1929), was strongly influenced by the Austrian philosopher Ludwig WITTGENSTEIN (1889 - 1951) who absolutely condemned recourse to any entity that is external to the perceptible world, eradicating from the new philosophy any metaphysical gloss. The members of this movement founded their philosophy on the ideas of the Austrian physicist and philosopher Ernst MACH (1838 - 1916) who, at the turn of the 20th century, applied himself to the definition of a terminology of physics that was based mainly on the mathematization of empirical observations. His aim was to give philosophy a scientific status, based on the experiment, eliminating the speculative theories of metaphysics and the revealed truths of theology. His radical physicalism meant that the natural sciences as well as the human and social sciences could be reduced to physical realities alone. When Germany annexed Austria in March 1938, several members of the Vienna Circle emigrated to the United States, where they had a strong influence on American epistemologists and logicians. In phase with the neopositivists, the British philosopher and mathematician Bertrand RUSSELL (1872 - 1970) wrote, in *My Philosophical Development* (1959), "Philosophy cannot be fruitful if divorced from empirical science. And by this, I do not mean that the philosopher should get up some science as a holiday task. I mean something much more intimate: that his imagination should be impregnated with the scientific outlook and that he should feel that science has presented us with a new world, new concepts and new methods."

Vitalism was still alive in the first decades of the 20th century, with the support of well-known scholars such as the chemist BERZELIUS, the physiologists TIEDEMANN and GMELIN (1788 - 1853) and the cytologist SCHWANN. The vitalist theory fitted in with speculation concerning the colloidal nature of cell cytoplasm, the organization of which was unknown at this time, and was based on the tricky **protoplasmic theory** of Thomas Henry HUXLEY (1825 - 1895) and Ernst HAECKEL (1834 - 1919), according to which the nutrients supplied to a cell organize themselves into a mysterious living macromolecule. Based on experiments on reflexes, Eduard PFLÜGER concluded that the retreating movement of the foot of a decapitated frog is a defense movement, related to a specific finality. He inferred from this that there is a "soul" in the spinal chord. Although support for vitalist theories was crumbling away, the embryologist Hans DRIESCH remained a fervent convert, and he was an aggressive adversary of physicalism, which he considered to be unable to account for the formation and evolutionary dynamics of germ layers. Henri BERGSON was to be the last eulogist of the vitalist movement.

Both vitalism, which was in decline, and an overly-simplistic physicalism gave way to a new concept, **organicism**. One of the first converts to this idea was the English embryologist Joseph NEEDHAM. From the radically neomechanistic attitude that he showed in *The Sceptical Biologist* (1928), while working as a young biochemist at Cambridge, Joseph NEEDHAM moved towards a more reserved attitude, which could be called organicism, when he entered the domain of embryology, some time later. Stripped of any subjective constraints, and sticking solely to the observation of facts and a strict experimental objectivity, organicism postulates that the structures that make up living beings have different degrees of organization, depending on physicochemical constraints, and that new properties emerge from the "whole", with respect to the sum of the properties that individually characterize the parts of the "whole". In the *Logic of Living Systems* (1970), François JACOB gives these structures the name **integrons**, and puts forward the idea that integrons that are not very structured, of a "lower level", are assembled to form integrons of a "higher level". Thanks to structural modeling, the latter acquire functional properties that do not allow the structure of "lower level" integrons to be predicted. Here we move towards the proposal of the holistic theory, according to which "the whole is greater than the sum of the parts."

8.2. *NOVUM ORGANUM* REVISITED AND CONTESTED

In his *Novum Organum*, Francis BACON laid down the principle that reasoning by induction is the basis of the experimental method (Chapter II-6.1). In his *Treatise on Human Nature* (1748), David HUME discussed the validity of the inductive method and limited the extent to which its principle could be applied, for the simple reason that it is not possible to extrapolate conclusions taken from cases for which we have experience to other cases for which we do not have experience. With the same idea in mind, the British philosopher John Stuart MILL (1808 - 1873) noted that in order for inductive reasoning to be conclusive, Nature must be uniform, in other words, the environmental circumstances must never change (*A System of Logic*, 1843). Reasoning on the principle of causality, MILL argued that while a given cause leads to a single effect, a given effect can result from different causes, and that several different causes can, by interaction, lead to unexpected effects. In short, it is necessary that broad generalizations arrived at by inductive reasoning be examined with care before being adopted, all the more so because MILL's causal relationship is found to have probabilistic characteristics in complex phenomena such as cancer, where the genetic determinant finds its effect modulated by environmental factors.

Adopting a radical attitude, the British philosopher of Austrian origin, Karl POPPER (1902 - 1994), who had been a member of the Vienna Circle but become a dissident, criticized inductive reasoning, developing the thesis that an accumulation of facts cannot be used to formulate a general statement, and that a single contradictory fact is sufficient to destroy this statement. This gave rise to the concept of **refutability (falsifiability)**, which is discussed in two of his major works, *The Logic of Scientific Discovery* (1935), with its *Post Scriptum: Realism and the Aim of Science* (1983) and *Objective Knowledge* (1971). POPPER states that Science should not aim to verify or confirm statements, but should rather aim to refute them, that the "scientificity" of a theory rests in the possibility of invalidating it and that what makes Science progress is that it is arguable. He writes that our certitudes only cover that which is false. A non-negligible advantage of this is that refutations which highlight our errors sometimes lead us to reveal the unexpected. In a certain sense, error is beneficial, and we "learn from our errors." For POPPER, a rational scientific process proceeds by **trial and error**. An idea is selected as an *a priori*, and, based on this idea, a model is theorized and submitted to experimentation. If this model resists falsification, it is kept. This process is analogous to that of the **Darwinian selection** of living species during evolution; species that do not withstand adverse environmental conditions are eliminated. POPPER illustrates the intelligence of proceeding by Popperian falsification, which is often used in the life sciences, by means of an allegory: "The difference between the ameba and EINSTEIN is that, although both make use of the methods of trial and error or elimination, the ameba dislikes erring while EINSTEIN is intrigued by it. He consciously searches for his errors in the hope of learning by his discovery and elimination. The ameba, cannot be critical with respect to its expectations or hypotheses, and it cannot be critical because it cannot confront its hypotheses; they are part of it." Nevertheless, while falsification clearly indicates the pathways to be avoided, it is not, in and of itself, sufficient to make science advance. POPPER introduces a radical demarcation between science, the results of which can be subjected to the criterion of falsification, and non-science or pseudo-science, for which he cites the examples of astrology, psychoanalysis or certain political doctrines such as Marxism.

The Austrian philosopher Paul FEYERABEND, who was considerably more provocative than POPPER, writes in his controversial work, *Against Method: Outline of an Anarchistic Theory of Knowledge* (1973), that "the idea that science can and should be organized according to fixed and universal rules is both unrealistic and pernicious." Such an idea is "detrimental to science, for it neglects the complex physical and historical conditions which influence scientific change." Taking as a basis the fact that often major discoveries are made under circumstances in which chance has intervened, FEYERABEND maintains that progress does not obey fixed rules and that in the scientific process "anything goes". He recommends **counter-induction**, which involves introducing hypotheses that do not necessarily agree with established theories, or may even be in opposition to them. He writes that "we must not hesitate to liberate ourselves from the shackles of the education received in schools, which stifles the imagination and mutilates by compressing, like the binding of the feet of Chinese women, that which is remarkable in human nature." In contrast to inductivity, which proceeds with no *a priori* conceptions, we have the hypothetico-deductive method; a speculation is made, a hypothesis is made, and experiment is carried out to test this hypothesis. This is a method that is commonly used in the biology of today. Nevertheless, it is necessary to ask the right question. Often this right question is asked when something completely unexpected happens during an experiment. Here again, we see the principle of serendipity.

In the original text of *Novum Organum Scientiarum* (1620), Francis BACON's major work, the term *"Instancia crucis"*, which has been translated, in a reductive manner, as "crucial experiment" , may be considered as a metaphor for the moment when a traveler, faced with a cross marking a fork in a road, must chose which of the two directions to follow. To sum up, in BACON's mind, the crucial experiment is not only an experiment whose validity is attested by a rigorous methodology, it is also an experiment that makes it possible to decide between two hypotheses, or even two theories. It is the use of the crucial experiment, with the intent to choose between two theories, which raises a polemic. The chemist and scientific historian Pierre DUHEM (1861 - 1916), in his *Theoretical Physics: its Objectives and its Structure* contrasted not only reasons of epistemological holism but also the possibility that two hypotheses to be decided between could both be wrong with the "crucial experiment". If an experiment can, by its elegance and its rigor, win over an outside observer, then its use for prospective purposes must be approached with circumspection. We can therefore appreciate the prudence with which Louis PASTEUR commented upon the experiment that he had carried out which demonstrated the absence of the spontaneous generation of germs. He was careful to specify that the conclusion that he drew from the results of this experiment was only valid under laboratory conditions and that his aim had been mainly to disprove the allegations of adversaries who had carried out similar experiments without surrounding themselves with the indispensable technical precautions (Chapter III-2.2.2).

8.3. RE-EXAMINING THE PROCESS OF THE EXPERIMENTAL PROCEDURE

Is progress in science continuous or discontinuous? Are there obstacles working against the progress of ideas? How is technical progress situated with respect to the birth of ideas? Is there a logic to the experimental method? Such are the questions with which contemporary epistemologists have been confronted.

Two of them, the American scientific historian Thomas KUHN (1922 - 1996) and the French philosopher Gaston BACHELARD (1884 - 1962) have provided particularly constructive answers. POPPER's concept of falsification and FEYERABEND's anarchic caprices inevitably lead to **situations of crisis**. In his *Structure of Scientific Revolutions* (1962), KUHN considers these situations as inescapable events that are scattered throughout the normal course of science. The idea that is developed is that any scientific theory contains a key idea, a **paradigm**, which imposes itself as a model. The creation of a paradigm in a scientific community follows on from a state of crisis that illustrates the observation of an anomaly or the impossibility of resolving a problem on the basis of a previous paradigm. A newly-created paradigm contains a theory, the postulates of which must be subjected to the approval of experimentation. KUHN considers the paradigm to be a "**disciplinary matrix**" that serves in the manufacture of **so-called normal science**. Most of the time, the new paradigm passes through an embryonic phase before being reinforced and asserted in opposition to the old paradigm, and, finally, being accepted. Thus, much experimentation was necessary before the validity of the theory of oxidation by oxygen formulated by LAVOISIER was accepted. This theory finally took the place of STAHL's phlogistics theory from the moment when it was realized that the negative weight of a substance postulated in the phlogistics theory was unreal, if not absurd. In the same way, a long series of steps was necessary before cell theory was accepted instead of HALLER's fiber theory. Cell theory achieved its final status with REMAK's enlightened formulation of the idea that any cell arises from the division of a mother-cell and with VIRCHOW's postulate that a pathological state arises from a cellular malfunction. The mission of the researcher in a situation of **normal science** is to reinforce the validity of the starting paradigm and to increase its scope, by arousing original ideas. There is therefore a dialogue between a paradigm and normal science that is derived from it. In other words, as written by the Hungarian epistemologist Imre LAKATOS (1922 - 1974), who was close to POPPER and FEYERABEND, "a research program progresses as long as its theoretical growth, built upon a certain thought structure (paradigm), anticipates its empirical growth, that is, for as long as it continues to predict new facts." Abandoning one paradigm for another can be compared to a revolution in the sense that it leads to a destructuring of acquired ideas and a reconstruction around new ideas, as well as changes in the methods of the experimental approach. POPPER noted that this is why a new paradigm in a particular specialty is often initiated by someone who is either very young or newly arrived in the specialty, and who, in either case, has not become impregnated with a doctrine that is the subject of consensus in the scientific community.

In *The Formation of the Scientific Mind* (1938), Gaston BACHELARD explains that experimental science makes progress by means of **breaks in dogma** resulting from the **beating down of obstacles**, including **technical** obstacles resulting from ill-adapted or non-existent instruments and low-performance or inappropriate methods, **epistemological** obstacles resulting from adherence to the erroneous theories that are accepted and spread by the teaching establishment, **verbal** obstacles arising from an uncontrolled vocabulary, **pragmatic** obstacles generated by research that has become utilitarian, and finally, **philosophical** or **religious** obstacles that impose prohibitions and oppose any deviation from a dogmatic norm. Epistemological obstacles are a consequence of the misadventures that are part and parcel of the scientific process. These misadventures become obstacles when they are taught as realities and perpetuated as dogma. In this context, we have seen how beneficial reasoning by analogy can be (Chapter III-2.2.4), but we have also seen how such reasoning can lead thinkers astray when it is used without discernment. The most formidable obstacles are philosophical and sociological ones. Thus, Auguste COMTE, in denouncing the cell theory put forward by a few biologists, caused a delay of several decades in the emergence of French cytology. In the middle of the 20th century in the USSR, the agronomist LYSSENKO (1898 - 1976) decided to oppose the principles of Mendelian genetics, for sociopolitical reasons, the result of which was a spectacular drop in agricultural yields and, in Universities, the eradication of any allusion to the laws of transmission of hereditary characteristics.

While the breakdown of dogma and changes in paradigms are features of scientific revolutions, exchanges between researchers and the trends that arise from these exchanges also play a determining role. In *Genesis and Development of a Scientific Fact* (1935), the Polish scientific historian and physician, Ludwig FLECK (1896 - 1961) develops the idea of the "thought collective" as a parameter that signifies how experimental science progresses. The thought collective is defined as the community of people who exchange ideas and interact intellectually. It is considered to be "the vector of historical development in a thought domain." The thought collective is more than a simple sum of the people involved. The thought that arises from it transcends this sum. Such behavior, in fact, was exclusive to the first academies in the western world. Operating in a liberal way, far from the constraints of the university, the academies acted as melting pots for new ideas and a springboard for an experimental science that was still building its foundations. Even when great individuals were at the origin of scientific revolutions, their ideas, coming up against those of their entourage in thought collectives, had to be backed up by irrefutable scientific proof before they could take over and be spread through society. This was the case for HARVEY's theory that blood flows through a circuit from the heart and back to the heart, and it was also the case for the refutation of LAVOISIER's phlogistics dogma, KOCH and PASTEUR's recognition of a relationship between microorganisms and infectious diseases and many other major discoveries that led to revisions of dogma.

In the second half of the 20th century, a new phenomenon appeared, the explosive development of technology and its sudden invasion of biology. Nowadays, biological research necessitates complex instruments that require professional expertise to run them, and, because of this, close collaboration between often disparate disciplines. This requirement leads to a new way of proceeding, accompanied by a style of thought in which the hunt for discoveries is organized around groupings of research teams, each team contributing its own expertise. Research on living beings has turned into a mass movement. What is more, certain domains in biology are so heavily dependent on the techniques of experimental physics that they cannot be distinguished from it. This is the case, for example, for everything having to do with imaging, the structural analysis of macromolecules or the electrophysiological exploration of the functioning of neurons. Nevertheless, conceptual approaches to the phenomena of biology and those of physics do maintain their own specificity. It is a fact that the specificity of physiological processes in living beings depends upon a fantastic network of macromolecular interactions, often associated with stochastic events, and upon the malleability of their genomes, which gives meaning to the movement of evolution through its multiple episodes.

Whether science proceeds by the breakdown of dogmas or by continuous progress, whether it involves strong individuals or multidisciplinary research consortiums, overall, it remains deterministic in nature, that is to say, it considers any phenomenon to be the consequence of a cause. Be that as it may, while this notion is acceptable from a statistical point of view when operating, for example, on a large number of cells or macromolecules, it is less acceptable when isolated cells or macromolecules that are taken individually are being looked at. In this case, stochastic hazards may mask a strictly determinist logic. As a consequence, where technology allows us to examine phenomena at the level of the infinitesimal, the way we evaluate relationships between cause and effect requires an outlook that is necessarily critical.

9. CONCLUSION - DETERMINISM AND THE EXPANSION OF THE EXPERIMENTAL METHOD - FROM THE ORGAN TO THE MOLECULE

In the 17th and 18th centuries, the experimental method led to remarkable breakthroughs in the understanding of living beings; the discovery of the circulation of blood by HARVEY, the refutation of the thesis of spontaneous generation of animals by REDI, and the demonstration of artificial fertilization by SPALLANZANI. Although these advances were remarkable, they remained few in number, as the naturalists of this period were mainly preoccupied with perfecting details in anatomical science and laying down a system that could organize the amazing diversity of living species in a coherent fashion.

In the 19th century, biology achieved the status of an experimental science, on the same level as physics and chemistry. Although biology was placed near the bottom of the scale, just before the social sciences, in the classification of the sciences that was proposed by Auguste COMTE, its avowed complexity, which was recognized but not yet explored, was an argument that justified this position. Nevertheless, by adopting a **resolutely determinist methodology**, physiology made rapid progress that gave rise to new concepts, such as those of the interior medium, endocrine secretions, cellular energetics and the cell metabolism, these concepts themselves generating original research. Within a few decades, thanks to the technical improvements in optical microscopy and the synthesis of refined dyes for endocellular structures, cytologists were able to define the different stages in the division of somatic and germ cells, a performance that signaled the birth of cell biology. Many other disciplines emerged during the same period, such as microbiology, immunology, genetics, enzymology and embryology, each of them establishing itself as an entity in its own right. This sudden blossoming was a revolution in the life sciences. It was underpinned by the **novel, fundamental theories** that germinated within a few years during the **middle of the 20th century**, the theory of evolution by natural selection, cell theory, and the theory of the transmission of hereditary characteristics, which fed ideas and spurred experimentation. Determinism, which had become the foundation stone of the experimental method, triumphed in all the disciplines of the study of living beings.

The complexification of knowledge in biology, increased specialization in the different domains of research and the **impact of applications on sectors of the economy** as varied as the pharmaceutical industry, agriculture and animal husbandry, have completely changed the way contemporary life sciences look at the experimental method. Vivisections carried out on dogs and other animals by 19th century physiologists helped to elucidate aspects of the functioning of organs that had previously been unknown. Conclusions from these vivisections, extrapolated to Man, were at the origin of significant, and sometimes decisive advances in the **diagnostic analysis of diseases** and the **elucidation of their pathogenesis**. Since this time, an immense amount of progress has been made. In the space of one century, we moved from the exeresis of organs to the elimination of proteins by invalidation of the genes that express them. Although the use of animal models continues to provide precious information about vital mechanisms, the almost-completely generalized practice of using cell cultures has become a preferred corollary or even alternative. Above all, the exploration of living beings, which was stuck at the tissue level, is now able to operate on the **cellular or even molecular level**, helped along by innovative technology.

In the 18th century, LAVOISIER had his balances, calorimeters and gasometers manufactured by specialist artisans. There was a dialogue between the man of science and the artisan-engineer, in order to perfect certain details of the apparatus so that once it was manufactured, it would be able to fulfill the experimental requirements as well as possible. In the 19th century, the **manufacture of scientific instrumentation became industrialized**. In the 20th century, it continued to grow and to specialize into diverse, targeted domains, such as spectrophotometry, radioactivity counting, ultracentrifugation, imaging, electrophoresis and chromatography. The structural details of thousands of the macromolecules of living beings can now be determined by means of increasingly powerful and effective instruments, such as cyclotrons, Nuclear Magnetic Resonance spectrometers and mass spectrometers.

Confocal microscopy, which makes it possible to explore the inside of the cell by means of "optical cross-sections", has revolutionized classical photonic microscopy. An original method for analyzing nanostructures, called near-field microscopy, has been shown to be capable of producing atomic scale images of the surface of macromolecules. In keeping with these unprecedented advances in the industrial domain, **drastic changes have revolutionized chemical analysis**. Until the middle of the 20th century, scientific researchers prepared the biomolecules necessary for their experiments, such as coenzymes, proteins of an enzymatic, hormonal or immunitary nature (antibodies), themselves. Nowadays, an order form sent to specialized companies is all that is necessary to obtain, in a short period of time, products whose preparation or synthesis once took weeks. While the amount of time saved is appreciable, it is nevertheless true that a certain distancing has come into being between the use of a product, and its history, i.e., the way in which it is manufactured and controlled in terms of purity, and possibly activity. However flamboyant the performance of contemporary biology may seem, it is in a direct line with the amazing way in which knowledge took off at the end of the 19th century. While the 20th century should be recognized for its remarkable ability to produce instruments that are adapted to the miniaturized exploration of living beings, it should be added that the ideas and concepts were already there; technical advances made them fruitful.

In the first decades of the 20th century, in keeping with the mechanistic view of living beings that was prevalent, based on themes of the animal-machine, experimental reductionism tried to disassemble the cogs of the living machine, from the cellular to the molecular levels, in the hope of understanding how they function, even if it meant checking the conclusions drawn from this type of study by means of reconstruction experiments. Helped along by continuous progress in instrumentation and analytical techniques, the **reductionist process** of the experimental method was crowned with an overwhelming success. We moved from the exploration of the organ to that of the cell and then to that of the macromolecules of the cell contents. This original way of proceeding made it possible to decipher the structural intimacy of cells and of the molecules that are present in the cells, and to disentangle the complexities of the networks of the metabolism and of cell signaling.

At the turn of the 21st century, the desire has grown to move on from molecular reductionism to an integrated biology of living systems, and to model the functioning of these systems. The genome, which has been explored up until now, is being manipulated. A new high-throughput technology is beginning to take over. Living entities are simulated using biorobots operating by means of highly-developed microelectronics. Technical feats, unimagined half a century ago, are heralding in a new era for the experimental method.

Chapter IV

CHALLENGES FOR EXPERIMENTATION ON LIVING BEINGS AT THE DAWN OF THE 21ST CENTURY

"We can talk endlessly about moral progress, about social progress, about poetic progress, about progress made in happiness; nevertheless, there is a type of progress that defies any discussion, and that is scientific progress, as soon as we judge it within the hierarchy of knowledge, from a specifically intellectual point of view."

Gaston BACHELARD
The Philosophy of No - 1940

In the last few decades of the 20th century, there was a change from a biology that was relatively poor in data to a biology that was overflowing with information. This sudden avalanche was triggered mainly by technological advances that were used profitably in the structural and functional study of genomes and proteomes. The need to organize these data in such a way as to draw from them an explanatory coherence has become an imperative that requires high-performance computing facilities. After the decoding of dozens of genomes of prokaryotes and eukaryotes, particularly that of Man, the first post-genomic era is currently opening up to a questioning of the use of knowledge that has been acquired. Man's health is at the heart of these preoccupations. Under pressure from the media and from a strong political current that leads to financial support being given to applications of so-called "high-visibility" fundamental research, society waits impatiently for quick, usable results, particularly in medicine. This can lead to a few paradoxical situations. Being subject to a mandatory finality in the short term, the experimental process comes into contradiction with academic research that is carried out over the long term, and is on the lookout for those contradictions that have often been and often remain at the origin of breathtaking findings. The interest of the pharmaceutical and agronomic industries in the economic development of discoveries made about living beings, with the necessity of making investments cost-effective and standing out from the competition, is contributing to a new image of biological research. Links that have been forged with the economic, social and political domains tend, nowadays, to have a preponderant affect on the application of the experimental method to living beings. One of the important parameters that has helped to modify the landscape of traditional biological research is related to the introduction of new exploratory methods such as biocomputing or bioinformatics and high-throughput screening, which involves the simultaneous processing of hundreds or even thousands of samples. This approach is in contrast with traditional biology, in which the research strategy is based upon the observation of effects obtained as a function of experimental parameters that are modified one by one. Another aspect of modern times is that, with the irresistible trend in genetic manipulation towards a focus on human beings, certain areas of fundamental research are finding themselves locked into philosophical dilemmas that are matter for ethical and sociocultural consideration, and the subjects of fierce debate.

1. THE ACCESSION OF BIOTECHNOLOGY
TOWARDS A NEW PARADIGM FOR THE EXPERIMENTAL METHOD

Instead of setting out to discover unknown mechanisms by analyzing effects that are dependent on specific causes, with some uncertainty as to the possible success of the enterprise being undertaken, which is the foundation stone of the Bernardian paradigm of the experimental method, many current research projects give themselves **achievable and programmable objectives** that depend upon the means available to them: sequencing of genomes with a view to comparing them, recognition of sequence similarities in proteins coded for by genes belonging to different species, with the aim of putting together phylogenetic trees, synthesis of interesting proteins in transgenic animals and plants, analysis of the three-dimensional structure of proteins, in order to find sites that are likely to fix medicinal substances, and synthesis of molecular species able to recognize pathogenic targets. The facilities that are called into play include instruments that are often sophisticated, the performance of which, in terms of miniaturization, computerization and robotization, is far beyond that of apparatus that was in use a few decades ago. These facilities, applied to research into living beings, have entered the framework of a methodology that has been given the label **biotechnology**. Proceeding hand-in-hand with applications that have become more and more meaningful in the domains of medicine, pharmacology, agronomy and animal husbandry, the **biotechnological process** has come to the fore as a **new paradigm** for the experimental method as applied to living beings. In addition to new discoveries, the driving forces behind biotechnologies are related to economic imperatives as well as the interest and support they receive from the political powers-that-be. The **academic spirit** that presides over fundamental science gives way to the **entrepreneurial spirit** that implements a rational programming of facilities and an efficient organization of scientific collaborations. As an example, the sequencing of the human genome, which includes three billion nucleotide base pairs, required the coordination of several dozen scientific teams around the world and the matching of several tens of thousands of results.

Research on DNA provides a typical illustration of the way in which research has become divided, over the last few decades, between an approach and an interest that had previously been purely academic, and the increasing role of technology, which can be justified by the results that arise in the life of society at large, but which, because of these results, also gives rise to questions concerning how well-founded some of these results are, particularly in the health domain. The experimental method, which had been confined to the laboratory, is now a matter for public debate.

1.1. THE GENOME EXPLORED

Before it won acclaim, DNA, which was isolated under the name of nuclein by Johann Friedrich MIESCHER (1844 - 1895), at the end of the 19th century, had to undergo a series of structural evaluation tests that were spread out over the first five decades of the 20th century. An overall conclusion then came to the fore. DNA is a polydeoxyribonucleotide that carries four cyclic bases, adenine, thymine, cytosine and guanine. Each base is involved in the structure of a mononucleotide where it is itself associated with a sugar, deoxyribose, which is associated with a phosphate residue. DNA was compared to a ladder, the rungs of which (mononucleotides) were linked by ester bonds between an acid group of a phosphate residue of a nucleotide and the free hydroxyl group of the deoxyribose of the following nucleotide.

1.1.1. From molecular biology to genetic engineering

"The very birth of molecular biology illustrates the impossibility of organizing research in a new domain, of scheduling it […]. This biology was born of individual decisions taken by a small number of scientists between the end of the thirties and the beginning of the fifties. Nobody pushed them in this direction. No administrator, no foundation, no Minister for Research committed them to this path. It was the curiosity of each, a new way of considering old problems, that led a few men and women to solve the problems of heredity."

François JACOB
Of Flies, Mice and Men - 1997

The middle of the 20th century saw an accumulation of experimental evidence showing that DNA carries genetic information, and because of this, that it controls the transmission of hereditary characteristics: the proof provided in 1944 by Oswald AVERY (1877 - 1955), Colin MACLEOD (1909 - 1972) and Maclyn MCCARTHY (1911 - 2005) of the **transforming power of DNA** in *Pneumococcus*, the highlighting by Alfred HERSHEY (1908 - 1997) and Martha CHASE (1927 - 2003) , in 1952, of the role played by **bacteriophage DNA** as an **infectious agent for bacteria**, the revelation by Erwin CHARGAFF at the beginning of the 1950s of the equivalence of molar concentrations of adenine (A) and thymine (T), on the one hand, and of cytosine

(C) and guanine (G), on the other hand, in DNAs arising from a multitude of sources, animal, plant and microbial, thus suggesting a **complementary pairing of adenine and thymine, and cytosine and guanine**. Based on the pairing of A/T and C/G bases, the model of the **double helix structure of DNA**, formulated in 1953 by James WATSON and Francis CRICK, made it possible to understand the **identical synthesis of double strands of DNA** by replication during cell division (Figure IV.1) and, as a consequence, the **conservation of hereditary characteristics** in descendants. Afterwards, it was found that the information contained in the DNA base sequence determines the amino acid sequence in proteins. Then the roles played by messenger RNA and **transfer RNAs** were elucidated, the former acting as a carrier of information between DNA and the proteins being synthesized and the latter acting as **double-headed adaptors**, able to recognize nucleotide triplets (codons) in messenger RNA and to specifically fix amino acids in order to position them on the ribosomes, the final result being the synthesis of a protein chain. In 1966, the genetic code was deciphered. The veil of mystery that had covered the mechanism of the synthesis of proteins was lifted, and the decisive role played by nucleic acids in this synthesis was shown.

Later on, there were a few adjustments. Although, **in bacteria, proteins are coded for by a continuous sequence of nucleotide triplets in DNA**, in the 1970s the surprising discovery was made that **in eukaryotic organisms, genes are discontinuous** and made up of coding DNA sequences (exons) interrupted by non-coding sequences (introns). From the end of the 1950s, François JACOB and Jacques MONOD had postulated the existence of a dual determinism for protein synthesis and shown that, next to **structural genes** expressed as proteins, there are **regulatory genes** able to control the expression of the structural genes. The importance of the differential regulation of gene expression in cell differentiation in higher organisms was quickly recognized. From this point on it was possible to explain why a particular species of protein is more specifically expressed in a given tissue and another species of protein is more particularly expressed in another tissue, each type of tissue finding its specificity in its molecular components. This fantastic framework of knowledge, which was built up over a couple of decades, has been used as the foundation stone for the so-called **central dogma of molecular biology**, which explains the **transcription of DNA sequences into messenger RNAs** and the **translation of messenger RNAs into proteins** and which, with only a few variants, is the same throughout the living world (Figure IV.1) [1].

---

1 The fascinating history of molecular biology was well described by James D. WATSON in *Molecular Biology of the Gene* (3rd edition 1976). There are now many works concerning this subject and how it has progressed, including two well-documented books in French: *The secrets of the gene* by François GROS (1986) and *Histoire de la biologie moléculaire* by Michel MORANGE (1994).

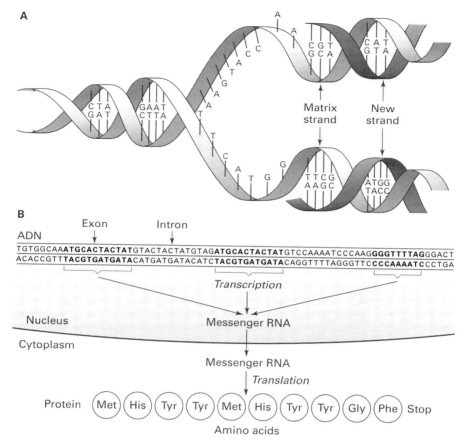

A - Double helix structure of DNA and simplified representation of its self-replication. Each strand of the parent molecule of DNA acts as a matrix for the synthesis of a daughter molecule of complementary DNA, in conformity with the rules of pairing: adenine (A) with thymine (T) and guanine (G) with cytosine (C). The double strands that appear are identical to each other and identical to the parent DNA molecule.

B - Transcription of DNA into messenger RNA and its translation into an amino acid chain. Diagram of gene expression in a eukaryotic cell. One of the strands of DNA (the coding strand) has coding sequences (exons) and non-coding sequences (introns). It is said that the gene is split. The transcription of the exons, accompanied by their splicing leads to the formation of a Messenger RNA, the codons (nucleotide triplets) of which are translated into amino acids that are linked to each other by covalent bonds in order to form a protein chain. In prokaryotic cells (bacteria), the genes do not contain introns and are not split.

A: adenine; C: cytosine; G: guanine; T: thymine; U: uracil. Met: methionine; His: histidine; Tyr: tyrosine; Gly: glycine; Phe: phenylalanine.

Figure IV.1 - The central dogma of molecular biology

Interlinked with the epic rise of molecular biology, there was a succession of technical innovations that led to the synthesis of DNA by chemical or enzymatic means, and to its being cleaved at specific locations, with the pieces that were obtained being joined together again [2]. In 1962, in Geneva, Werner ARBER (b. 1929) [3] and Daisy DUSSOIX highlighted the **restriction** phenomenon, which involves the degradation of bacteriophage DNA by a recipient bacterium. They discovered that an extract of *E. coli* has a restriction activity, and that this activity is of an enzymatic nature, caused by a nuclease that breaks the phosphodiester bonds in DNA. In 1970, the Americans Hamilton SMITH (b. 1931) and Kent WILCOX [4] purified the first restriction enzyme from a strain of *Hæmophilus influenzae*. In 1971, Daniel NATHANS (1928 - 1999) and Kathleen DANNA [5] (b. 1945) at Johns Hopkins University in Baltimore (USA) drew up the first restriction map based on the circular DNA of the monkey SV40 virus, using a restriction enzyme that was named *Hind*III and a follow-up of the sequential appearance of shorter and shorter fragments resulting from the partial digestion of DNA. In the following years, dozens of **restriction enzymes** were isolated, all of them endowed with a surprising specificity with respect to specific base sequences in DNA (Figure IV.2). These enzymes were to be indispensable tools in genetic recombination experiments.

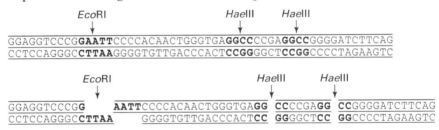

Certain restriction enzymes cleave DNA, leaving free cohesive (or sticky) ends (this is the case for the enzyme *Eco*RI), while others cleave DNA leaving blunt ends (this is the case for the enzyme *Hae*III).

A: adenine; C: cytosine; G: guanine; T: thymine.

Figure IV.2 - Mode of action of restriction enzymes

---

2 In order to take into account the contemporary nature of the facts discussed in this chapter, some references (necessarily limited in number) are given at the bottom of most pages, thus allowing a reader who wishes to know more about experimental details to refer to original articles, the list given in these footnotes being, of course, far from exhaustive.

3 W. ARBER and D. DUSSOIX (1962) *"Host specificity of DNA produced by* Escherichia coli. *I. Host controlled modification of bacteriophage"*. Journal of Molecular Biology, vol. 5, pp. 18-36 and 37-49.

4 H.O. SMITH and K.W. WILCOX (1970) *"A restriction enzyme from* **Hemophilus influenzae***. I. Purification and general properties"*. Journal of Molecular Biology, vol. 51, pp. 379-381.

5 K. DANNA and D. NATHANS (1971) *"Specific cleavage of simian virus 40 DNA by restriction endonuclease of* **Hemophilus influenzae***"*. Proceedings of the National Academy of Sciences, USA, vol. 68, pp. 2913-2917.

The transformation of RNA back into DNA was observed by Howard TEMIN (1934 - 1994) and S. MIZUTANI [6], in experiments on the ROUS sarcoma virus, a virus with RNA that, when it proliferates in host cells, is able to synthesize a DNA that is complementary to its RNA. The enzyme responsible, **reverse transcriptase**, was purified by both H. TEMIN and David BALTIMORE (b. 1938) [7]. Starting with a determined messenger RNA, it then became possible to work back to the DNA, i.e., to the gene, by a simple enzymatic reverse transcription operation. DNA that has been synthesized in this way is called **complementary DNA (cDNA)**. In eukaryotic organisms, reverse transcription has proved to be all the more useful as a technique in that all cDNA is coding, unlike the situation *in vivo* in which the genes are divided up into portions that are coding (**exons**) and portions that are non-coding (**introns**).

The ability to cleave DNA and to join together the fragments obtained in a deliberately chosen order, or, in other words, to manufacture previously unseen DNA sequences by making new combinations, led to the dawning of recombinant DNA technology and caused scientists to come to the sudden realization that the Pandora's box that contains the secrets of life had been opened, that uncontrollable catastrophes might arise from this, and that there was a potential danger of causing tumorigenic viruses to reproduce in commensal bacteria such as the enterobacterium *Escherichia coli*. In 1975, around a hundred molecular biologists gathered together at the Asilomar Conference Center [8] near Monterey in California, in order to discuss the dangers of the new DNA technology. They proposed strict regulation to govern genetic manipulation. Time and experience have shown that the risks being run were very low.

In 1977, DNA sequencing methods were published. One of them made use of chemical techniques [9], the other made use of enzymatic techniques [10]. Applications were not slow in appearing. From 1977, the team led by Frederick SANGER (b. 1918) in Cambridge (UK) determined the first sequence of a genome, that of bacteriophage PhiX174, which is 5 375 nucleotides long. This was the beginning of an audacious adventure, the apparently senseless challenge being met with unbelievable rapidity thanks to the innovative methods of bioengineering, resulting, during the first years of the 21st century, in the **sequencing of the human genome**. Analysis of the human DNA sequence involved the participation of two rival groups,

---

6   H.M. TEMIN and S. MIZUTANI (1970) *"RNA-dependent DNA polymerase in virions of Rous sarcoma virus"*. Nature, vol. 226, pp. 1211-1213.

7   D. BALTIMORE (1970) *"RNA-dependent DNA polymerase in virions of RNA tumour viruses"*. Nature, vol. 226, pp. 1209-1211.

8   P. BERG, D. BALTIMORE, S. BRENNER, R.O. ROBLIN and M.F. SINGER (1975) *"Asilomar conference on recombinant DNA molecules"*. Science, vol. 188, pp. 991-994.

9   A.M. MAXAM and W. GILBERT (1977) *"A new method for sequencing DNA"*. Proceedings of the National Academy of Sciences, USA, vol. 74, pp. 560-564.

10 F. SANGER, S. NICKLEN and A.R. COULSON (1977) *"DNA sequencing with chain-terminating inhibitors"*. Proceedings of the National Academy of Sciences, USA, vol. 74, pp. 5463-6567.

one of them being academic, coordinated by Francis COLLINS (b. 1950), and bringing together dozens of laboratories around the world, and the other being a private Californian company directed by Craig VENTER (b. 1946).

At the beginning of the 1980s, when everyone was persuaded that the RNAs could be placed into three well defined categories, messenger RNAs, transfer RNAs and ribosomal RNAs, it was with great surprise that it was learned that there were RNAs that have catalytic properties (see Thomas CECH [11] [b. 1947]). These RNAs, called **ribozymes**, have, in keeping with enzymatic proteins, structured catalytic sites that are able to catalyze RNA or DNA cleavage or ligation reactions. Recently, engineering techniques have been used to obtain artificial ribozymes that have been found to be able to catalyze reactions as varied as oxidations or the synthesis of peptides and nucleotides, thus opening up wide-ranging possibilities of applications in molecular therapeutics, and, in addition, reinforcing the famous theory of the "World of RNA" at the beginning of the appearance of life on Earth [12].

Another discovery of the 1980s was the role of methylation of DNA bases, cytosine and adenine, and its deregulation in a certain number of pathologies: fragile X syndrome, scapulohumeral dystrophy, certain forms of cancer [13]...

In the past decade or so, basic proteins known as histones that are associated with the nuclear DNA of eukaryotes in the form of a complex called **chromatin** and which had previously been assigned a structural role, have now acquired the status of functional partners. Thanks to specific modifications of certain amino acids (acetylation, methylation, phosphorylation), histones control the state of condensation of the chromatin and the efficacy of transcription of DNA contained in the chromatin, to such an extent that we now speak of the "histone code". The development of our understanding of histones is a good illustration of the complexification of a concept, the DNA code, into an entity that comes closer to living reality, **the DNA code in partnership with the histone code**. There has also been the discovery of **interfering microRNAs**, small polymers made up of around twenty nucleotide units, the role of which is to control protein synthesis (Chapter IV-1.2.2). Methylation of DNA, structural modifications of the chromatin histones, and blocking of transcriptional activity by interfering microRNAs are a few of the major areas of research in a scientific domain that is in full expansion, **epigenetics**, which could be said to have "pipped the science of genetics at the post," and which explains the plasticity of the functions of living beings.

---

11 T.R. CECH, A.J. ZAUG and P.J. GRABOWSKI (1981) *"In vitro splicing of the ribosomal RNA precursor of* **Tetrahymena**. *Involvement of guanosine nucleotide in the excision of the intervening sequence"*. Cell, vol. 27, pp. 487-496.

12 R. FIAMMENGO and A. JÄSCHKE (2005) *"Nucleic acid enzymes"*. Current Opinion in Biotechnology, vol. 16, pp. 614-621.

13 For review see: M.I. SCARRANO, M. STRAZZULO, M.R. MATARAZZO and M. D'ESPOSITO (2005) *"DNA methylation 40 years later: its role in human health and disease"*. Journal of Cellular Physiology, vol. 204, pp. 21-33.

1.1.2. DNA becomes a molecular tool

With the arrival of restriction enzymes and reverse transcription, the foundation stones of **genetic engineering** have now been laid, and are ready to be used, this being all the easier in that **synthetic chemistry** is now able to manufacture DNA chains that are several hundreds of nucleotides long, and progress in **robotics** and **computing techniques** made it possible for chemists to avoid carrying out tedious routine tasks by using completely-programmable machines. The hope that it would be possible to experiment on living beings by means of the manipulation of DNA became a reality when the American researchers Paul BERG [14] (b. 1926), Stanley Norman COHEN [15] (b. 1937), and Herbert BOYER [15] (b. 1936) succeeded in incorporating foreign DNA into a bacterial cell and making it express itself as a specific protein. To make a foreign DNA penetrate into a bacterium, it is often inserted into a **bacterial plasmid**, i.e., a circular molecule of extrachromosomal DNA that acts as a **vector** (Figure IV.3). The plasmid-foreign DNA chimera is replicated inside the bacterium at the same time as the bacterium divides. Viruses are also used as cloning vectors. Other techniques for making DNA penetrate into cells are now available: bombarding cells with tungsten microbeads covered with DNA, electroporation of cells submitted to rapid, high-voltage electrical pulses in the presence of DNA, and direct injection of DNA into mammal cells by micromanipulation (Figure IV.4).

In the first work carried out on the expression of foreign genes, the use of plasmids as vectors was preferred, particularly that of plasmid pBR322, because of its considerable replication capacity. In 1978, a first success was obtained by Herbert BOYER and his co-workers, with the **expression of the gene for somatostatin**, a peptide hormone comprising twelve amino acids that negatively regulates the secretion of growth hormone, **in the bacterium** *E. coli*. Because of its small size, the somatostatin gene was synthesized by chemical means. The expression of somatostatin in *E. coli* was verified using immunological and physiological criteria, thus demonstrating the validity of the procedure that was used.

The following year, **human insulin was produced in** *E. coli*. Fairly soon, yeast was substituted for this bacterium because, as a eukaryotic organism, it has enzyme systems that are able to carry out chemical finishing operations on neosynthesized proteins that bacteria are unable to do, for example the formation of disulfide bridges in insulin.

---

[14] D.A. JACKSON, R.H. SYMONS and P. BERG (1972) *"Biochemical methods for inserting new genetic information into DNA of simian virus 40: circular SV40 DNA molecules containing lambda phage genes and the galactose operon of* Escherichia coli*"*. Proceedings of the National Academy of Sciences, USA, vol. 69, pp. 29054-2909.

[15] S.N. COHEN, A.C.Y. CHANG, H.N. BOYER and R.B. HELLING (1973) *"Construction of biologically functional bacterial plasmids in vitro"*. Proceedings of the National Academy of Sciences, USA, vol. 70, pp. 3240-3244.

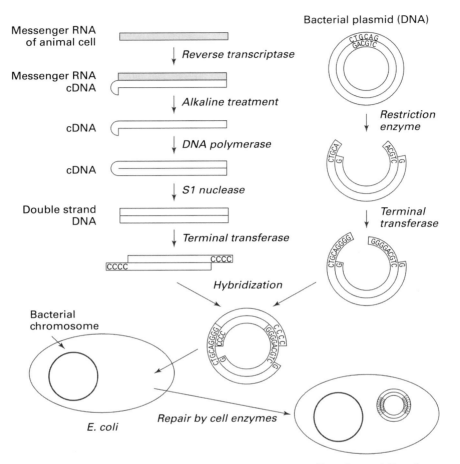

Genetic recombination is used in order to cause bacteria to manufacture a foreign protein of animal or plant origin. This involves the insertion of the fragment of animal or plant DNA that codes for this foreign protein into a plasmid. The plasmid, a small ring of bacterial DNA, acts as a vector for the foreign DNA. In order to carry out insertion, plasmid DNA is cleaved by an appropriate restriction enzyme. The foreign DNA is obtained by reverse transcription from a useful messenger RNA. Its duplication is catalyzed by a DNA polymerase. The S1 nuclease makes it possible to break the covalent bond between two strands of DNA. In the following step, a terminal transferase is used to add four nucleotides for which the base is a cytosine (C) to each of the two DNA strands. The same lengthening operation is carried out on the bacterial plasmid, but, in this case, the addition involves a sequence of four nucleotides for which the base is a guanine (G) (complementary to the cytosine C). The bacterial plasmid is hybridized *in vitro* with foreign animal or plant DNA and then introduced into the bacterium which, using its own machinery, perfects the junction between the integrated DNA and the plasmid DNA.

Figure IV.3 - Genetic recombination technique

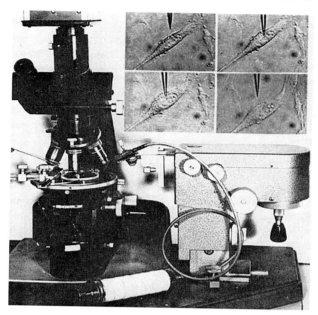

A very small volume of DNA (2.10^{-9} ml) is injected under the microscope into eukaryotic cells (HeLa cells in inset) using a micropipette with a very fine end that pierces the cell membrane. The swelling of the cells at the moment of injection can be seen (inset).

Figure IV.4 - Injection of DNA into individual cells by micromanipulation under the microscope (reproduced from J.E. DARNELL, N. LODISH and D. BALTIMORE - *Molecular Cell Biology*, Scientific American Books, W.H. Freeman and Company, New York, USA, 1986, p. 207, with permission of A. GRÄßMANN, Ph.D. thesis, 1968, Freie University, Berlin)

Supported by these successes, genetic engineering started to come to the fore as an application-oriented discipline. Levels of performance that would never have been imagined half a century before were achieved, such as the production of growth hormone, interferons, blood coagulation factors and vaccines. In the final decades of the 20th century, phenotype transformations using genetic modifications that had previously been carried out in bacteria and yeasts were successfully attempted in animals and plants. It was observed that a mutated DNA integrated into a plasmid and introduced into a fertilized mouse egg (by micromanipulation) modifies the mouse's genetic inheritance, which affects first the embryo and then the adult mouse with phenotype modifications. Such mice, which are said to be **transgenic** because of the stable integration of a foreign DNA into their genome, are now widely used as animal models in studies that aim to understand the mechanisms involved in high-incidence human pathologies such as cancer, diabetes, and rheumatoid conditions. In 1982, two American researchers [16], Ralph BRINSTER (b. 1932)

---

16 For review see: R.D. PALMITER and R.L. BRINSTER (1986) *"Germ line transformation in mice"*. *Annual review Genetics*, vol. 20, pp. 465-499.

and Richard PALMITER (b. 1942) carried out a spectacular transgenesis experiment in mice. Using microinjection, they introduced the growth hormone gene (obtained from the rat) into oocytes of mice from a "little" germ line. Once the transgeneic mice had reached adulthood, they were giants. At present, the transgenesis technique is being applied both to the animal kingdom and to the plant kingdom. In the 1980s, Kary MULLIS [17] (b. 1944) perfected an ingenious technique, the **Polymerase Chain Reaction (PCR)**, which makes it possible to produce several tens of thousands of copies of a fragment of DNA. Using this technique, it is possible to detect traces of a fragment of DNA of a given sequence down to an attomolar concentration, i.e., one billion billion (10^{-18}) times smaller than molar concentration.

By the end of the 20th century, genetic engineering had become well-established and wide-spread, thanks to a mastery of techniques involving the manipulation of DNA such as the **accurate cleavage of a gene into fragments using** commercially available **restriction enzymes**, the **covalent assembly of two fragments of DNA by ligases**, the **automated chemical synthesis of fragments of DNA** of more than one hundred nucleotides and the possibility of manufacturing a **complementary DNA (cDNA)** from a messenger RNA by using a reverse transcriptase and **automated DNA sequencing**. Given this particularly well-equipped toolbox, the molecular biologist is now able to manipulate DNA, that is to say, the chemical material that contains the information that is central to the functioning of living structures (microorganisms, plant and animal organisms), and thus to modify, at will, the genotype of these structures that the selective pressure of evolution has previously favored.

Genomics has produced enormous quantities of data that are stored in databanks. Automated procedures have been invented to make the information contained in these data intelligible, these procedures forming the basis for a new discipline, biocomputing, or bioinformatics, which develops programs, or algorithm-based strategies, that are able to solve specific problems, of which the **annotation of genomes**, i.e., the identification of coding and non-coding sequences. While the annotation of the prokaryotic genomes is relatively easy, because of the absence of introns, that of the eukaryotic genomes is considerably more difficult because of the alternating exons and introns and the small proportion of coding exons (fewer than 2% in the case of the human genome). This explains why, at the time of writing, several hundred genomes of prokaryotes (around 300 at the beginning of 2006) have been sequenced, as opposed to only a few dozen genomes of eukaryotes. Annotation was carried out manually at first, but it has become automated and it is now possible to analyze thousands of items of genomic data.

The comparison of nucleotide base sequences in DNAs and of amino acids in proteins of different origins involves biocomputing. The identification of similar or

---

[17] K. MULLIS (1990) *"The unusual origin of the polymerase chain reaction"*. Scientific American, vol. 262, pp. 56-65.

identical regions that provide information about functional similarities and phylogenetic proximity involves the use of **alignment methods**. One of these methods, which is in current use, is called BLAST (Base Local Alignment Search Tool). Comparison of protein sequences has been particularly instructive in the science of evolution. It has highlighted evolutive processes in the **phylogenesis** of proteins and linked these processes to precise functions. It has been possible to deduce that, over time, different families of proteins with similar functions appeared independently and evolved along different routes. This is the case for membrane proteins whose polypeptide chain crosses the thickness of the membrane six or twelve times; thus, the mitochondrial membrane proteins that transport metabolites are formed by triplication of an element with two transmembrane segments, while proteins located in other membranes of the cell are derived from duplication of an element with three transmembrane segments. From this academic context arose the study of Paleogenetics, a new discipline that compares DNA sequences extracted from fossils and amplified by PCR with DNA sequences of current species. In addition to being of immense interest to fundamental biology, genetic bioengineering has led to innumerable industrial applications making use of **genetically modified microorganisms** that are able to synthesize molecules with a high added value that can also be used in xenobiotic depollution operations.

So much data has already been deposited, equivalent to the sequencing of more than one hundred billion nucleotides, that it is inevitable that there have been errors, some of which might prove prejudicial for future use (comparison of sequences, screening of drugs...). Nevertheless, the ever-increasing numbers of genome sequencing projects for animal, plant and microbe species show the interest that is shown in understanding the genetic information present in different types of cells, in order to be able to exploit their potential.

1.1.3. DNA chips and protein chips - From genomics to proteomics

DNA chips appeared in the last decade of the 20th century, and came to the fore as part of a new technical revolution, the "high throughput" revolution (Figure IV.5). An article written by the group headed by Ronald DAVIS and Patrick BROWN (b. 1954) at the University of Stanford [18] gives a precise description of the hybridization technique used in DNA chips. Thus, around a hundred short DNA strands, corresponding to portions of genes of the plant *Arabidopsis thaliana*, commonly known as mouse-ear cress, a small plant in the brassicaceae (formerly cruciferae) family, are synthesized. A robot is used to deposit microquantities of these DNAs in solution in a dot pattern on a small glass slide coated with poly-L-lysine, thus comprising a "chip" on which the covalently fixed DNAs act as **probes** for specific

---

[18] M. SCHENA, D. SHALON, R.W. DAVIS and P.O. BROWN (1995) *"Quantitative monitoring of gene expression patterns with a complementary DNA microarray"*. Science, vol. 270, pp. 467-470.

molecules. A later step involves both the use of reverse transcription to produce **complementary DNAs (cDNAs)** from messenger RNAs arising from the expression of genes in the same plant and the labeling of these cDNAs with fluorescent ligands for use in screening. Once these fluorescent cDNAs have been denatured, i.e., after separation into single strands, they are brought into contact with the DNA chip. The unhybridized molecules are removed by washing.

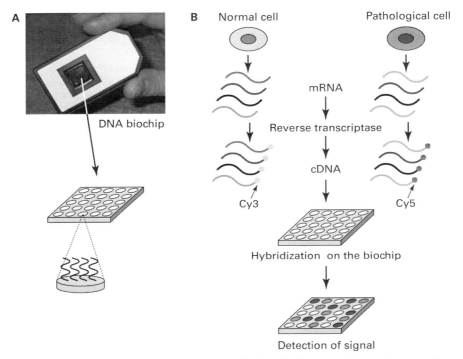

A - The term DNA chip corresponds to a small, chemically-treated glass (or sometimes silicon) plate on which a robot has deposited DNA strands of known sequence in a pre-determined order.

B - The DNA chip may be used for different types of diagnostic procedures. In the differential diagnosis experiment represented here, messenger RNAs are prepared from two cell samples that have been treated in parallel, the control sample (normal cell) and the experimental sample (pathological cell). These messenger RNAs are reverse transcribed into complementary DNAs (cDNAs) by means of a reverse transcriptase. Each of these two types of cDNA, corresponding to the two types of messenger RNA, is labeled by a chemical reaction with a specific fluorescent ligand (Cy3, which emits at 568 nm and Cy5, which emits at 667 nm). They are then hybridized with chip DNA strands. After hybridization and then washing, the fluorescence emitted at 568 nm and 667 nm under laser irradiation are analyzed using an appropriate detection system. The differential expression of the genes in the control cell sample and the experimental sample (shown by a color difference) can thus be analyzed.

Figure IV.5 - Technology of DNA chips

Hybridization between the fluorescent DNAs called **targets** and the complementary nucleotide **probes** fixed to the DNA chip is detected by means of an automated fluorescence detection system. At the beginning of the 1990s, Stephen FODOR [19] and his group, who were working at the Affymax Research Institute (Palo Alto, California), developed an ingenious procedure for microphotolithography that led to the synthesis of a network of a thousand peptides on chemically pre-treated glass microscope slides. The resolution of the network was shown by epifluorescence microscopy after fixation of specific antibodies labeled with fluorescent probes. Soon after this, microphotolithography was used for the manufacture of DNA molecule networks on solid supports. From then on, two competing techniques for the preparation of DNA chips became well-established, either the depositing on a solid support of cDNA obtained by gene amplification (technique used by DAVIS and BROWN), or the synthesis *in situ* of oligonucleotides carried out directly on a solid support (technique used by the American Affymetrix company, arising from Affymax).

One considerable advantage of DNA chip technology is that it provides information on the **level of transcription of thousands of genes into messenger RNAs (mRNAs), in a simultaneous manner and in a relatively short lapse of time**. Experiments that previously required weeks, months or even years to be completed can now be carried out in a matter of hours. We therefore have a sort of instantaneous, precise, freeze-frame picture of the state of a cell at a given moment, with a great number of parameters explored in a semi-quantitative manner. DNA chips are a typical example of the application of high throughput technology to the study of living beings.

The panoply of mRNAs produced by the transcription of DNA is called the **transcriptome**. The method by which transcriptomes are obtained is called **transcriptomics**. It should be added here that there is not necessarily a correlation between the abundance of a mRNA, evaluated on a DNA chip, and the functionality of the corresponding protein, which depends on multiple factors that particularly involve post-translational modifications: phosphorylation, glycosylation, hydroxylation, and so on.

At the end of the 1990s, DNA chips were being used extensively in the research programs of many biology laboratories. They are used in a variety of domains: human pathology, to differentiate between forms of cancer linked to multiform mutations; microbiology, to identify pathogenic germs; comparative genomics, to look at model eukaryotic organisms that have in common a certain number of genes; or even in populations genetics, to detect polymorphisms linked to a change in a single base in a DNA sequence.

---

19 S.P.A. FODOR, J.L. READ, M.C. PIRRUNG, L. STRYER, A.T. LU and D. SOLAS (1991) *"Light-directed spatially addressable parallel chemical synthesis"*. Science, vol. 251, pp. 767-773.

As a complement to the DNA chip technique, the FISH (Fluorescent In Situ Hybridization) method holds a key position in the study of **cytogenetics**. This method is based on hybridization between fluorescent nuclear probes of known nuclear sequence and of complementary motifs located in the DNA of the chromosomes. It allows the detection of chromosomal modifications with a gain or loss of genetic material, such as those that are found in certain tumors. It is used in prenatal diagnostics to diagnose such modifications.

Protein chips, which are used to characterize the reactivity of proteins with respect to specific ligands, are another example of the application of high throughput technology. Dozens of proteins of different types (antibodies, for example), as well as derivatives of nucleic acids or even molecules capable of being ligands of proteins that might arise from combinatorial chemistry (Chapter IV-3.3), are arranged in a network on small glass plates that are chemically treated to act as hooks to entrap specific proteins present in a tissue extract or serum (Figure IV.6A). This procedure, which is essentially analytical in nature, is complemented by a functional study in which proteins that have been isolated in their native form, i.e., those that are capable of expressing the same functions that they posses *in vivo*, are deposited on a glass microplate. This type of biochip makes it possible to analyze the reactivity of the proteins that are fixed to it with respect to a multitude of targets (proteins, nucleic acids or pharmaceutical substances) (Figure IV.6B).

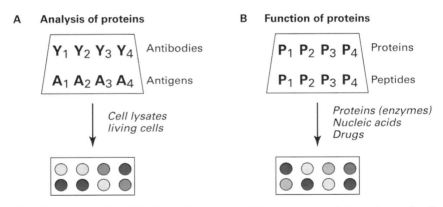

A - Biochip for the identification of proteins. Different types of ligands, antibodies, antigens, DNA or RNA, small molecules with a high affinity and specificity, are deposited on a reactive surface. These biochips can be used to determine the level of expression of proteins and the type of proteins expressed in cell extracts. They can be used for clinical diagnostics.

B - Biochip for the functional study of proteins. Native proteins or peptides are arranged in micronetworks on a reactive medium. Biochips produced in this way are used to analyze the activities of proteins and their affinities as a function of post-translational modifications. They are useful for identifying drug targets.

Figure IV.6 - Use of biochips in proteomics

Analytical proteomics, which was still in its infancy in the last decades of the 20th century (Chapter III-6.2.3) has become a vigorous discipline. The association of liquid nanochromatography and of mass spectrometry allows the identification of peptides obtained by the trypsin hydrolysis of samples of proteins of around one picomole in size. Applied to fundamental and pathological cell biology, the aim of analytical proteomics is not only to decipher the list of proteins on the scale of the cell, but also to highlight variations in the abundance of synthesized proteins as a function of environmental conditions. It also aims to determine the post-translational modifications that are undergone by the proteins inside the cells, which, for a large part, control the specificity of their operation. In parallel with the study of proteomics, the study of **peptidomics**, or the study of all of the peptides (peptidome) present in animal and plant cells and in the fluids that bathe these cells, has developed. For example, several hundred different peptides have been found in the cerebrospinal fluid.

Structural proteomics, which deals with the three-dimensional structure of proteins, has also become a domain in which the activity has been increasingly dominated by high-throughput techniques. This is partly due to the fact that the pharmaceutical industry has given considerable, sustained attention to the understanding of the structure of proteins that could play the role of therapeutic targets. This is the case for protein kinases that catalyze, *via* ATP, the phosphorylation of endocellular proteins, of proteases involved in hydrolysis reactions, or even of cell surface receptors that are able to combine with ligands such as hormones. The use of automated crystallization systems has become common in structural biology. From the classical technique of the hanging drop, in plates comprising 24 wells, we have moved on to plates with 96, 384 and, recently, 1536 wells. Obviously, this increase in dimensions requires the use of an automated system that includes a robot in charge of transferring microaliquots of the protein solution into the wells and adding media that differ according to their pH, ionic force and molecular composition to these wells. The crystallization process is followed by an automated microscopic examination coupled with video photography.

Making use of the recent development of genomics and of proteomics, and a detailed inventory of the structures and functions of the different protein species of living beings, contemporary biology is now able to sketch out a scheme of **molecular systematics**, including a classification into phylla, families, and classes that echoes those of the zoological and botanical systematics of the 17th and 18th centuries. However, modern systematics does not tell us how protein macromolecules interact within dynamic networks.

There is still an enormous amount of work to be done in order to achieve an understanding of the meaning of the dialogue between macromolecules in a normal or pathological cell context. This work will require a detailed analysis of metabolic pathways and of how they are controlled, and their evaluation in kinetic and thermodynamic terms. It will be accompanied by modeling (Chapter IV-4). There is no doubt that it will be successful. Making use of subtle differences in the qualitative and quantitative expression of genes, it will become possible to understand the molecular principles that modulate differences in morphology and in function between neighboring animal, plant or microbial species. The science of evolution should benefit from this. In medicine, the forecasting of predispositions for certain diseases should be made easier (Chapter IV-3.2), opening up the perspective of prevention strategies. Using recombinant DNA technology, metabolic engineering applied to microorganisms and plants should make it possible to improve the production of molecules that are of economic interest or can be used for drugs.

1.1.4. From genomics to metagenomics

The diversity of bacteria is amazing, much greater than might be supposed by looking at the number of bacterial species identified by culturing on appropriate media. In fact, the number of bacterial species that can be cultivated only represents 1% of the total number of existing bacterial species on the surface of the Earth. There are two major reasons for this:
- We do not know the appropriate conditions for culturing these bacteria;
- A certain number of environmental bacteria live in symbiosis, acting as commensal organisms that benefit from the products secreted by other organisms.

Nevertheless, the study of the bacterial genome, without any clonal culture, has been carried out, and comprises a branch of genomics known as **metagenomics**. Instead of looking at an isolated, well-identified bacterial species, in order to analyze the sequence of its DNA, as has been done traditionally, researchers look at a heterogeneous bacterial sample from which the DNA is extracted, amplified, and then sequenced by high throughput methods. Computer processing of the data provides information about individual germs. Craig VENTER, who had already gained notoriety with the sequencing of the human genome, recently applied "metagenomic" procedures to the study of the sequence of the "metagenome" of the bacterial species of the Sargasso Sea [20]. He came up with nucleotide sequences corresponding to approximately 1 million kilobases of non-redundant nucleotides, attributable to more than two thousand different genomes.

The challenge to be met by metagenomics is to connect a function to its phylogenetic source and to extend this information to specific species within a bacterial community. The functional analysis of metagenomics banks has already led to the identification of new antibiotics and of new proteins equipped with various enzymatic activities. For example, analysis of the metagenome of symbiotic bacteria hosted by a marine sponge, *Theonella swinhoei*, has shown that the genes it contains are at the origin of the synthesis of antitumoral substances of the polyketide

---

[20] J.C. VENTER *et al.* (2004) *"Environmental genome shotgun sequencing of the Sargasso Sea". Science*, vol. 304, pp. 66-74.

group [21]. In human biology, a metagenomics approach has been applied to the study of the population of bacteriophages present in the intestinal flora. Approximately 1 200 genotypes have been identified, a number that greatly exceeds the 400 bacterial species of this flora [22]. This result leads us to think that the luxuriant community of bacteriophages which cohabits with that of the intestinal bacteria may influence the diversity of the latter by selective bacterial lysis and also by promoting the exchange of genes between bacteria.

A rapid overview of the history of the exploration of genomic DNA over the last fifty years shows the rapidity with which a traditional experimental paradigm can move thanks to modern computing and robotics procedures. In less than twenty years, we have moved from the manual sequencing of DNA that was developed at the end of the 1970s to automated high throughput sequencing. At the turn of the 21st century, the sequencing of communities of genomes (metagenomics) has been substituted for the sequencing of individual genomes. DNA and protein chips have become objects of everyday use in fundamental and applied biology. Transgenesis is widely practiced. DNA, a molecule that remained mysterious for a long time after it was discovered, delivered some of its secrets during the second half of the 20th century.

1.2. THE MANIPULATED GENOME

The purpose of the first experiments on DNA was to understand how DNA, detector of the genetic code, transmitted its message. After having questioned DNA, researchers moved on to manipulating it. The current aim is to use oligonucleotides to build nanoscale constructions with original and, if possible, useful, properties. In addition, the possibility that has recently become available of being able to interfere with the expression of the genome in living cells, with the intervention of small RNA molecules, allows the programmed manipulation of the genome. Another challenge, the extending of the coding power of the genetic code, now appears to be achievable.

1.2.1. DNA used as a construction material

The production of nanomachines made of DNA is no longer just a dream. Such constructions have been built recently by the American Nadrian SEEMAN [23] (b. 1945) using fragments of DNA with cohesive or "sticky" ends, which result

---

21 J. PIELS et al. (2004) *"Antitumor polyketide biosynthesis by an uncultivated bacterial symbiont of the marine sponge* Theonella swinhoei*". Proceedings of the National Academy of Sciences, USA,* vol. 16, pp. 16222-16227.

22 M. BREITBART et al. (2003) *"Metagenomic analyses of an uncultured viral community from human feces". Journal of Bacteriology,* vol. 185, pp. 6220-6223.

23 N.C. SEEMAN (2005) *"From genes to machines: DNA nanomechanical devices". Trends in Biochemical Sciences,* vol. 30, pp. 119-125.

from the fact that each of the strands of the double helix overhangs in one direction, and in the other the strand with which it is paired, thus leaving a few bases free (Figure IV.7). If two strands of DNA with sticky ends are brought into contact, when the bases of these ends are complementary, a branched structure will appear spontaneously. Using this principle as a basis, cube-shaped nanometric constructions that make it possible to encage molecules of interest have been built. The opening of the cage by appropriate devices liberates the encaged molecules, which can act as substrates in specific reactions.

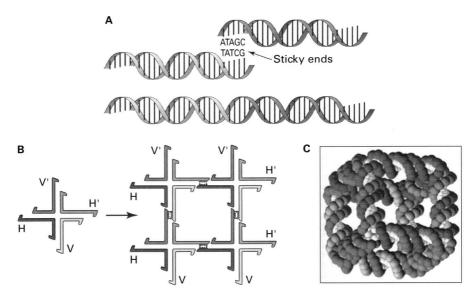

The cutting of a double strand of DNA using restriction enzymes able to create fragments with cohesive ends (**A**) has been used to "build" an artificial construction (**B**), which, in this case, is a cube (**C**), but which could be an object of a different geometrical type.

Figure IV.7 - Building a cubic construction from DNA double helices with sticky ends
(reprinted from N.C. SEEMAN © (2003) "DNA in a material world", *Nature*, vol. 421, pp. 427-431, by permission from N.C. SEEMAN and Macmillan Publishers Ltd)

A **DNA nanomachine** that is capable of movement is becoming a reality. One DNA nanomachine, which is admittedly still rudimentary, has been put together based on the structural difference that exists between B-DNA, the classical double helix that twists to the right, and Z-DNA, a double helix that twists to the left. A propensity to adopt the Z-form is triggered when there is an alternating sequence of cytosine (C) and guanine (G) (CG sequence) in the DNA. The experiment illustrated in figure IV.8 makes use of a duplex formed of B-double helices.

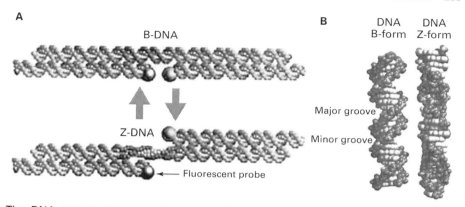

The DNA nanomachine constructed by N.C. SEEMAN comprised a duplex of double strands of DNA. One of the double strands, of the classical B-form of DNA (right-hand twist), has been cleaved in such a way as to fix fluorescent molecular probes onto the cleavage zones. Facing this cleavage zone, a short nucleotide sequence, in which the cytosine (C) guanine (G) motif is repeated, can be found in the other DNA double strand, which is also of B-form. The addition of cobaltihexammine induces the transition of the CG segment from a right-hand twist (B-DNA) to a left-hand twist (Z-DNA), which leads to a rotation of this segment and to a rotation of the assembly, which can be detected using FRET (Fluorescence Resonance Energy Transfer) spectroscopy.

Figure IV.8 - Construction of a DNA nanomachine
(reprinted from N.C. SEEMAN (2003) "DNA in a material world", *Nature*, vol. 421, pp. 427-431; C. MAO, W. SUN, Z. SHEN and N.C. SEEMAN (1999) "A nanomechanical device based on the B-Z transition of DNA", *Nature*, vol. 397, pp. 144-146, by permission from N.C. SEEMAN and Macmillan Publishers Ltd)

One of the double helices has a short CG segment. Facing the CG segment, the other double helix is interrupted, and its ends, where the interruption is, carry fluorescent molecular probes. The simple fact of adding a cationic substance such as cobaltihexammine, which neutralizes the negative charges of the phosphate groups, triggers a conformational transition, with the CG segment taking the Z-form, causing a rotational movement of the assembly that is detected by the movement of the probes.

There is no doubt that the use of DNA strands in order to build nanomolecular constructions that are capable of programmed movement marks the beginning of an adventure that we may imagine will be rich in outlets for domains such as computer technology, nanomechanics and even the life sciences. In addition, the discovery that DNA conducts electrical current gives rise to dreams of a revolutionary technology in which DNA may be used in the design of electrical circuits, in competition with classical electronics [24].

---

[24] V. BHALLA, R.P. BAJPAI and L.M. BHARADWAJ (2003) "DNA electronics". *EMBO Reports*, vol. 4, pp. 442-445.

1.2.2. *RNA interference: a new frontier in the manipulation of the expression of the genome*

Interfering RNAs are non-coding RNAs of around twenty nucleotides that control gene expression at post-transcriptional level. As with many discoveries, that of RNA interference was the result of serendipity. It began during the 1980s with observations made by two American research groups, that of Victor AMBROS [25] now at Darmouth Medical School, Hanover, and that of Gary RUVKUN [26] (b. 1951) at Boston's Massachusetts General Hospital, that a gene named *lin*-4, which is involved in the post-embryonic development of the nematode *C. elegans*, did not code for a protein, but for a small size RNA that played an antisense role.

This odd discovery was supported and made more explicit a few years later by the research groups of Andrew FIRE (b. 1959) at Baltimore's Carnegie Institute and Craig C. MELLO (b. 1960) at the University of Massachusetts in Worcester [27]. In order to block the production of certain proteins in the nematode *C. elegans*, the researchers used synthetic antisense RNAs. The control involved the use of sense RNAs according to a classical protocol. Unexpectedly, protein synthesis was blocked in both cases, suggesting that a contaminant was present in the sense and antisense RNA preparations. This contaminant was identified as a double strand RNA (dsRNA – double strand) that is, an RNA that is folded back on itself in a "hairpin" loop because of the pairing of complementary bases (adenine *vs* uracil and guanine *vs* cytosine). In order to verify the mechanism by which the translation of messenger RNAs into proteins is silenced, the nematode *C. elegans* was injected with a synthetic dsRNA, part of the sequence of which was complementary to that of the gene *unc*-22, known to code for a protein involved in muscular contraction. Within a few hours, the worm was making disordered movements, suggesting that the dsRNA interferes with the production of proteins in the process of muscular contraction.

The mechanism of action of dsRNA was quickly unraveled: dsRNA gives rise to two single strand RNAs after cleavage by a specific enzymatic mechanism. One of the single strand RNAs (siRNA – small interfering RNA) is paired thanks to a complementarity of bases with a short sequence of messenger RNA transcribed from the gene *unc*-22. The result is a blockage of the translation of messenger RNA into a protein, followed by the destruction of messenger RNA. This phenomenon was named **RNA interference** (Figure IV.9). These results shed light on the obser-

---

25 R.C. LEE, R.L. FEINBAUM and V. AMBROS (1993) *The* C. elegans *heterochromatic gene* lin-4 *encodes small RNAs with antisense complementarity to* lin-14. *Cell,* vol. 75, pp. 843-854.
26 B. WIGHTMAN, I. HA and G. RUVKUN (1993) *"Posttranscriptional regulation of the heterochronic gene* lin-14 *by* lin-4 *mediates temporal pattern formation in* C. elegans*". Cell,* vol. 75, pp. 855-862.
27 A. FIRE, S. XU, M.K. MONTGOMERY, S.A. KOTSAS, S.E. DRIVER and C.C. MELLO (1998) *"Potent and specific genetic interference by double stranded RNA in* Caenorhabditis elegans*". Nature,* vol. 391, pp. 806-811.

vations made ten years earlier at the University of Arizona by Richard A. JORGENSEN (b. 1951) on purple petunia plants treated with an excess of copies of the genes involved in the synthesis of the purple pigment. Completely unexpectedly, instead of their becoming brighter in color, the treated petunias lost their color and became white. The treatment triggered a phenomenon in which the gene responsible for the pigmentation was silenced, no doubt by the implementation of interfering RNAs.

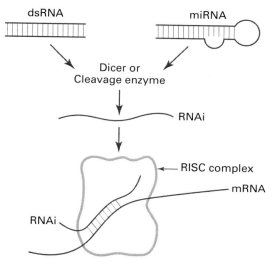

The DICER cleavage enzyme, which has a ribonuclease activity, cuts the double strand RNA into two strands. In the presence of the RISC (**R**NA-**I**nduced **S**ilencing **C**omplex) protein complex, one of the RNA strands finds a complementary nucleotide sequence in a messenger RNA (mRNA) and associates itself with this RNA, making it unable to be translated into protein.

Figure IV.9 - Silencing of messenger RNA (mRNA) by interfering RNAs from double strand RNA that is either synthetic (dsRNA) or natural (miRNA)

We now know that eukaryotic cells from animals and plants produce and host **interfering RNAs** that are said to be "natural" [28]. Natural interfering RNAs, of around twenty nucleotides, are called **microRNAs (miRNAs)**. Although a few details differ between the modes of formation and action of natural interfering RNAs and those of synthetic interfering RNAs, in particular the fact that messenger RNAs are not destroyed by miRNAs but blocked in their translation, the effect of negative regulation on the production of specific proteins comes to the same thing (Figure IV.9). There is a far from negligible number of genes that code for miRNAs. Already, several hundred miRNAs have been identified in the genomes

---

[28] A.E. PASQUINELLI, S. HUNTER and J. BRACHT (2005) *"MicroRNAs: a developping story"*. Current Opinion in Genetics and Development, vol. 15, pp. 200-205. (review)

of plants and animals. The amount of interest that they arouse, and the feverishness of the research being carried out on them, are in keeping with the major mechanisms that they control: embryogenesis, hematopoiesis, neuronal differentiation, etc.

Given an understanding of the genome sequence in Man, the rat and the mouse, trials have begun that aim to achieve an understanding of how the expression of mammal genes of known sequence might be manipulated by the interplay of interfering RNAs (Chapter IV-3.1). The treatment of viral infections such as AIDS or hepatitis B, which are worrying public health problems, could benefit from this new technology. It appears that interfering RNAs have much more to give to us in the near future than they have taught us up until now.

1.2.3. The experimental transgression of the genetic code

The deciphering of the genetic code in the middle of the 1960s was the end of a first step in elucidating the mechanism by which a sequence of nucleotides in DNA is translated into a sequence of amino acids in a protein (Chapter IV-1.1). During the years that followed, the subtleties of the transcription of DNA into messenger RNA and of the translation of messenger RNA into protein *via* transfer RNAs were explored in hundreds of laboratories around the world. Particular attention was paid to the understanding of how a given amino acid is activated and bound to a transfer RNA (tRNA) after being picked up by an aminoacyl-tRNA synthetase. Nevertheless, the idea remained of a code in which triplets of purine and pyrimidine bases of messenger RNAs are translated into natural amino acids.

Recently, methods have been developed that give more flexibility to the action of the aminoacyl-tRNA synthetases, or, in other terms, relax their specificity [29]. Synthetases that have been manipulated in this way are able to recognize non-natural amino acids and to incorporate them into proteins by working together with the ribosomal machinery. It is in this way that, at the time of writing, around thirty non-natural amino acids, obtained by insertion of different types of chemical residue (photoactivable, fluorescent or radioactive residues capable of acting as probes for structural and functional analyses) (Figure IV.10) have been incorporated into protein structures. With such an innovation, an unexpected field of exploration has opened up to research in domains as far apart as pharmacology and the science of evolution, giving rise to burning questions: could such non-natural proteins have therapeutic properties? could they give a selective advantage to the organisms that host them? With the addition of non-natural amino acids to the genetic code and the demonstration that proteins containing such amino acids can function in living cells, in sum, with the transgression of the potentialities of the natural genetic code, the experimental method appears to challenge the order of living beings.

---

29 J.W. CHIN, T.A. CROPP, J.C. ANDERSON, M. MUKHERJI, Z. ZHANG and P.G. SCHULTZ (2003) *"An expanded eukaryotic genetic code"*. Science, vol. 301, pp. 964-967.

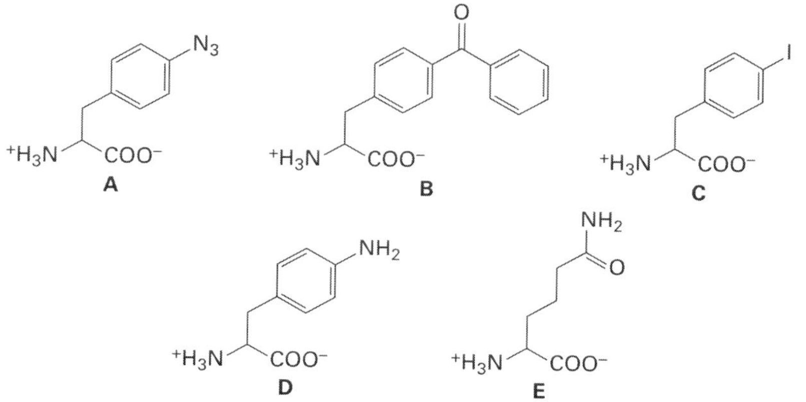

Examples of amino acids added to the genetic code: **A**: *p*-azido-*L*-phenylalanine; **B**: benzoyl-*L*-phenylalanine; **C**: *p*-iodophenylalanine; **D**: *p*-aminophenylalanine; **E**: homoglutamine.

Figure IV.10 - Expansion of the genetic code
(adapted from T. Ashton CROPP and Peter G. SCHULTZ (2004) "An expanding genetic code", *Trends Genet.*, vol. 20, pp. 625-630, with permission from Elsevier)

The triumph of genetic engineering *via* the study of DNA is not unique to biology. Many other sectors are undergoing changes in their type of experimental approach, dictated by the technosciences and making use of computer sciences, robotics and high-throughput screening. However, given the many questions that its operation continues to raise, and its central position at the heart of scientific ethics, the study of DNA remains a typical example of the way in which the experimental life sciences and the techniques that underlie them are evolving nowadays.

2. TOWARDS A MASTERY OF THE FUNCTIONS OF LIVING BEINGS FOR UTILITARIAN PURPOSES

"I perfectly agree that when physiology is sufficiently advanced, the physiologist will be able to make new animals or plants, as the chemists produces substances that have potential, but do not exist in the natural order of things."

Claude BERNARD
Principles of Experimental Medicine - 1877

More than a century after Claude BERNARD predicted a genetically manipulated world, it has come to pass. The molecular biologist, having original, high-performance methods for "tinkering" with DNA, has moved on to the application and use of his technical expertise for utilitarian ends. During the 1970s, with transgenesis, research on bacteria (Chapter IV-1.1.2) opened up a new biological domain, that of genetically modified organisms (GMOs). Results led to predictions that it would be possible to transfer a fragment of DNA corresponding to a gene of a certain species into the genome of another species and have this foreign gene express itself as a protein in the host cell. In 1983, the successful trans-genesis of a gene for resistance to an antibiotic, kanamycin, in tobacco plants, signaled the beginning of the technology of the first plant-type genetically modified organisms, still called GMPs (genetically modified plants). In 1996, the birth of Dolly the ewe unveiled the era of reproductive cloning in mammals, i.e., the identical reproduction of an already-existing organism. An additional step was taken with the first tests on the differentiation of embryonic stem cells towards different types of lines that are characteristic of well-defined tissues, such as nerve tissue, cardiac tissue or the hepatic parenchyma, thus opening up promising perspectives in regenerative medicine. The frontiers of the experimental method continue to be pushed back to the limit of what is feasible and sometimes into the realm of fiction, as in immunology, for example, with the idea of xenotransplantation, using "humanized" animal organs.

2.1. MANIPULATIONS OF PLANT DNA
THE CHALLENGE OF GENETICALLY MODIFIED PLANTS

Given the universality of the genetic code, any gene that is introduced into the genome of a plant, whether that gene is of animal, plant or microbial origin, is able to replicate itself and be expressed as a specific protein. Thus plant GMOs or genetically modified plants (GMPs) are able to express specific foreign proteins from another plant, a bacterial microorganism or an animal organism. In the 1990s, a short time after fundamental research had revealed the feasibility of plant transgenesis, the first transgenic plant, the Flavr Savr tomato, was marketed in the USA. Since this time, numerous other plant GMOs have been cultivated on a large scale and become available on the world market, including corn, soya, rice, cotton and the poplar. One of the desired aims is to produce modified plants that are able to **resist destruction by the herbicides** that are commonly used to eliminate weeds, while another is to prevent **predation by harmful insects**. In the first case, transgenesis involves the insertion of a herbicide-resistant gene, and in the second case, the inserted gene codes for an insecticidal toxin. Recently, plant GMOs that **produce proteins with a therapeutic effect** have appeared, ranging from antibiotic peptides to antibodies or proteins as unexpected as human hemoglobin. Current projects aim to create plants that are resistant to adverse conditions such as the dryness of arid climate zones.

The preferred procedure for producing a plant GMO is to **use a bacterium**, *Agrobacterium tumefaciens*, a microorganism that is able to insert fragments of its own DNA into plant cells (Figure IV.11). The useful gene that we wish to transfer into the plant may be a gene for resistance to a pesticide such as glyphosate, marketed under the name of *Roundup*, phosphinothricin (*Basta*) or glufosinate (*Liberty*).

IV - CHALLENGES OF EXPERIMENTATION ON LIVING BEINGS AT THE DAWN OF THE 21ST CENTURY 267

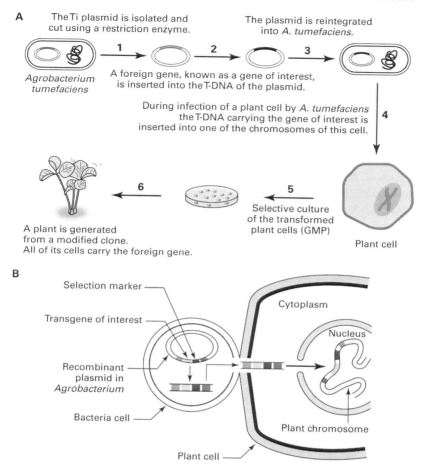

A - Use of the Ti (Tumor-inducing) plasmid as a vector for genetic engineering in plants. The Ti plasmid resides naturally in the bacterium *Agrobacterium tumefaciens*. This bacterium infects certain plants. During infection, the part of the Ti plasmid called T-DNA is integrated into the genome of the infected plant. The T-DNA fragment can be modified by insertion of a selective marker such as the gene for resistance to kanamycin, which will allow selective isolation, in the presence of kanamycin, of modified clones and a gene of interest (gene for resistance to a herbicide or gene expressing an insecticidal toxin). When *A. tumefaciens*, carrying the plasmid modified T-DNA, infects a plant cell, it transfers the T-DNA to it in linear form after its excision from the plasmid (step 4). After this, the T-DNA is integrated into the chromosome of the plant cell.
B - Enlarged view of step 4: transfer of DNA from the *Agrobacterium* to the plant cell.

Figure IV.11 - Creation of a plant GMO by genetic engineering

In the case of the fight against insect predators, the useful gene is carried by a fragment of DNA contained in the genome of the bacterium *Bacillus thuringiensis*. This gene, called Bt, expresses a toxin responsible for the insecticidal capability of

B. thuringiensis. A current application involves the protection of Bt corn with respect to the corn-borer, a devastating insect whose caterpillars are particularly destructive. Another, more direct, gene transfer method, known as **biolistics**, involves bombarding plant cells with tungsten microbeads covered with modified DNA.

With the implementation of large-surface-area experimental fields and the first marketing of GM soya, in 1996, the question of whether or not the advantages achieved with respect to crop yields are counter-balanced by risks for the environment and for consumers came to the fore. Food risks could arise from the toxicity or allergenic power of artificially synthesized proteins. At the time of writing, this question remains unanswered, due to the lack of epidemiological studies carried out rationally over several years.

When the first creations of GMOs took place, the transfer of the gene of interest was carried out by means of the co-transfer of an antibiotic resistance gene. The transformed cells were selected according to the criterion of their resistance to this antibiotic, which involved a risk of dissemination of the resistance gene. This selection technique has been abandoned. In practice, it is difficult to evaluate the theoretical ecological risk of wild plants being invaded by genes that have been inserted artificially into GMOs. As a precaution, zones used for experimentation of plant GMOs are now surrounded by refuge zones, i.e., fields in which the same species of plants, in non-GMO form, are cultivated. There has been a much fiercer and completely legitimate debate concerning the presence of the Terminator gene in seed from the first GMOs marketed by the Monsanto company in the USA. The Terminator gene blocked germination of the seed from the cultivated plant, so it was necessary for the farmer to buy more seed from the company each season, thus creating a state of dependency. This technique is no longer in use, but the fact remains that most transgenic seed is patented, and therefore farmers who use such seed are dependent on the companies that posses this genetic know-how.

The culture of plant GMOs has spread around the world, covering more than a billion hectares of our planet, more than half of which are in the United States of America. This type of culture is used on a large scale for soya and in a less extensive way for corn, rape seed and cotton, but there are many other applications of plant transgenesis. Among the countries that are actively involved we may mention Argentina, Brazil, Canada and China, and more recently India, Paraguay and South Africa. While the policies of these countries are based on the fact that GMO products do not differ fundamentally from non-GMO products with respect to checks carried out *a posteriori*, and that there is thus no reason to prohibit them, **European policy** has taken refuge behind a principle of precaution, and it remains **basically restrictive**. Although the moratorium on the culture and marketing of plant GMOs that was put in place in 1999 was lifted in 2004, mandatory labeling for any consumable product containing more than 0.9% GMO remains dissuasive. The United States of America has refused to use such labeling.

The worries that are aroused by plant transgenesis, which are often exacerbated by the diktats of **ecology groups**, must be analyzed in a reasoned manner. Common sense and lucid thought dictate that the debate should be situated within a scientific perspective in which the main role is played by the experimental method in long-term applications. Simple reflection leads us to think that with time parasites and self-propagating plants will develop a resistance to the most drastic treatments, as was the case for bacteria confronted with antibiotics. The perspective of an acquisition of uncontrolled resistances, which gives rise to so much passionate debate, is, in fact, only the first stage of a technology with promising applications. The mastery of plant transgenesis that was acquired through the first experimentations should, in fact, allow the emergence of plant GMOs that are assigned to the production of **molecules with therapeutic effects** (drugs, vaccines, human proteins, vitamins...). In this domain, there have already been creations that include golden rice, which carries β-carotene, the precursor of vitamin A, banana plants that express a vaccine against hepatitis B and tobacco that produces human lactotransferrin and hemoglobin. If we just look at the production of golden rice as a palliative for vitamin A deficiency, it should be remembered that, in certain countries of our planet, this deficiency affects people's sight and is a frequent cause of blindness, that it generates problems with development and the immune response to infections, that it affects more than a hundred million children around the world and that it is responsible for the death of three million of them each year. If these plants are considered to be a material of choice for the production of proteins with a therapeutic effect, this is partly due to the yield of such crops over large surface areas, and also partly due to the low risk of transmission of viral pathogens to Man, because of the species barrier, a risk that is less negligible when animal productions are involved. Genetically modified plants are also potential factories for the manufacture of **chemical products with an industrial impact**, for example lubricants, perfumes and aromas. Given the unpredictable outlets that plant GMOs may have in human medicine and the different domains of the economy, plant GMO technology should be considered in a manner that is free of any pressure or passion, and, as far as the political authorities are concerned, it should be subject to appropriate measures to surround and protect certain strategic experiments.

When looking at the worries being expressed by the European society, it should be remembered that the genetic inheritance of plants has never ceased changing, not only in the most of natural of manners, over millions of years, particularly with the mobility of transposable elements located in the genome, but also artificially, at the hands of farmers from ancient times onwards, with their methods of hybridization and selection. The nervousness of European authorities, showing an ignorance of basic scientific ideas, with the pretext of a principle of precaution, and sometimes political compromises that are exemplified by fluctuating and contradictory positions, runs the risk, in the short term, of causing their countries to lag disadvantageously behind the United States of America, which holds the majority of plant biotechnology patents.

2.2. MANIPULATIONS OF HUMAN DNA AND HOPES FOR GENE THERAPY

The principle of **gene therapy** is simple: the introduction of an appropriate gene into the cells of a patient who carries a mutation can correct the phenotypical consequences of this mutation, or, in other terms, cure the disease affecting the patient, or at least slow down its evolution. The technical difficulty involved in gene therapy is that of finding an appropriate vehicle or vector for the transfer of the gene and addressing it to an appropriate location in the genome of the host cell. The most commonly used **vectors** in human gene therapy are **viral**. A certain number of criteria are necessary for a transfer to be efficacious, including a high concentration of viral particles carrying the gene to be transferred (more than a billion viral particles per milliliter) and a good capability on the part of the foreign gene to be integrated into the host's genome. The patient's immune response remains a major worry in the use of viral vectors: at cell level it often leads to a proliferation of cytotoxic lymphocytes and, especially at humoral level, to the synthesis of antibodies directed against the viral proteins. In order to minimize its immune response, the genetic material of the viral vectors is modified.

For **ethical reasons**, gene therapy is currently only applied to **somatic cells**, germinal gene therapy being rejected. Somatic gene therapy has been experimented in the treatment of hereditary illnesses linked to hematopoiesis. One of the technical reasons for this choice is easy access to the progenitor cells of the bone marrow, with the aim of transfection. It was with this in mind that mouse gene therapy models were developed a few years ago. The **sickle cell mouse** is one of these models. Human drepanocytosis (sickle cell anemia) is a serious disease that is caused by a mutation in the β protein chain of normal human hemoglobin A. The molecules of sickle cell hemoglobin S tend to aggregate and form fibers that obstruct the blood capillaries of the microcirculation. Somatic gene therapy has been applied to these sickle cell mice. This involves an autograft of bone marrow hematopoietic cells transfected with a retrovirus hosting the gene coding for the β subunit of normal hemoglobin. Encouraging results have shown the validity of this approach.

In 2000, a gene therapy protocol that had been applied with success to Man was described by the group of Alain FISCHER (b. 1949) and Marina CAVAZZANA-CALVO at the Necker hospital in Paris [30] (*Science*, vol. 288, pp. 669-672). The purpose of this therapy was to bring about a long term remission in the case of an immune disease known as SCID-X1 (Severe Combined ImmunoDeficiency linked to a mutation on the X chromosome). Because of their susceptibility to microbial and viral infections, babies who are affected can only survive in sterile rooms. They are known as **bubble babies**. In this illness, the hematopoietic progenitor cells of the bone marrow

---

[30] M. CAVAZZANA-CALVO et al. (2000) *"Gene therapy of human severe combined immunodeficiency (SCID)-X1 disease"*. Science, vol. 288, pp. 669-672.

are unable to differentiate into T and NK (Natural Killer) lymphocytes because of a mutation that affects a cytokine receptor. Previous experiments carried out on model mice show that SCID can be corrected by *in vivo* transfer of the cytokine receptor gene into hematopoietic progenitors. The transfer of the gene of interest paired with a retroviral vector was carried out first in March 1999, in two babies, one of them eleven months and the other two months old. Progenitor cells from their own bone marrow, cultured and modified genetically, were injected into them. These were therefore autografts, without any risk of immune rejection. A remission of symptoms over a period of nearly a year, shown by the almost normal behavior of the babies' immune cells, encouraged the application of the same therapy to other babies. In total, ten babies were given this therapy. The enthusiasm that greeted the successes that were recorded was nevertheless tempered by fact that in the spring of 2002, and again in the following year, a child who had undergone the gene therapy developed a leukemia characterized by an anarchical proliferation of lymphocytes, necessitating chemotherapy. These two occurrences were explained by the random character of the insertion of the gene of interest into the patients' genomes: insertion into a site close to a proto-oncogene had led to activation of this proto-oncogene and the proliferation of the lymphocytes. While the trial carried out at the Necker hospital gave rise to great hopes, it nevertheless showed that there is still a long way to go before we achieve a **targeted transfection of genes** so that no undesirable consequences follow. Here we have a typical example of the limits of an experimental method that is based on an in-depth technological know-how, but also on a still imperfect understanding of the complex arcana of the mechanisms that regulate the positioning and interaction of genes in the chromosomes of eukaryotic cells. This example highlights a harrowing ethical dilemma: should we not treat a patient whose illness is likely to be fatal, or attempt a therapy that may save the patient, without having any formal assurance of its success?

An experimental medicine that has the power to modify the human organism *via* its genetic material is now able to take over from the experimental method that up until now operated on animals and plants. We can easily understand, given the progress that has already been accomplished and that which is to come in the domain of gene therapy, that the temptation will be great, in the future, to consider manipulations of the human germ cell genome as being licit, insofar as such manipulations make it possible to eradicate a handicapping defect in our descendents. At present, the idea of any attack on the **germinal genetic inheritance** has been rejected unconditionally on the basis of ethical considerations. Nevertheless, the history of science shows that prohibitions that were once considered to be untouchable finish by being contravened. This was the case for abortion. In a text entitled *Why genetic engineering should continue its battle* [31], James WATSON writes of

---

31 Published in *A Passion for DNA. Genes, Genomes and Society*, 2000.

his confusion when faced with a choice that is likely to become more and more insistent over the years: "Dare we be entrusted with improving on the results of several million years of Darwinian natural selection? or do the human germ cells represent on the contrary Rubicons that geneticists will never dare to cross?"

2.3. STEM CELLS AND CLONING

A mastery of the differentiation of stem cells and of cloning are two essential weapons in the biotechnological arsenal, the use of which for utilitarian ends, particularly in human medicine, gives rise to hope and disquiet, agreement and disapproval.

2.3.1. The hope of stem cells

At the beginning of the 1960s, experiments carried out by the Canadian biologists Ernest MCCULLOCH (b. 1926) and James TILL (b. 1931) attracted attention to the particular properties of cells in the bone marrow, the **stem cells**, which would subsequently be found in other tissues [32]. The experimental protocol is simple. Bone marrow cells from a mouse are injected into another mouse that has previously been irradiated in order to destroy its stem cells. The injected cells go to the spleen where they divide and form colonies that take the form of nodules of different sizes. The researchers realized that the cells of these nodules present differences in their potential for renewal, which is more or less rapid. They reinjected the nodule cells into mice from a second batch. The reinjected cells showed themselves capable of multiplying and generating several types of blood line.

These observations suggest the presence in the nodules of progenitor cells that have a strong potential for self-renewal and self-differentiation. In the following years, these observations were confirmed and explained by two characteristic criteria of stem cells; self-renewal and differentiation into multiples cell lines with specific characteristics. From this point on, it was possible to understand the enigma of the amputated hydra in the experiments carried out by TREMBLEY, two centuries beforehand (Chapter II-3.4.2). We now understand why, like the **hydra**, organisms like the **flatworm**, the **salamander**, the **starfish** and the **zebrafish** are able to recreate an amputated or damaged part of their bodies. The hydra mobilizes stem cells that it has preserved since its birth. In the case of the salamander, regeneration involves the reprogramming of cells that have already been differentiated.

---

[32] E.A. MCCULLOCH and J.E. TILL (1960) *"The radiation sensitivity of normal mouse bone marrow cells, determined by quantitative marrow transplantation into irradiated mice"*. Radiation Research, vol. 13, pp. 115-125.

A.J. BECKER, E.A. MCCULLOCH and J.E. TILL (1963) *"Cytological demonstration of the clonal nature of spleen colonies derived from transplanted mouse marrow cells."* Nature, vol. 197, pp. 452-454.

Like all stem cells, **embryonic stem cells** (or **ES cells**) are able to self-renew and differentiate into the different types of known adult cell line, giving rise to different types of cell such as neurons, cardiac cells that are able to contract, or hepatocytes (Figure IV.12). This potential has led to the hope that ES cells could be used in **regenerative medicine**.

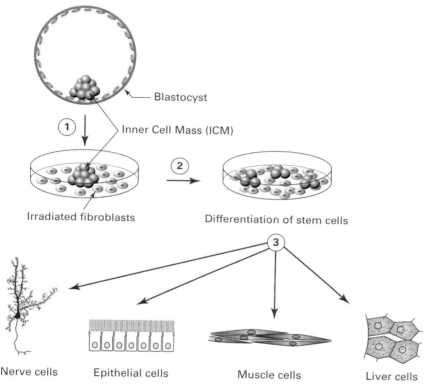

In a fertilized egg that has developed to the blastocyst stage, it is possible to distinguish a cell mass (inner cell mass, ICM) which protrudes inside the blastocyst. The ICM cells are removed and placed on a mat of irradiated (and thus unable to divide) fibroblasts that provide them with a support and nutrients (steps 1 and 2) so that they can proliferate. The stem cells arising from the ICM cells, placed in a medium that has been specifically conditioned to provide cytokines and other biomolecules, are able to differentiate into various cell types (step 3).

Figure IV.12 - Diagram illustrating how to obtain differentiated cells from stem cells

At what stage of embryo development is it possible to remove ES cells for experimental purposes? After fertilization by a sperm cell, the ovum undergoes a series of divisions that give rise to a microstructure, the **blastocyst**, the cells of which are called **blastomeres**. Each isolated blastomere remains capable of producing an entire organism of fetus and placenta, by division and differentiation. At this stage, blastomeres are totipotent. Five days after fertilization, the embryo has the form of a hollow sphere. An external layer of cells, the **trophoectoderm**, surrounds a cavity, the blastocele, inside which a small mass of cells, the **inner cell mass**, protrudes. From the beginning of the implantation of the blastocyt in the uterus, the trophoectoderm evolves to form the placenta. The cells of the inner cell mass take part in the process of differentiation that generates all of the tissues of the future adult organism. These are called **embryonic stem cells (ES cells)**. ES cells are said to be **pluripotent**. Isolated, they have lost their ability to give rise to a complete individual, but they have maintained the possibility of differentiating, according to their environment, into any of the two hundred cell types that make up animal tissues. During their division, ES cells evolve from a stage of being **pluripotent** to a stage of being **unipotent**, passing through a stage of **multipotency** beyond a hundred cells. A state of multipotency characterizes cells that give rise to a restricted number of cell lines in the tissues in which they nest. This is the case for of the hematopoietic stem cells of the bone marrow that form the red blood cells and the white blood cells. The term unipotent refers to the progenitors, which give rise to a single type of cell, for example the hepatocyte of the liver or the cardiomyocyte of the heart.

When ES cells are cultivated for 4 to 7 days in a conventional nutritive medium, they multiply and aggregate. If the culture medium is supplemented with certain biomolecules such as insulin, retinoic acid, transferrin or fibronectin, **the differentiation** of the ES cells **is oriented** towards cells of different types, such as neuron cells, glial cells or muscle cells. There are many publications about experiments concerning the grafting of differentiated stem cells in the mouse or the rat. For example, neuron precursors derived from the spinal cord or the brain are grafted into rats whose spinal chords have been injured. Five weeks after the grafts are carried out, the transplanted cells have filled the area of the injury and differentiated into oligodendrocytes, astrocytes and neurons. What is more, after about twelve weeks, locomotive function has been partially restoredIrradiated [33]. Other experiments involving the grafting of differentiated stem cells have been carried out on rats in which the dopaminergic neurons of the "substantia nigra" of the brain that secrete the neurotransmitter dopamine have been selectively destroyed by injection of 6-hydroxydopamine. The problems found in the rat as a result of this neuronal degenerescence mimic those found in Man in patients suffering from PARKINSON's disease. Dopaminergic neurons obtained by the differentiation of mouse ES cells are grafted into the striatum of each of these rats, a region of the brain whose neurons communicate with those of the substantia nigra and play a fundamental role in the control of movement. This results in a significant improvement in the motor deficit, coupled with the establishment of functional

---

33 K. WATANABE et al. (2004) *"Comparison between fetal spinal-cord and forebrain-derived neural stem/progenitor cells as a source of transplantation for spinal cord injury"*. Developmental Neuroscience, vol. 26, pp. 275-287.

synapses between the injected neurons and those of the host [34,35]. A recent publication bringing together the results of two French research teams, that of Michel PUCÉAT (b. 1961) in Montpellier and that of Philippe MENASCHÉ (b. 1950) in Paris, provides interesting information about how mouse embryonic stem cells, grafted into sheep cardiac tissue where an infarctus has been artificially induced, are able to colonize the infarct zone and regenerate cardiac contraction in a functional manner. Moving from the mouse to the sheep constitutes a considerable species leap, and the absence of any immune rejection leads us to say that embryonic stem cells have an "immune privilege" [36].

The use of ES cells in regenerative medicine necessarily requires that their differentiation be regulated in an exhaustive manner into well-defined pathways, in order to produce homogeneous cell lines with a view to implanting them in damaged tissues. In fact, contamination with non-differentiated ES cells is likely to cause tumors (**teratomas**) over the long term. The mastering of the use of ES cell culture and differentiation, as well as of cloning, in such a way as to overcome problems of histocompatibility, is still in its infancy.

For a long time, the **mouse** was the preferred animal model for experimental studies on the differentiation of ES cells. In 1981, the first ES cells from mouse blastocysts were isolated and successfully cultured by two groups of researchers in Great Britain and the USA. It was only in 1998 that **human embryonic stem cells (hES)** were isolated for the first time [37] and held in culture, on a nutritive layer of fibroblasts from irradiated mice. This delay with respect to the ability to culture animal ES cells can be explained by the fact that the molecular machinery that activates replication and cell differentiation programs is not completely identical in Man and the mouse [38]. For example, a cytokine called LIF (Leukemia Inhibitory Factor), which is indispensable for the renewal of ES cells in an undifferentiated state in the mouse, has no effect on human ES cells. There are several other differences concerning the control of proliferation and differentiation in human and murine ES cells by growth factors. Briefly, the conclusions obtained from experiments carried

---

34 J.H. KIM, J.M. AUERBACH, A. RODRIGUEZ-GOMEZ et al. (2002) "*Dopamine neurons derived from embryonic stem cells function in an animal model of Parkinson's disease*". Nature, vol. 418, pp. 50-56.

35 T. BARBERI, P. KLIVENYI, N.Y. CALINGASAN et al. (2003) "*Neural subtype specification of fertilization and nuclear transfer embryonic stem cells and application in Parkinsonian mice*". Nature Biotechnology, vol. 21, pp. 1200-1207.

36 C. MÉNARD, A.A. HAGÈGE, O. AGBULUT, M. BARRO, M.C. MORICHETTI, C. BRASSELET, A. BEL, E. MESSAS, A. BISSERY, P. BRUNEVAL, M. DESNOS, M. PUCÉAT and P. MENASCHÉ (2005) "*Transplantation of cardiac-committed embryonic stem cells to infarcted sheep myocardium: a preclinical study*". Lancet, vol. 366, pp. 1005-1012.

37 J. THOMSON, J. ITSKOVITZ-ELDOR, S.S. SHAPIRO et al. (1998) "*Embryonic stem cell lines derived from human blastocysts*". Science, vol. 282, pp. 1145-1147.

38 G. GUASH and E. FUCHS (2005) "*Mice in the world of stem cell biology*". Nature genetics, vol. 37, pp. 1201-1206. (review)

out on murine embryonic stem cells cannot be transposed automatically to human embryonic stem cells. This simple observation should encourage politicians to **think constructively** about the use of human embryonic stem cells for fundamental research studies and applications in therapeutics.

While there is a highly promising future for the use of ES cells, this future is littered with obstacles, and rigorous checks and balances need to be put in place. Nevertheless, research on such cells is mandatory if we wish to move on to a regenerative medicine that aims to be a new frontier in the art of healing. After specific differentiation, hES cells could provide unlimited quantities of the tissues needed to replace damaged tissues responsible for handicapping illnesses (dopaminergic neurons in PARKINSON's disease, cardiomyocytes in myocardial infarction, pancreatic islets of LANGERHANS cells in diabetes, fibroblasts in skin grafts, chondrocytes in rhumatoid arthritis). In addition, metabolic analysis of hES cells carrying defective genes whose phenotypical expression is known in human pathology should improve our understanding of the perturbed mechanisms, and could lead to pharmacological advances. As well as the technical difficulties involved, which have not yet been adequately overcome, the handling of hES cells is subject to much ethical debate in many countries, with those who object to it holding to their prejudices, which are linked to religious or cultural traditions. This is the case in France, where, nevertheless, a few timid dispensations had begun to appear at the time of writing. In contrast, in Great Britain, the law authorizes the isolation of hES cells for therapeutic purposes, using embryos of less than one hundred cells, produced by *in vitro* fertilization, and surplus to requirements. The British response to the burning question of whether an isolated hEs cell may be considered as a potential human embryo is clearly "no", for, in order to be able to develop *in utero*, such hES cells would need to have the placental progenitor cells.

An alternative to the use of ES cells is to make use of **adult stem cells**. However, the proliferation capacity of adult stem cells is considerably lower than that of their embryonic homologues. The hematopoietic stem cell is the paradigm of the adult stem cell. It can differentiate into all known types of cells. In the last decade of the 20th century, several publications concerning the plasticity of the adult stem cell awakened a hope that these cells could transform the treatment of degenerative illnesses. Certain of these publications stated that adult bone marrow stem cells, implanted into different types of tissues, differentiate into hepatocytes, cardiomyocytes or neurons, depending on the specific environment. Careful re-examination of the techniques used revealed that, in certain cases, interpretation of the results as showing cell transdifferentiation was an erroneous one, and that the fusion of the bone marrow stem cells with cells from other tissues was a more plausible explanation.

In any case, while not ignoring the use of adult stem cells, experimentation on hES cells remains a judicious choice, given our current state of understanding. In France, the 2004 law application decree that was issued on the 7th of February 2006, revising the restrictive bioethical standards of 1994, opens up the possibility of using human embryonic stem cells for scientific purposes, with certain ethical reserves being maintained.

2.3.2. *The specter of cloning*

One of the obstacles to the stabilization over time of a stem cell graft in a receiver involves the phenomenon of rejection for reasons of histocompatibility. Considered to be foreign by the receiver (host), grafted stem cells coming from a donor are rejected. This obstacle could be overcome by using the technique of **cloning**. Based on experiments on several animal species, it is now accepted that the transfer of the nucleus of an adult somatic cell from a host into an enucleated oocyte makes it possible to obtain from this oocyte, which is once again nucleated, and which is the equivalent of a zygote and able to divide, ES cells whose genome is identical to that of the host. Because of this, the ES cells are immunologically compatible with the tissues of the host. In Man, such cells could be directed by differentiation towards stable cell lines creating well-defined tissues and organs (liver, muscle...) that could be used in regenerative medicine. This is the principle of **therapeutic cloning**. In March 2004, Korean veterinary researcher Woo Suk HWANG (b. 1953) and his co-workers, who were recognized experts in animal cloning, announced in the American review *Science* [39] that they had succeeded for the first time in obtaining around thirty human blastocysts by cloning, i.e., by the transfer of nuclei of somatic cells into enucleated ova. This first experiment involved autologous cloning (ovum nuclei and enucleated somatic cells taken from the same woman). HWANG and his team used 176 ova, and the yield from the experiment was close to that obtained at that time for the cloning of mammals. Using the inner cell mass of one of the blastocysts, they isolated a line of embryonic stem cells able to maintain a normal karyotype after several dozen divisions. This publication, which appeared in a highly prestigious scientific review, triggered an enthusiasm in the media that was in keeping with the spectacular nature of the team's exploit, tempered here and there by a few comments that were mainly linked to questions of medical ethics. In 2005, there were numerous other articles by the same team on the same subject, reinforcing the first results with a heterologous cloning technique (ovum nuclei and enucleated somatic cells taken from different people), thus giving rise to great hopes that the era of regenerative medicine was near. At the beginning of 2006, Professor HWANG's retractation of all his work, and a public confession of a spectacular fraud, were even more dramatic, offering certain media an occasion for a disproportionate level of fury against therapeutic cloning. However, despite such rear-guard combats, it is obvious that one day these technical difficulties will be overcome. Human cloning, in order to obtain stem cells for therapeutic purposes,

---

[39] Woo Suk HWANG *et al.* (2004) *"Evidence of a pluripotent human embryonic stem cell line derived from a cloned blastocyst"*. *Science*, vol. 303, pp. 1669-1674.

cannot escape the future. Once this aim has been achieved, it will be spoken of as the outcome of a long story.

The adventure of **animal reproductive cloning** began in 1960. In *Developmental Biology* [40], two American researchers, Robert BRIGGS (1911 - 1983) and Thomas KING (1921 - 2000) described experiments involving the transfer of cell nuclei of embryos from a frog (*Rana pipiens*), at the blastula and gastrula stages, into enucleated eggs of the same species. A high percentage of the clones obtained in this way were able to reach the tadpole stage when the transferred nuclei came from the early blastula stage, but only mediocre success was achieved when the nuclei came from the later gastrula stage. These experiments emphasized both the totipotency of the embryo somatic cells and the equivalency of the somatic cell nucleus and the nucleus of the fertilized egg in cell division and differentiation. BRIGGS and KING's publication did not arouse any particular interest. It is true that the 1960s were dominated by the saga of molecular biology, which would reach its culmination in the deciphering of the genetic code.

From the 1980s onward, the first attempts to clone mammals (rat, mouse, pig) began. Moving from the amphibian egg, which was a millimeter wide, to a mammal egg that was one hundred times smaller, presented a technical difficulty that would be overcome by a technique of cell-to-cell electrofusion. Cloned embryos were thus obtained by nuclear transfer and then implanted into the uterus of a surrogate female. However, in all cases, the nucleus came from embryo cells. In February 1997, the announcement made by Ian WILMUT (b. 1944), Keith H. CAMPBELL (b. 1954) and their collaborators at Edinburgh's Roslin Institute [41] of the birth of the cloned lamb Dolly had an immediate effect in the media. In fact, this was not only the cloning of a higher mammal, but, above all, the cloning by insertion of an **adult somatic cell**, in this case a mammary tissue cell, into an enucleated oocyte. This went far further than the experiment carried out by BRIGGS and KING, which essentially involved the transfer of embryo cell nuclei into enucleated frog eggs. The trick that gave WILMUT and CAMPBELL their success was to bring the cells providing the nuclei to a quiescent state corresponding to the interphase stage of the cell cycle, by impoverishing their culture medium, before electrofusion with enucleated oocytes. Although we should be aware that 434 attempts were made before a positive result was achieved, this does not make it any less astonishing that the nucleus of a cell in its adult state, i.e., completely differentiated, was able to behave as if it were **totipotent**. Despite being committed to a program of differentiation that is considered to be more-or-less irreversible, and which will give it a specific identity, **the nucleus of an adult cell can be reprogrammed** and become totipotent. Since Dolly, many other mammals have been cloned from nuclei of adult cells; mice,

---

40 R. BRIGGS and T.J. KING (1960) *"Nuclear transplantation studies on the early gastrula* (Rana pipiens). *I. Nuclei of presumptive endoderm"*. Developmental Biology, vol. 2, pp. 252-270.

41 I. WILMUT et al. (1997) *"Viable offspring derived from fetal and adult mammalian cells"*. Nature, vol. 385, pp. 810-813. Erratum in: *Nature* (1997), vol. 386, p. 200.

cows, goats, pigs, rabbits, cats, dogs, rats and horses. As far as ethical discussion about cloning is concerned (Chapter IV.5), it is essential to note that the demarcation line between reproductive cloning and therapeutic cloning is situated where decisions are made concerning the **destiny of the cloned blastocyst** (Figure IV.13).

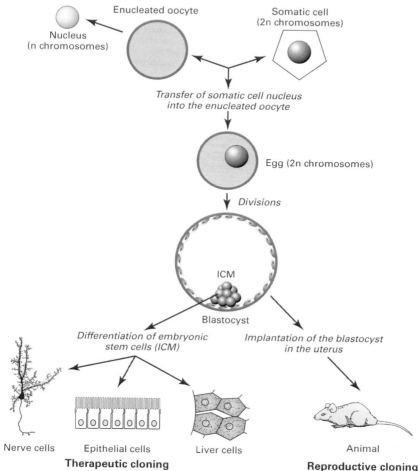

Figure IV.13 - Therapeutic cloning *versus* reproductive cloning

The transfer of the nucleus of a somatic cell (liver, epidermis, muscle) containing 2n chromosomes into an enucleated oocyte gives rise to an egg (2n chromosomes) that is able to divide and to produce a blastocyst. The cells of the blastocyst inner cell mass (ICM) can be used as stem cells that can differentiate into different types of cell line (therapeutic cloning). On the other hand, if the whole blastocyst is implanted into a uterus, it will produce an embryo which, after birth, will grow into an adult animal (reproductive cloning). Reproductive cloning and therapeutic cloning therefore differ because of the fact that in reproductive cloning, the whole blastocyst is used, while in therapeutic cloning, only certain cells, corresponding to the inner cell mass (ICM) of the blastocyst, are used.

Its implantation into a uterus determines reproductive cloning, while the use of the cells of the inner cell mass, with the aim of making them differentiate towards different cell lines, constitutes the basis of therapeutic cloning.

The structural and functional identity of the cells of a given tissue in an adult organism involves a basic mechanism: while each cell has the same set of genes, only some of the genes are expressed as proteins and the genes that are expressed differ according to the tissue involved. The key to the mechanism responsible is in the **epigenetic type** chemical modifications of cell DNA, for example methylations, which repress the expression of certain genes without altering the expression of others. These modifications of the DNA, which control cellular specificity (muscle, liver, brain...) are not very reversible, but, in certain circumstances, they can become so. This is what happens from time to time when the nucleus of an adult cell is inserted into an enucleated oocyte. We are thus able to assume that in the molecular arsenal of the oocyte cytoplasm there are substances that can cancel the epigenetic modifications of the DNA present in the nucleus of an adult somatic cell and recreate a state of pluripotency in this nucleus, or, in other words, provoke the **reprogramming** of the somatic cell nucleus. In the long term, it is to be hoped that biochemical technology will be able to find and purify the molecules responsible for the nuclear reprogramming of somatic cells.

The use of human oocytes for the purpose of therapeutic cloning is still subject to severe criticism. Certain groups wish it to be prohibited, because of a fear of a drift towards reproductive cloning. To obviate this risk, the idea has been to make use not of human oocytes but of those of animals, transferring the nuclei of human somatic cells into them. Even supposing that the technical difficulties involved could be overcome, the cells that would result, a sort of Man-animal chimera, would also be the subject of an ethical debate, even if the purpose of this type of cloning were to be solely therapeutic.

2.3.3. *The bias of parthenogenesis in cloning*

Some Japanese researchers [42] have succeeded in creating mice according to a parthenogenetic process that involves adding the nucleus of an oocyte that is haploid (1n chromosomes) to another haploid oocyte, the result being the equivalent of a fertilized egg (2n chromosomes). This exploit is achieved by the invalidation of one of the genes (H19) involved in the control of the parental imprint. It is known that sexual reproduction in mammals involves a phenomenon called the parental imprint, which, by means of the methylation of DNA and perhaps also of histones, allows the expressing or silencing of certain genes in male and female gametes. A single copy of a given gene, originating either from the oocyte or from the sperm cell, is therefore expressed, while the other is inactive. In the Japanese experiment,

---

42 T. KONO et al. (2004) *"Birth of parthenogenetic mice that can develop to adulthood"*. Nature, vol. 428, pp. 860-864.

if the mouse H19 gene had not been invalidated, the result would have been an anarchical development of the responsible genes involved in the parental imprint with overexpression in the case of some and an absence of expression in the case of others. These disturbances would have been incompatible with the viability of the embryo. However limited its application might be, the manipulation of the germinal genome poses the problem of the mechanism by which the parental imprint intervenes in the viability of the egg, a parameter that at the time of writing is still not completely understood, but is being actively explored.

2.4. THE "HUMANIZATION" OF ANIMAL CELLS FOR PURPOSES OF XENOTRANSPLANTATION

In Boston, Massachusetts, in 1954, a kidney was transplanted from a healthy boy into his twin brother, who was suffering from a fatal renal anomaly. The success of this graft ushered in the era of transplantations of such organs as the heart, liver and kidney in Man. In order to try to prevent the rejection of grafts, caused by an immune incompatibility between the receiver and the graft from the donor, different immunosuppressing treatments were tried, one by one, involving corticosteroids or cytostatic agents such as 6-mercaptopurine. In the 1980s, a decisive step forward was made with the fortuitous discovery of the powerful immunosuppressive effect of the cyclosporin A, a cyclic polypeptide isolated from the mold *Tolypocladium inflatum*.

Each year, human organ transplants into patients make it possible to save many lives. However, for some time now, organ transplantation has been suffering from penury of donors. One **alternative to the homograft** is the grafting of animal organs, or the **xenograft**, and, at the dawn of the 21st century, this type of graft has entered an active, promising phase, with the creation of pigs that have been partially "humanized" and are thus, as a consequence, **immunocompatible**. For reasons of genetic and physiological similarity, the first choice for such grafts was to use apes or monkeys. However, this idea was quickly abandoned, for several reasons; a non-negligible risk of viral infection due to the phylogenetic kinship of human and simian species; a slow growth rate; a low reproduction rate and, finally, laws that protect primates. These disadvantages are not found, or are at least minimized, in the **pig**: the risk of a viral infection passing from the pig to Man should be low because of the species barrier (but nevertheless, it ought to be evaluated), the pig growth rate is relatively rapid, pig litters are large and pig organs are of a size close to those of Man.

Hyperacute rejection of grafts is the critical obstacle that must be overcome before it is possible even to envisage the feasibility of xenotransplantation. Hyperacute rejection is caused by the presence in Man of natural antibodies (**xenoantibodies**) that accumulate throughout a lifetime and are directed against antigenic motifs carried by the products of the digestion of food or dust that is breathed in.

Xenoantibodies are mobilized when a xenograft occurs, and when they combine with **xenoantigens** brought by the graft this activates immune proteins such as the **complement** proteins. The catastrophic effect of this xenoantibody/xenoantigen combination is a vascular thrombosis followed by necrosis and rejection of the graft. The pig xenoantigen that is considered to be the one mainly responsible for the phenomenon of rejection in Man is a sugar molecule, galactose α-1,3-galactose, located on the plasma membrane of endothelial cells. Synthesis of this molecule requires the enzyme α-1,3-galactosyltransferase, which is present in most mammals, but absent in Man and the primates. This enzyme disappeared in Man around twenty million years ago, following a double mutation of the gene. In 2002, cloning by nuclear transfer, associated with the invalidation of the gene coding for galactosyltransferase, made it possible to create pigs without galactose α-1,3-galactose [43]. This performance shows that the xenotransplantation objective, although it can only be envisaged over the long term, is not based on false hopes.

Plant GMOs, gene therapy, embryonic stem cells, therapeutic cloning, and xenotransplantation are a few of the many examples that show how far experimentation on living beings has progressed in just a few decades, from inquiries into the operating mechanism of an organ or a cell, in the interests of pure understanding, to a programmed process, planned with an objective in mind, the chances of success of this objective being analyzed and counted in terms of impact and cost-effectiveness.

During the Renaissance, ecclesiastical authorities, worried by the libertarian forces that were assailing them, applied the brakes to audacious questioning of dogma such as the circumterrestrial revolution of the Sun that had, since Ancient times, placed Man at the center of the Universe. Nowadays, civil authorities, conscious of the potential but also of the possible misuses of genetic manipulation, insist on having the right to oversee such procedures. In truth, since the 19th century, governments have been interesting themselves in research on living beings and encouraging it, as long as its applications have allowed improvements in human health. This has been the case for vaccinations against infectious diseases or for prevention of microbial infections by means of aseptic or antiseptic methods. With the breakthroughs made in genetic manipulation at the end of the 20th century, it was more than just the results of experiments on living beings that attracted the attention of the political authorities, it was, above all, the **manner in which** the experimental method, with all its hazards, made use of living material, sometimes of human origin, in order to unlock mysteries. Conscious of the **social impact** of emerging discoveries that are subject to considerable media coverage and are

---

43 Y. DAI et al. (2002) *"Targeted disruption of the α-1,3-galactosyltransferase gene in cloned pigs"*. Nature Biotechnology, vol. 20, pp. 251-255.
See also: L.M. HOUDEBINE (2006) *"The interventions de la transgenèse sur le xénotransplantation"*. Biofutur, vol. 264, pp. 32-37.

sensationalized in both the written and audiovisual media, the State, with the help of researchers and philosophers, has laid down a **code of bioethics**, applied through strict or even restrictive legislation. It remains to be seen whether the rules of this code will continue to be an inviolable absolute or will be modified according to the evolution of the moral codes and the cultures of nations. University teaching and the education of a society must now take into account not only the content of successive discoveries, but also the fallout of these discoveries, insofar as they concern Man, and even the ethical justification of the methods that have allowed these discoveries. In "remaking" living beings according to imposed norms, and in scheduling, in a certain fashion, the manufacture of life according to new codes, certain questions move from the "how" to the "why", i.e., from the scientific domain that is accessible to human thought to the metaphysical sphere, with its problems of the **limit of what is surmountable and tolerable in terms of ethics**.

3. THE PROGRESS OF MEDICINE
FACE TO FACE WITH THE EXPERIMENTAL METHOD

In his *Birth of Predictive Medicine*, Jacques RUFFIÉ (1921 - 2004) reminds us that medicine has evolved through three stages over the course of time: **curative medicine**, which has been practiced since Ancient times and is still being practiced; **preventive medicine**, which is more recent, and is designed to prevent people from falling ill, either by vaccinating them, in the case of infectious diseases, or by recommending an appropriate diet and medication in the case of metabolic disorders such as diabetes or arterial hypertension that have been detected by means of systematic examination; and, finally, **predictive medicine**, a branch of medicine that is still in its early phases, and which is based on modern technology and is able to predict situations of risk because of anomalies detected in the genetic inheritance or because of exposure to environments that are reputed to be dangerous (for example, carcinogenic smoke, asbestos).

About one and a half centuries ago, the publication of the introduction to *The Study of Experimental Medicine* (1865) provided proof, based on scientific arguments, that the time had come to transfer the experience that had been acquired through the experimental method practiced on animal models to the ill person. After Claude BERNARD, attentive to the progress made by ideas and techniques in the physical and chemical sciences, and making use of its own advances in the understanding of the living cell, both normal and pathological, experimental medicine was to live through a development that was without precedent in the history of Humanity. To understand the causes of epidemics, nutritional deficiencies, metabolic deviations of hereditary origin and degenerative illnesses, and then to translate these causes in cellular and molecular terms, this was the process undertaken by medicine once it began to use the experimental method. In fact, for several decades, from the beginning of the 19th century, medicine had already undergone some major revisions of outdated practices and had inaugurated a new era in diagnosis. For example, the differential diagnosis of pulmonary ailments became possible because of the invention of the stethoscope by René LAENNEC (1781 - 1826) and the practice of auscultation and percussion by Joseph SKODA (1805 - 1881), the uncontested master of the Vienna school. In France, Pierre LOUIS (1787 - 1872) used statistical methods to evaluate the efficacy of different treatments. Armand TROUSSEAU (1801 - 1867), a pupil of Pierre BRETONNEAU (1787 - 1862) wrote a famous treatise on the Hôtel-Dieu Medical Clinic in Paris. In Great Britain, chronic nephritis, with its identifying symptoms, was described by Richard BRIGHT (1789 - 1858), paralysis agitans by James PARKINSON (1755 - 1824) and ADDISON's disease, which affects the adrenal glands, by Thomas ADDISON (1793 - 1860). During the 19th century, many other famous names signaled the arrival of a medicine that was resolutely anatomo-clinical in nature, in line with Bernardian doctrine.

3.1. FROM EMPIRICAL MEDICINE TO EXPERIMENTAL MEDICINE

"Experimental medicine is thus a medicine that claims to understand the laws of the organism in sickness and in health, in such a way that it not only predicts phenomena, but also in such a way that it can regulate and modify them, within certain limits."

Claude BERNARD
Introduction to the Study of Experimental Medicine - 1865

In the introduction to *The Study of Experimental Medicine,* Claude BERNARD stigmatizes the relics of an empirical medicine that was still being practiced in his day and was forgetful of rationalism. The terms that he uses are without leniency: "I have often heard doctors who, when asked the reason for a diagnosis, reply that they don't know how they recognize such a case, but it is obvious, or who, when asked why they administer certain remedies, reply that they don't really know how to put it exactly, and that anyway they are not required to give a reason, because it is their medical tact and their intuition that guides them. It is easy to understand that doctors who reason in this way are denying science. What is more, **it is impossible to be too forceful in rising up against such ideas, which are bad not only because they stifle any scientific seed in the young, but also, above all, because they favor laziness, ignorance and charlatanism.**" In order to evaluate the meaning of these words, it should be remembered that in Claude BERNARD's time, the medical profession was far from considering the microscope as a useful instrument for the study of cell structures and that the cause and effect relationship between bacterial germs and infections was still to be shown.

With the development of increasingly effective instruments for exploration, and of methods for microanalyses concerning a wide range of blood and humoral constants, throughout the 20th century, medicine, which was once **empirical**, has now become **scientific**. Claude BERNARD's dream, experimental medicine, is now operative. This medicine is no longer content simply to determine the cause of an illness and to locate the affected organ, which was the major objective of clinical medicine, but it **seeks to detect the mechanisms of pathological processes** by means of **histological and physicochemical explorations**. This medicine is no longer willing to passively monitor the evolution of an infectious disease. After having identified the responsible germ, it tries to **target** this germ with the chemical weapon that is able to selectively destroy it. This medicine is no longer content simply to find **remedies**, it aims to understand the **mode of action**. It sets itself the goal of meeting challenges such as finding the **genetic cause** of degenerative illnesses or of cancers and developing **appropriate therapies**. It is supported by **statistical data**. When a new drug is implemented, the results are now evaluated by the double blind method: neither any of the patients (treated and non-treated) nor any of the investigators are aware of who has been administered with the drug and who has been administered with a placebo. In the surgical domain, audacious techniques have also led to considerable progress, particularly in neurosurgery and in cardiovascular surgery. Thanks to robotics and to computer technology, remote surgery or telesurgery has become practicable, although up until not that long ago, it was only to be found in fiction.

Faced with emerging problems in public health, the task undertaken by experimental medicine is immense. In the middle of the 20th century, the spectacular recovery from high-incidence infectious diseases such as pneumococcal pneumonia, meningococcal meningitis or acute forms of tuberculosis, which that was brought about by antibiotics, gave rise to the idea that medicine had won a battle against the microbial world and that, from then on, it would be able to control the evolution of infectious diseases and to offer rational treatments. The gradual appearance of a microbial resistance to antibiotics has brought an end to this euphoric era. Penicillin, for example, which was put on the market at the end of the 1940s, was active on practically all strains of *Staphylococcus aureus*. Sixty years later, more than 90% of the strains of this same microbe are resistant to penicillin. The incidence of **nosocomial infections**, which are contracted in health care facilities, never ceases to rise. At present, around 10% of the hospitalizations that take place are complicated by the patient developing a nosocomial infection. Equally worrying are the re-emergence of diseases that were once considered to be under control, such as tuberculosis or poliomyelitis in Africa, and the emergence of new diseases such as AIDS, whose HIV virus (Human Immunodeficiency Virus), which was identified at the beginning of the 1980s, has generated a pandemic that has spread throughout the planet. Infectious diseases are currently responsible for more than a quarter of human deaths. The KOCH bacillus responsible for tuberculosis and the pneumococcus kill three to four million people a year, around the world. In 2004, HIV killed more than three million people, and more than forty million people are infected. One person is infected every 30 seconds.

In viral diseases, the role of vectors (insects, various animals) as well as the notions of contagiousness and aggressivity have been emphasized. We have only to remember the dreadful contagiousness and aggressivity of the Spanish flu virus (*Influenzavirus* AH1N1) which, in 1918 - 1919, killed more human beings around the world than the First World War that preceded it. In contrast, the SARS (Severe Acute Respiratory Syndrome) epidemic of 2003, the vector of which was doubtless the civet, a small carnivore raised in China and desired for its meat, was rapidly contained because of its low contagiousness and also because of the isolation measures that were taken. Human behavior is not without its effect on the emergence of viral diseases. The growth in intercontinental travel and human migration, as well as intensive deforestation in Africa and South America, which bring virus vectors into contact with Man, are factors concerned in the emergence of viral diseases that risk being explosive and devastating. In this context, the history of the Ebola virus and of the Marburg virus, which cause violent hemorrhaging, is edifying. In 1967, in the German village of Marburg, an epidemic of unknown origin broke out, the illness manifesting itself with brutal suddenness by vomiting, diarrhea, a high fever and an increased tendency to bleed. This pathology, which was contained rapidly by means of drastic isolation measures, was found to be of viral origin. The pathogen concerned was a filovirus (filiform virus). A brief enquiry showed that the origin of the epidemic was contact between technicians of a pharmaceutical company and monkeys that had been imported from Uganda and that were carrying the virus. In 1976, two other epidemics, characterized by severe and often fatal hemorrhagic fevers, were reported in the Sudan and the Republic of the Congo. Here again, the illness was caused by a filovirus, the Ebola virus. At the time of writing, only public health organizations, including the NIH (National Institutes of Health) in the USA, have attempted to set up vaccination and therapeutic strategies. Research on these dangerous viruses requires high security installations that are particularly costly, so that private companies are reticent about investing in work that is only targeted on poor regions and which concerns epidemics that have so far been contained successfully, although one day the Ebola and the Marburg virus could quite well escape their African niches.

Experimental medicine must also understand the colossal challenge of the five thousand **hereditary diseases** that are currently listed, the most handicapping of which are myopathies and neuropathies. Given the means that are available to the contemporary clinician in order to assign each of these diseases to a genetic defect, one can only be amazed by the mass of information about them that has accumulated over a century, since the first diagnosis of a hereditary disease, **alcaptonuria**, which was made in 1902 by Archibald GARROD (1857 - 1936), a doctor at London's St Bartholomew's hospital. Alcaptonuria is a non-serious genetic flaw that can be detected easily by a blackening of the urine. It is the result of a blockage caused by the mutation of an enzyme involved in the catabolism of an amino acid, tyrosine, this blockage leading to the accumulation of homogentisic acid, the polymerization of which gives rise to a brownish color. The patient examined by GARROD was a young boy. Investigation of the family history revealed that transmission of the flaw was correlated to cross-cousin marriages and followed MENDEL's laws for recessive traits. GARROD demonstrated other hereditary-type anomalies, cystinuria, porphyria and pentosuria. In 1909, these observations were published in a work that became a classic: *Inborn Errors of Metabolism*.

In 1956, the specific molecular defect of a metabolic anomaly linked to a mutation was identified for the first time by the German-born British biochemist Vernon INGRAM. This was the hemoglobin defect responsible for drepanocytosis or sickle cell anemia: a glutamic acid in the β chain is replaced by a valine. The consequence of this simple change is a modification of the structure of the hemoglobin, leading to a sickle-shaped deformation of the red blood cells, the increased fragility of these cells and also a tendency towards cell lysis. This discovery made use of the electrophoresis and chromatography techniques that had just been introduced in biochemistry (Chapter III-6.2.2): such a discovery would not have been possible without these techniques. Because of the progress made in molecular biology, the nosological framework of hereditary diseases has been greatly enriched over the last twenty years. For example, at present, more than one hundred hereditary-type myopathies have been identified by accurately locating molecular lesions in the genomic DNA and characterizing the structural and functional modifications of the mutated proteins.

Certain health problems present real challenges for experimental research. This is the case for the spongiform encephalopathy caused by a prion (proteinaceous infectious particle) , which has all the more impact on the imagination because its etiology remains a mystery. It is also the case for degenerescence of the central nervous system correlated with aging, ALZHEIMER's disease being a striking example, although, as far as familial forms of this illness are concerned, i.e., those of the hereditary type, it has been possible to link the invasion of the brain by a so-called amyloid peptide, which accumulates in plaques, on the one hand, and, on the other hand, the absence, due to a mutation, of an enzyme, a peptidase, which normally degrades the amyloid peptide.

Contemporary scientific medicine sometimes acquires a revolutionary aspect. Here again, as with other disciplines involved in the study of living beings, it arises from discoveries resulting from the principle of **serendipity** (Chapter III-2.2.3). This was the case when, in January 1987, a team in Grenoble, France [44], led by the neurosurgeon Alim-Louis BENABID (b. 1942) and the neurologist Pierre POLLACK (b. 1950) discovered by accident that in patients affected by PARKINSON's disease a beneficial effect was achieved by deep, high-frequency electrical stimulation of the brain. The

---

44 P. LIMOUSIN, P. KRACK, P. POLLAK, A. BENAZZOUZ, C. ARDOUIN, D. HOFFMANN, A.L. BENABID (1998) *"Electrical stimulation of the subthalamic nucleus in advanced Parkinson's disease"*. The New England Journal of Medicine, vol. 339, pp. 1105-1111.

three major symptoms of PARKINSON's disease are muscular rigidity, a tremor when at rest and a slowing down of the execution of movements. In the 1960s, the Swedish team of Arvid CARLSSON (b. 1923), who won the Nobel prize for Physiology and Medicine, demonstrated a relationship between the PARKINSON syndrome and a deficit in the secretion of a neurotransmitter, dopamine. A group of neurons that is limited to half a million (of the 100 billion contained in the brain) produces this neurotransmitter in a small structure located in the midbrain, called the **substantia nigra**. The neurons of the substantia nigra have elongations that interact with different nerve formations (called nuclei) including the subthalamic nucleus. In 1990, BERGMAN et al. [45] published an article that describes a curious relationship between a provoked lesion of the subthalamic nucleus and the disappearance of the signs of PARKINSON's disease in a monkey which had been made Parkinsonian by chemical treatment. This publication led the team in Grenoble to target their electrical stimulation on the subthalamic nucleus. This was completely successful. This electrical stimulation procedure, which is now well-codified, involves using stereotactic neurosurgical techniques, controlled by Magnetic Resonance Imaging, to implant an electrode into the subthalamic nucleus. The electrode is connected to a generator that is implanted under the patient's clavicle. The generator sends brief electrical impulses of frequencies from 100 to 200 Hz. Under the effect of this stimulation, the characteristic symptoms of the illness, particularly the static tremor and bradykinesis, regress in a spectacular manner. The mechanism by which this stimulation acts is not yet understood. No doubt this has to do with complex phenomena involving the inhibition of certain neuron relays near to the substantia nigra, which remain to be deciphered. Here we have a typical case of a progression from an experimental fact, discovered by accident, towards the analysis of its cause. From the point of view of the experimental method, it is interesting to make parallels between this discovery by serendipitous means of the beneficial role of electrical stimulation of the midbrain in PARKINSON's disease and Cartesian style programmed research that aims to graft into the brain of PARKINSON's disease sufferers embryonic stem cells differentiated into dopaminergic neurons [46].

Civilian society and its armed force, the political authorities, have understood that experimental science has the tools, the method and the thought processes necessary to develop strategies for prevention and healing. Every year a few new antibiotics are isolated and tested and new vaccines are developed. This is the case for DNA vaccines obtained by inserting a bacterial or viral immunogenic DNA sequence into a bacterial plasmid. The modified plasmid, amplified by bacterial culture and then injected into an individual, becomes incorporated into the

---

45 H. BERGMAN, T. WICHMANN et al. (1990) *"Reversal of experimental parkinsonism by lesions of the subthalamic nucleus"*. Science, vol. 249, pp. 1436-1438.
46 A.L. PERRIER, V. FABER, T. BARBERI et al. (2004) *"Derivation of midbrain dopamine neurons from embryonic stem cells"*. Proceedings of the National Academy of Sciences, USA, vol. 101, pp. 12543-12548.

immune cells of this person and induces the synthesis of immune proteins. Finally, it seems relatively certain that hopes concerning gene therapy for hereditary diseases will be fulfilled within before too long (Chapter IV-2.3.1).

One of the traits that is characteristic of the period we live in, and which arises partly from the economic stakes involved, is the **shortening of the time that elapses between a discovery being made and the application of that discovery**. For example, Interfering RNAs, which were discovered in the 1990s (Chapter IV-1.2.2) are already the subject of therapeutic investigation. More than a hundred biopharmaceutical companies around the world are using them with a view to producing drugs from them [47]. In mice, a certain number of synthetic interfering RNAs have proved their efficacy in silencing genes which, following mutation, have acquired carcinogenic potential. However, the use of interfering RNAs as therapeutic agents requires them to be stabilized, because they are fragile molecules. The group headed by Achim AIGNER [48] (b. 1965), at the school of Medicine in Marburg, Germany, managed to stabilize a synthetic interfering RNA by complexing it with polyethyleneimine, and this interfering RNA is able to block the expression of a receptor involved in cancerization (c-erbB2/neu(HER-2) receptor). Used in mice, such a drug appears promising.

Despite the undeniable progress that has been made, experimental medicine is still some way from finding solutions to some of the enigmas that it meets along the way, and which underline the complexity of living beings. Some time ago, it was thought that after having invalidated a gene coding for a protein that is indispensable to a function, we would discover the secret of a cause-and-effect relationship. Experimental practice has shown that, generally, this is far from being the case. Another example of the complex relationships that exist in living beings is the interference of the mental and the organic. One experiment that suggests this interference was carried out on mice who had acquired a form of pathology similar to HUNTINGTON's chorea, by transgenesis. Mice from the same line were separated into two batches, one acting as a control, and the other being subjected to daily mental stimulation, including memorization tests. Unexpectedly, the appearance of symptoms was noticeably slowed down in mice who had been subjected to mental gymnastics [49], as if the brain, by intentionally mobilizing its neuron activity, was able to secrete substances able to alleviate its own defects. In short, by means of possible retroactive mechanisms that are called upon by the mind, the brain appears to act as actor and spectator.

---

[47] F. DE RUBERTIS (2005) *"ARN, cible et agent thérapeutiques. Un nouveau secteur prometteur pour l'investissement"*. Biofutur, vol. 255, pp. 22-23.

[48] B. URBAN-KLEIN, S. WERTH, S. ABUHARBEID, F. CZUBAYKO and A. AIGNER (2005) *"RNAi-mediated gene-targeting through systemic application of polyethylenimine (PEI)-complexed siRNA in vitro"*. Gene Therapy, vol. 12, pp. 461-466.

[49] A. VAN DELLEN, C. BLAKEMORE, R. DEACON, D. YORK and A.J. HANNAN (2000) *"Delaying the onset of Huntington's in mice"*. Nature, vol. 404, pp. 721-722.

3.2. CONTEMPORARY ADVANCES IN BIOTECHNOLOGY
THE EXAMPLE OF MEDICAL IMAGING

At the turn of the 21st century, experimental medicine was being nourished by techniques inherited from experimental physics, chemistry, and even mathematics and computer technology, in the same way as the other sciences of living beings. The progress made in medical imaging techniques has been particularly impressive since the time, at the end of the 19th century, when the X-rays discovered by Wilhelm RÖNTGEN made it possible to view the structure of the human skeleton. The saga of X-radiation continued through the 20th century (Chapter III-2.6.1). For the last few decades, new imaging techniques have come to the fore. They have spread rapidly, and been refined.

Ultrasound imaging, which is based on the principle of the reflection of ultrasound waves off of different kinds of surfaces, has become an everyday technique for viewing blood flow in blood vessels and the heart. However, it is mainly in the study of the brain that medical imaging has benefited from technical advances in the domains of physics and computer technology, and it has been innovative in assigning cognitive activities to well-identified anatomical structures. This functional neuroanatomy makes it possible, in a non-trauma-inducing manner, to monitor and locate the operation of neuron networks with great temporal and spatial precision, during various cognitive tasks such as reading and the written or oral expression of thought.

The middle of the 20th century saw the gradual development of two methods for exploring zones of cerebral activity, **electroencephalography** and **magnetoencephalography**. At present, these techniques are being taken over by **MRI** (Magnetic Resonance Imaging) (Figure IV.14). The principle of MRI is based on the detection of hydrogen nuclei and their differentiation according to their environment. **Functional** MRI leads to the location of the areas of the brain that are active during calculation exercises, the perception of sounds, language and objects, and memorization, with a resolution of just a few millimeters. Its power of exploration is such that it has been possible to analyze the brain response, in sleeping or awake babies who are only three months old, to auditory stimuli from language that either makes sense or does not make sense [50]. The response, located in the left hemisphere and the prefrontal cortex, leads to the conclusion that, from the first months of life, there are zones of the brain that are potentially active before the first attempts at language appear.

Both in France (CEA-Saclay and the Frederic JOLIOT Hospital at Orsay) and abroad, recent MRI performance has encouraged projects concerning the manufacture of instruments able to produce magnetic fields of around ten teslas, which allows an

---

50 G. DEHAENE-LAMBERTZ, S. DEHAENE and L. HERTZ-PANNIER (2002) *"Functional neuroimaging of speech perception in infants"*. Science, vol. 298, pp. 2013-2015.

IV - CHALLENGES OF EXPERIMENTATION ON LIVING BEINGS AT THE DAWN OF THE 21ST CENTURY   291

unequaled definition in the identification of areas of the brain assigned to specific cognitive functions and in the highly accurate determination of the location of pathological lesions.

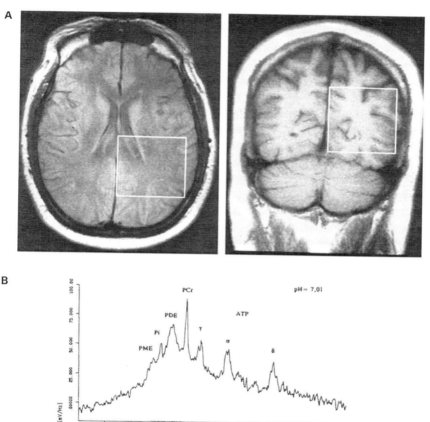

A - The two images of the magnetic resonance of protons (H) correspond to two virtual cross-sections of the brain along the axial plane, in T mode$_1$ (interaction of the protons with the environment).

B - The profile corresponds to the NMR spectrum of phosphorus ^{31}P (framed region). The pH is determined as a function of the position of the peak for (inorganic) mineral phosphate, Pi. PME: phosphomonoesters; PDE: phosphodiesters; Pcr: phosphocreatine; ATP: adenosine triphosphate with the specific resonances of phosphate groups in the α, β and γ position.

Figure IV.14 - Application of NMRI (Nuclear Magnetic Resonance Imaging) to the study of the neuron activity in a normal human brain
(reproduced from D. MAINTZ et al. (2002) "Phosphorus-31 NMR spectroscopy of normal adult human brain and brain tumors", NMR in Biomedicine, vol. 15, pp. 18-27, John Wiley & Sons Ltd, with permission)

A technique that is complementary to MRI is **Positron Emission Tomography (PET)**. This generally uses water labeled with oxygen 15 (^{15}O), a radioactive isotope of natural oxygen that has a very short lifetime (123 s), produced extemporaneously in a cyclotron by bombardment of an ^{14}N target with protons. The radiolabeled water is injected into the blood flow of the patient. It is found in greater concentration in the zones that are the most irrigated by blood capillaries. The positrons that it emits collide with the surrounding electrons and give rise to photons that can be detected by the appropriate apparatus. Affected by a stimulus (whether this stimulus results from talking, writing or listening), the blood irrigation of the zones of the brain that have been specifically excited increases noticeably. The location of the positron emission provides information about the location of these zones. Within a few dozen minutes, it is possible to locate a highly vascularized cerebral tumor. PET can use molecules other than water, such as organic molecules labeled with positron-emitting atoms, (^{18}F) fluorine (half life 110 min) and (^{11}C) carbon (half life 20 min). Around twenty years ago, in Canada, an analogue of L-dopa, the precursor of dopamine in the brain, ^{18}F-6-L-fluorodopa, was synthesized, and was found to be an excellent probe for determining the capture capability of the endings of the dopaminergic neurons in the striatum. In patients suffering from PARKINSON's disease, this capture capability is noticeably reduced. At present, PET involving ^{18}F-6-L-fluorodopa is being used to evaluate the survival of dopaminergic cells grafted into the striata of PARKINSON's disease sufferers [51,52].

Nowadays, brain imaging techniques can be used to explore the electromagnetic anomalies of neurological or neuropsychiatric illnesses such as HUNTINGTON's chorea, the different forms of ALZHEIMER's disease or even autism, the genetic origin of which is in the process of being deciphered. A bridge has now been built between the molecular defects identified by genetics and the electromagnetic anomalies that result, analyzed by functional cerebral imaging. It was not so long ago that DESCARTES considered that human thought was unconnected to a material support (Chapter II-3.4.3). We are not far from the era when BROCA located the language area in a specific zone of the brain after the autopsy of an aphasic patient (Chapter III-3.1), thus opening the door to another scientific domain, neuropsychology, which had previously only been the subject of speculation.

---

51 P. REMY, P. HENTRAYE and Y. SAMSON (1999) *"La tomographie par émission de positons, un outil de recherche fondamentale devenu indispensable à la recherche clinique. L'example des greffes neuronales dans le maladie de PARKINSON"*. Medicine/Sciences, vol. 15, pp. 490-495.

52 The PET technique is applicable in many pathologies other than degenerative pathologies of the brain. It is in everyday use for oncologic imaging. In order to locate the origins of a cancer, certain specialized centers use ^{12}F-2fluorodesoxyglucose (FDG). Given the highly marked glycolytic activity of tumor cells, FDG accumulates in them in its phosphorylated form, which cannot be metabolized, and is easily shown by the PET technique. To review: J. MAUBLANT, J. LUMBROSO, F. CACHIN, J.L. RAOUL, A. SYROTA and J.P. VUILLEZ (2001) *"Médecine nucléaire en oncologie: nouvelles modalités diagnostiques et thérapeutiques"*. Bulletin du Cancer, n° 88, pp. 35-44.

The consequences, from the societal point of view, were far from being insignificant. Thus autism, which was once suspected of being caused by errors in the mother's behavior with respect to her child, has been shown to be a disturbance in the development of the fetal nervous system, in the temporooccipital region.

While the neurosciences occupy a preponderant position in the medicine of the beginning of the 21st century, because of the development of techniques that aim to analyze even the functions of thought, emerging methodologies of another order, such as gene therapy (Chapter IV-2.2), are in the process of completely modifying our ways of treating and curing a range of previously incurable human diseases, from incapacitating immune disorders to cardiovascular diseases and cancer.

3.3. FROM EXPERIMENTAL MEDICINE TO PREDICTIVE MEDICINE

"It is in the domain of thought about the future that Man is singled out. We are beings who have an imagination. Not content to live in the present, to profit from past experience, we remain haunted by a future that we are conscious of constantly entering. This obsession with the future has been a powerful driving force in cultural evolution. We seek to predict in order to avoid the worst and to better prepare for our tomorrows."

Jacques RUFFIÉ
Birth of Predictive Medicine - 1993

By predicting potential dangers in subjects who are in good health, predictive medicine aims to provide the means of avoiding these dangers. These dangers can be intrinsic in nature, being written, for example, into a certain genome DNA sequence, or they can be extrinsic in nature, linked to an unsuspectedly deleterious environment. In each generation, mutations occur, certain of which can lead to so-called genetic diseases; between 3 and 4% of newborns are affected. Besides these spontaneous mutations, there are also mutations arising from the genetic inheritance of the parents. The purpose of **genetic counseling** is to warn parents when the existence of a potentially serious genetic flaw is suspected.

The highlighting of genes that give a predisposition for cancer (proto-oncogenes) is a convincing illustration of the power of predictive medicine. This involves genes that control the synthesis of growth factors, the activity of which is essential to embryogenesis and to the repair of damaged tissue. While they are normally subject to strict control by anti-oncogenes, proto-oncogenes are able to become active in an anarchical manner, under different influences, and to transform themselves into cancer-generating oncogenes. Recently, mutations have been found in two genes, BRCA1 and BRCA2, these mutations giving a predisposition for cancers of the breast and of the ovary. Thanks to genetic exploration, it will soon be possible to predict whether a cancer of the breast will have a rapid progression leading to uncontrollable metastases or a slow progression. Depending on the case patients will be subject to heavy chemotherapy or to a less aggressive treatment. In this context, targeted therapy with monoclonal antibodies is a source of great hope. While genetic inheritance has a role in cancer, the environment plays a not-insignificant role as well. This is the case, for example, in lung cancer sufferers who smoke tobacco, cancer of œsophagus in those who drink alcohol and job-related cancers in those working in factories producing colorants or materials derived from asbestos or tars.

Cardiovascular diseases are the primary cause of death in the more developed countries, involving either an infarctus, or a stroke. Many risk factors for these diseases are known, i.e., metabolic deviations affecting cholesterol or the blood serum proteins involved in the transport of lipids. These metabolic anomalies result in a syndrome known as atherosclerosis, which is characterized anatomically by the deposit of fats in the form plaques in the arteries. While genetic factors are at the origin of these metabolic problems, the latter are clearly amplified by an inappropriate diet. The role of predictive medicine is to recognize the genes that are responsible, warn individuals of the risks they are running and to advise them about the types of lifestyle and diet that do not increase these risks.

Being **able to predict**, predictive medicine should be **able to prevent** by means of targeted drugs. Within this context, it gives rise to reflection upon **polymorphism linked to variation in a single nucleotide** in the DNA of the genome of an individual. Known as SNP (Single Nucleotide Polymorphism), this polymorphism has proved to be a very useful auxiliary in molecular medicine. Hundreds of thousands of SNPs are present in the human genome and several tens of thousands in genes coding for the proteins. Where they are located differs according to ethnic backgrounds. Among these SNPs, some appear to be linked to certain pathologies, such as certain forms of cancer or degenerative illnesses such as ALZHEIMER's disease. In addition, in a small number of patients, the location of certain SNPs has been connected with previously-inexplicable drug incompatibilities. In line with these observations, **pharmacogenomics**, a branch of pharmacology that deals directly with genome sequence data, is trying to evaluate the impact of "SNP variants" on the efficacy and toxicity of drugs and to understand the genetic bases that explain the differences that are observed in the responses of different individuals to the same medication [53]. Rather than using a standard drug that is not very efficacious or causes adverse side effects, it might be possible, depending on the genetic profile of the patient, to use a drug that is more appropriate to his or her **genetic map**. It is doubtless not just a fantasy to imagine that, in 20 or 30 years' time, a patient visiting the doctor will be offered a genetic map thanks to cells taken from the buccal mucosa. Finding SNP variants that are known to be responsible for drug incompatibilities in such a map will make a targeted prescription possible. It will allow the detection of **genes for susceptibility to an illness**, at the same time

---

[53] T. REISS (2001) *"Drug discovery of the future: the implications of the human genome project". Trends in Biotechnology*, vol. 19, pp. 496-499.

uncovering targets for new drugs. Pharmacogenomics, which is still called **new pharmacogenetics**, contrasts with old pharmacogenetics in which, having found an adverse clinical response to a certain therapy, an attempt was made to identify the protein target of the incriminated drug, and then to go back to the gene coding for this protein, and to look for the mutation responsible for the aberrant response to the drug.

The existence of **customized predictive medicine**, which would read the destinies of individuals in their genes, would not be without its consequences in the life of a citizen. By registering each citizen with a genetic map, matched with a named identity card, predictive medicine might begin to take on the aspect of a Janus, with his beneficent face warning subjects of potential risks of metabolic problems, and guiding them towards the actions to be taken to lower the risks, but also with his evil face delivering each individual's intimate details to the indiscrete inquisitiveness of investigators who are operating towards their own ends (insurance companies, employers…). No less worrying would be the sly but predictable transformation of the individuality of the repaired or even doped human being within a system of imposed, docilely-accepted assistance.

3.4. THE DRUG LIBRARY OF THE FUTURE

In the 19th and 20th centuries, the methodology for biological experimentation underwent a revolution caused by the progress made in the domain of chemistry, both analytical chemistry, with the deciphering of increasingly complex molecular structures, and also in synthetic chemistry, with the large-scale production of tens of thousands of new molecules. The effects of these molecules, which might eventually be used as drugs, were tested directly on animals. It was thus that in 1910 the German chemist Paul EHRLICH discovered Salvarsan, a derivative of arsenic, which was active against a type of treponeme, the agent of syphilis. This was the result of a systematic analysis of the effect of synthetic products, aromatic derivatives of arsenic acid, on syphilis in rabbits. Salvarsan was the 606th derivative that was tested, and this is why it was called 606 for a long time before it was given the name Salvarsan. Sometimes, lucky chance shows surprising and unexpected properties in synthetic molecules. This was the case for chlorpromazine, which was initially used as an antihistamine. It was luck that led to its antipsychotic activity being discovered in 1950. A new era opened up in psychiatry with the arrival of synthetic narcoleptics like chlorpromazine.

A new chemical science known as **combinatory chemistry**, which dates from the 1990s, has aroused an increasing amount of interest in pharmacology. This involves making two or more species of organic molecules that carry reactive functional residues react in solution or in the solid phase in such a way as to synthesize, by means of all possible combinations, a number of final and intermediary products that is situated in the hundreds or even the thousands, and which makes up

chemical library or drug library. We can directly test all of the products formed on a sample of eukaryotic cells, in order to verify their effects (for example the inhibition of an anarchical proliferation of cancerous cells), or on microorganisms in order to evaluate an antibiotic capability. We can also proceed straight away with the fractioning of the reaction products and the testing of each of the fractions. If the response is positive, fractioning is continued until the molecular species responsible for the desired effect is obtained. Other evaluation parameters for this molecule, such as its absorption, its toxicity and its metabolic future (distribution in the organs, chemical modifications and excretion) are then explored, first in cells, and then in animals (rats, mice), thus comprising pre-clinical tests. These **screening** operations, which are said to be **high-throughput**, require **automation** and **robotization** aided by powerful **computer technology**. Each year, pharmaceutical companies screen tens of thousands of different molecules on hundreds of targets.

Complementary to combinatory chemistry, *in silico* chemistry works by **molecular modeling** and uses **computer programs** for the rational design of new drugs that are able to fix onto specific protein targets. The purpose of this is to provide a virtual follow up to modifications in the reactivity of a given drug molecule as a function of the modifications imposed on its structure, for example, the addition of residues that differ according to their electrophilic or hydrophilic properties, or according to the length of their side-chain. Provided there is a chemical library and we know the three-dimensional structure of a macromolecule, for example an enzyme, as well as the nature of the residues that define its active site, we can hope to select and chemically modify a substance that is able recognize the active site of this enzyme and to make an almost perfect ligand out of it which is able to efficiently block the operation of the target enzyme. This method, which is based on computer-aided chemistry, is called "**Structure-based drug design**", and has had some notable successes. It has made it possible to develop an inhibitor capable of blocking a protease involved in the replication of the AIDS virus.

However, both in combinatory chemistry and in molecular modeling, the many successes that have been achieved remain modest in number compared to the means that have been deployed to achieve them. In terms of statistics, out of ten thousand molecules that are recognized as being efficacious for a given target *in vitro*, around one hundred are chosen for preclinical trials on animals, around ten are chosen for preclinical trials in Man and only one will come out as a drug. The financial and economic effect is far from being negligible. It has even become a preoccupation in a system where merciless competition is the rule.

In addition to synthetic chemistry, preparative chemistry, which is based on the isolation of natural molecules, is now the subject of renewed interest, due to the introduction of high-throughput techniques. High-throughput screening, which is an essential tool in combinatory chemistry, is also carried out to ensure **the systematic detection and isolation of natural substances** having interesting pharmacological activities such as antibiotic activities or anti-cancer activities, based on marine animals, microscopic fungi, prokaryotic organisms and various plants. For example, among the substances that have been isolated recently are cibrostatin, a specific cytostatic of melanoma cells, from a marine sponge, mannopeptimycin, a bacterial antibiotic from an actinobacterium *Streptomyces hydroscopicus* and a whole set of alkaloids with a cytostatic activity with respect to human tumor cells from an exotic plant of the genus *Daphniphyllum*. The molecular diversity of the living world is such that the reserves of natural products having pharmacological activities are far from being exhausted. So far, only a small percentage of the microbial species populating the Earth have been listed. The depths of the oceans harbor many unknown species. Thousands of insect species remain to be discovered in the canopies of tropical rainforests. Exploration of the plant kingdom is far from being complete. The listing of natural molecules having a therapeutic activity has only just begun. The hunt promises to be a fruitful one, all the more so because the high-throughput screening methods that can now be used greatly increase the efficiency of the search.

High-throughput screening, applied to natural molecules, has overturned the methodological procedures that were in use until recently, which progress through logical steps, using relatively simple artisanal analytical methods, from observation, often resulting from **serendipity**, to the isolation of the active substance. Thus, in the 19th century, using inherited traditional knowledge that a decoction of *Cinchona officinalis* bark calms malaria crises, Pierre Joseph PELLETIER (1788 - 1842) and Joseph Bienaimé CAVENTOU (1795 - 1877) decided to isolate the active substance of this bark. From the raw extract, they purified an alkaloid, quinine, which proved to be the anti-malarial substance they were looking for. More recently, the starting point of FLOREY and CHAIN's isolation of penicillin from the microscopic fungus *Penicillium notatum* was the fortuitous observation made by FLEMING that this *Penicillium* secretes an antibiotic factor (Chapter III-2.2.5).

There are many examples in which serendipity has been the principle factor involved in the discovery of a drug, and this will no doubt continue to be the case. The appearance of a lucky chance, after all, is not incompatible with high-throughput practices. Also, it is not impossible that in the future there will be a conjugation of the discovery of new natural substances and the use of combinatory chemistry, with the aim of manufacturing derivatives having a much greater power of action and quality of specificity from these substances [54].

To sum up, the experimental method has caused contemporary medicine to take a giant leap forward, with the discovery of increasingly high-performance functional exploration techniques, the development of therapies using molecules that are already present in Nature or are manufactured by synthesis and the more and more advanced understanding of molecular mechanisms that takes into account

---

54 A. GANESAN (2004) *"Natural products as a hunting ground for combinatorial chemistry"*. Current Opinion in Biotechnology, vol. 15, pp. 584-590.

the functions of the cell and makes it possible to predict, if not to prevent, pathological malfunctions.

4. TOWARDS A GLOBAL UNDERSTANDING OF THE FUNCTIONS OF LIVING BEINGS

The basic idea of the pioneers of molecular biology was that the function of a macromolecule depended on its structure. Thus, PERUTZ's elucidation of the tetrameric three-dimensional structure of hemoglobin, and of its modifications depending on the degree of oxygenation, shed a considerable amount of light on the cooperative mechanism of the transition from the hemoglobin state to the oxyhemoglobin state (Chapter III-6.2.1). In the same way, an understanding of the structure of many enzymes, receptors and transporters of metabolites has shed light on their mechanisms. In a parallel manner to the exploration of the structures and functions of proteins, that of genomes has made remarkable progress. The subtle entanglements of genomics and proteomics that have become accessible to the experimental method are the order of the day. One major challenge for **post-genomics** is to understand how proteins, expressed by genes, interact with one another to generate functions that characterize cellular specificity. Even more ambitious are attempts to understand the operation of organs or even of living organisms, based on mechanisms that are implemented at molecular level. These attempts lead straight to an **integrated biology**, that is, a biology that aims to understand the overall functioning of living beings. Taking as its purpose the access to emerging functions, resulting from interactions between macromolecules, integrated biology first tries to invent methods that make it possible to detect these interactions. Strengthened by the information obtained, it tries to integrate this information with a mathematized language into modules that attempt to simulate living beings.

4.1. EXPERIMENTAL DEMONSTRATION OF PROTEIN INTERACTIONS

From the simplistic procedures of the middle of the 20th century, which were justifiable within the reductionist context of this period, and which involved considering each species of proteins as an autonomous functional entity, we have moved on to the idea that the different species of protein that inhabit a cell have a dialogue with one other, and that they may move from one endocellular organelle to another, depending on post-translational modifications (for example, phosphorylations) that change their conformation and, at the same time, their reactivity and their behavior. Thus, an enzyme protein is not only defined by its catalytic performance with respect to a given substrate, but also by its place in a metabolic network where **it interacts in a dynamic and transitory manner** with a multitude of other protein species (Figure IV.15A).

The concept of **cell signaling** has also evolved. Instead of considering that a cell membrane receptor, activated by fixation of an extracellular ligand (a hormone, for example), addresses the information received to an endocellular effector protein *via* a linear cascade of individual proteins, it has come to be postulated that communication between an activated receptor and its effector is mediated by proteins organized into **interactive networks** (Figure IV.15B). This machinery provides more flexibility in the addressing of messages to effectors.

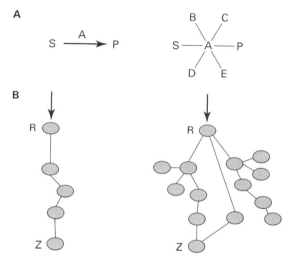

A - Case of enzyme systems. The diagram on the left refers to the classical idea of enzyme transformation of a substrate S into product P, catalyzed by enzyme protein A. The diagram on the right shows that besides its catalytic function, protein A interacts with other proteins in the cell.

B - Case of the transduction of a signal that is external to the cell (a hormone, for example). The diagram on the left refers to the classical idea of signaling from a receptor R according to a linear cascade of protein-protein interactions inside the cell, leading to an effector Z. The diagram on the right shows that the signal is spread through proteins organized into interactive networks.

Figure IV.15 - Evolution of ideas concerning the transfer of information between endocellular proteins

Another subject to be considered is the density of macromolecules of all types, such as proteins, nucleic acids, lipids and polysaccharides, contained by a microorganism or a eukaryotic cell, which reaches values of 300 to 500 g/liter, denoting a **semi-solid state** or a considerable degree of **compacting**. However, for technical reasons, kinetic studies carried out *in vitro* on isolated enzymes have been carried out with solutions that are 1 000 or 100 000 times more dilute. Conscious of this difference in scale between information obtained from *in vitro* studies and the *in vivo* reality, the biology of today is trying to re-evaluate **molecular dynamics** within the context of a cell. This is why we are seeing the birth of an **integrated (or integrative) biology of functions**, which, using modeling procedures, aims to achieve an understanding of the temperospatial dynamics of the interactive components inside cells. This **holistic conception of biological systems** ("systems biology") has been made possible by progress in **technological expertise** in domains as varied as biochemistry, molecular biology, physical optics, electronics, nanomechanics, physical and mathematical modeling and computer technology. It is a necessary complement to the classical experimental method based on Bernardian determinism which, in order to connect an effect with a cause, explores living beings in a manner that is often monoparametric and is inevitably reductionist. This signals a **change in paradigm** in the experimental approach to living beings.

A particularly effective investigative method used to explore the dialogue between proteins is the **double-hybrid** method described in 1989 by Stanley FIELDS (b. 1955) and Ok-Kyu SONG [55] (Figure IV.16).

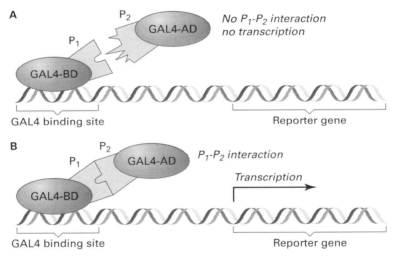

By genetic construction, two proteins, P_1 and P_2, whose interaction is to be tested, are expressed in the form of fusion proteins in yeast. Protein P_1 is fused with the binding domain (GAL4-BD) to the DNA of GAL4, a protein that regulates the transcription of the β-galactosidase gene. Protein P_2 is fused with the activation domain of GAL4 (GAL4-AD). Insofar as P_1 interacts with P_2 (B), the GAL4 transcription regulation activity is re-established, which is verified by the transcription of the reporter gene. If the opposite occurs, i.e., in the absence of any interaction between the two domains of GAL4 (A), the reporter gene is not transcribed.

Figure IV.16 - Principle of highlighting protein-protein interactions using the double hybrid system

---

[55] S. FIELDS and O.K. SONG (1989) *"A novel genetic system to detect protein-protein interactions"*. Nature, vol. 340, pp. 245-246.

The principle of this method is based on the modular nature of numerous transcription factors in eukaryotes. These factors contain both a DNA-binding domain that includes a specific DNA-**binding** site and a transcription **activation** domain that starts up the machinery for transcribing DNA into messenger RNA. These two domains can be dissociated and then re-associated in a functional manner, by forming hybrids with interacting proteins. A first protein, P1, is fused with the DNA-binding domain of a transcription factor by genetic manipulation, and a second protein, P_2, is fused with the activation domain of the same transcription factor. If P_1 is able to interact with P_2, the transcription factor is reconstituted and the reporter gene upon which it depends can be expressed.

The **trapping technique**, which is complementary to the double-hybrid system, was developed to make it possible to identify a set of **interactive proteins** within a cell. A protein that is included in this set (protein of interest) is fused by genetic engineering techniques to a short polyhistidine chain (called a tag). Using this tag, the protein of interest is fixed to a solid medium containing nickel ions, a material that is reactive with respect to the polyhistidine chain. In the presence of a soluble cell extract, the **protein of interest** binds the **cognate proteins** of this extract, making it possible to retrieve a complex whose components, corresponding to interactive proteins, can be resolved after denaturing gel electrophoresis and then characterized (Figure IV.17).

The techniques described above are backed up by **cell imaging** techniques that make use of **confocal microscopy**, which is more directly in keeping with living reality. The optical performance level of confocal microscopes has improved lately, with the arrival of biphotonic and multiphotonic lasers that illuminate precise points of the cell. As we have seen previously (Chapter III-6.2.1), it is possible to create a protein chimera made up of a protein of interest fused with a **fluorescent protein**, in this case GFP (Green Fluorescent Protein), inside a cell. There are currently several variants of GFP that are able to emit fluorescent light at different wavelengths. This has allowed the development of a technique known as FRET (Fluorescence Resonance Energy Transfer) which explores the interaction between two fluorescent proteins. In practice, two GFP variants that have neighboring emission spectra are fused, by genetic engineering inside the cell, to two proteins of interest, P_1 and P_2, that are suspected of being interactive. If this is the case, the fluorochromes that they carry are sufficiently close that the result is a modification in the intensity of the emission fluorescence of the donor fluorochrome (decrease) and of the receiver fluorochrome (increase), which is readily detectable.

All of these studies, taken together, have given rise to the idea that **endocellular proteins are organized into networks, that these networks are interactive** and that **their location in defined compartments of the cell is dependent on epigenetic events such as phosphorylations**. Two attributes can be found in integrated systems: firstly, the presence of modules, i.e., interactive motifs, which, like the pieces of a jigsaw puzzle, fit together to produce a complex, coherent structure, and, secondly, the **emergence** of functional properties due to the newly created interactions.

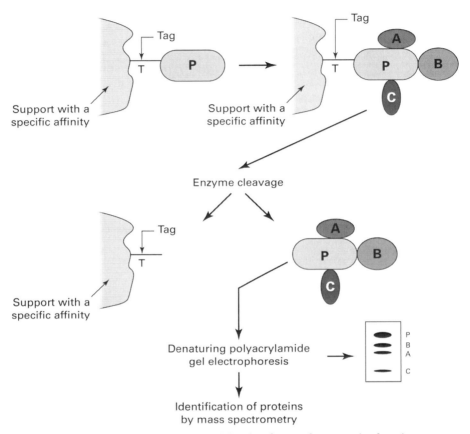

The protein of interest P is expressed in the form of a protein fused to a protein "tag" T that is able to bind to a solid support with a specific affinity. The assembly is brought into contact with a cell extract. Certain proteins of this extract, A, B and C, which are able to interact with the protein P, become fixed to the latter. In a second step, the tag T is freed from its attachment to protein P by means of a specific cleavage enzyme. The PABC complex that is recovered from the solid medium in soluble form is subjected to polyacrylamide gel electrophoresis, in order to separate and identify its components.

Figure IV.17 - Principle of isolating proteins that are interactive with respect to a protein of interest immobilized on a solid medium

Given an analytical description of the basic building blocks that are used to construct living systems, and an understanding of their modes of association in defined circumstances, it is normal to try to **reconstruct**, in their entirety, mechanisms that show the functioning of these systems. This idea was first applied to the yeast *Saccharomyces cerevisiae* for different reasons, such as cell homogeneity (in principle, and, in any case, statistically speaking, all yeast cells have the same genome and the same proteome), an in-depth understanding of the genome and the proteome and the presence of a vast directory of well-characterized mutants.

The use of techniques for the detection of interactions between proteins has revealed the existence of a potential dialogue of unexpected richness between a multitude of proteins (Figure IV.18), in the yeast cell. It is necessary to reflect upon this evidence, which leads to the postulate that, for a given protein, there are mechanisms that restrict and select the many partners able to react with it at a precise moment. Faced with a situation in which chance has the upper hand, leading to uncontrollable anarchy, it is necessary to have regulation, which is underpinned by Darwinian logic.

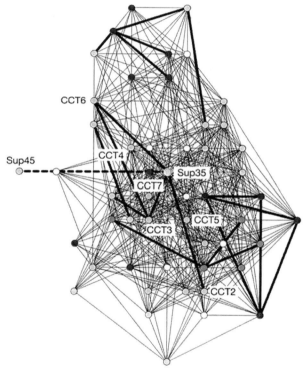

Example of an interaction network involving the yeast Sup45 prion protein. The line of dashes refers to experimental data concerning the interaction of Sup45 with another protein, Sup35. The lines in bold refer to interactions taken from experimental data; while the fine lines refer to predictions, particularly phylogenetic ones.

Figure IV.18 - Network of protein interactions in yeast
(reprinted from D. EISENBERG, E.M. MARCOTTE, I. XENARIOS and T.O. YEATES (2000)
"Protein function in the post-genomic era", *Nature*, vol. 405, pp. 823-826,
by permission from Macmillan Publishers Ltd, D. EISENBERG and *Nature*)

This logic arises from a choice of the most efficient reaction path, which is first of all dictated by the speed constants involved in the association and dissociation of molecular partners, without, however, neglecting any stochastic events that may arise. Chemical modifications of proteins participate in this regulation, such as phosphorylation, glycosylation and acylation. The result at cell level is a **coherent channeling** of the information that is carried from a molecular signal. Thus, fixation of a hormone onto a receptor induces a series of modifications to the intracellular proteins that channel information towards an effector terminal, for example an enzyme responsible for the production of a metabolite with a strategic function.

4.2. MATHEMATICAL MODELING OF THE COMPLEXITY OF LIVING BEINGS

How can the sum of the scattered **experimental data** that we have concerning the catalytic capabilities of a multitude of enzymes of cellular origin be integrated into the operation of a cell? How can we envisage the gene-enzyme relationship according to current evidence concerning the complexity of the genetic message? Biocomputing, or bioinformatics, a science that emerged towards the end of the 20th century, proposes to try to answer these questions.

At the turn of the 21th century, with the development of increasingly powerful computer microprocessors that are able to carry out complex operations with amazing rapidity, the hope arose that it would be possible to simulate processes as varied as the regulation of the cell cycle, molecular flow in metabolic pathways and the reception of molecular signals, for example from hormones by living cells, as well as the transmission of the messages that result. The dream of an *in silico* **virtual biology** has become achievable.

The first mathematical theory of simple **enzyme reaction kinetics** was put forward approximately a century ago, by Victor HENRI (1872 - 1940). Born in Marseilles to Russian parents, Victor HENRI studied in Saint Petersburg and then spent time at the universities of Göttingen and Leipzig before becoming established in Paris. Having an eclectic mind, studying both psychology and physicochemistry, he had the wonderful intuition that enzyme catalysis arises from a specific mechanism, different from that implemented in a chemical reaction. The study carried out by HENRI concerned the cleavage of sucrose (table sugar) into fructose and glucose by the action of an enzyme called invertase (sucrase). The term invertase was used because during reaction there was a change in the rotatory power of the sucrose solution, shown by a polarimeter. Analysis of the reaction suggested that an **enzyme-substrate complex** is formed, which breaks down to regenerate the enzyme and liberate the product of the reaction. This analysis gave rise to an equation for the speed of the enzyme reaction as a function of substrate concentration. HENRI published the results of his experiments both in his thesis, which he presented to the Paris Faculty of Sciences in 1902, and in two articles that appeared in the *Reports of the Academy of Sciences* [56,57]. In 1913, in *Biochemische Zeitschrift* (vol. 49, pp. 333-369), Leonor MICHAËLIS and Maud MENTEN (1879 - 1960) confirmed the results of Victor HENRI and formulated an equation that became a classic, describing the speed of formation of a product from a substrate in enzyme catalysis. Since the period of these first studies, the concepts involved in enzyme kinetics have evolved considerably. The first metabolic pathway be deciphered was that of the degradation of glucose (glycolysis), either into ethanol in yeast, or into lactic acid in muscle tissue.

After this, researchers became aware that the **activity** of enzymes could be **modulated** as a function of **covalent modifications** of amino acid residues of their protein chain (phosphorylation, dephosphorylation, acylation…). Metabolic flow analysis led to the idea of the limiting reaction. In the 1970s, the idea that there is a single limiting reaction in a chain or a cycle of reactions gave way to the idea that metabolic control is distributed over several reactions, and that each reaction has its own, more-or-less intense **control force**. Another complexity factor came to light with the discovery of **allostery** [58]. Allosteric enzymes have the particularity that they can fix reversibly onto a site that is different from the active site (allosteric site) molecules that are often the terminal products of a chain of reactions: the consequence of this is a conformational modification of the structure of the enzyme that has repercussions on the geometry of the active site and modifies its reactivity with respect to the substrate.

In the 1980s, faced with the complexity of the tangle of listed metabolic and signaling networks, attempts were made to use **mathematical modeling** to show the progress of the traffic of molecules inside a cell in relatively simple metabolic pathways such as glycolysis. In the modeling procedure, the concentrations of the different molecular species are considered to be variables whose variations over time depend on their speed of production and their speed of disappearance, which leads to a set of paired differential equations. With this procedure, we entered the domain of **virtual biology**.

One of the earliest examples that illustrates the scope of this virtual biology was the modeling of a bacterial infection by a bacteriophage, which was carried out at the University of California Berkeley campus in the USA [59]. The simulation encom-

---

56 V. HENRI (1902) *"Recherches sur la loi de l'action de la sucrase"*. Comptes Rendus de l'Académie des Sciences, vol. 133, pp. 891-894.

57 V. HENRI (1902) *"Théorie générale de l'action de quelques diastases"*. Comptes Rendus de l'Académie des Sciences, vol. 135, pp. 916-919.

58 J. MONOD, J.P. CHANGEUX and F. JACOB (1963) *"Allosteric proteins and cellular control systems"*. Journal of Molecular Biology, vol. 6, pp. 306-329. (classic article on the concept of allosteric regulation)

59 D. ENDY, L. YOU, J. YIN and I. MOLINEUX (2000) *"Computation, prediction and experimental tests of fitness for bacteriophage T7 mutants with permuted genomes"*. Proceedings of the National Academy of Sciences, USA, vol. 97, pp. 5375-5380.

passed the different phases of the infection of the enterobacterium *E. coli* by the bacteriophage T7. This infection involves the translocation of the bacteriophage DNA into the bacterium, the replication of this DNA in the body of the bacterium and the bacterium's own protein synthesis machinery being diverted to the production of bacteriophages (Chapter III-6.1). In order to model the infection of *E. coli* by phage T7, the genome of the phage was divided into 73 numbered segments, each segment representing one part of the genome. The modeling took into account the translocation of the viral DNA into the bacterium, the transcription of the viral DNA into messenger RNAs (mRNA) and the translation of the messenger RNAs into viral proteins. It used measurement values taken from experiments carried out *in vivo*, such as those concerning the kinetic constants of various enzyme reactions involved in virus proliferation. It was observed that, depending on whether the RNA polymerase is of bacterial or viral origin, the speed of transcription is 40 or 200 base pairs per second. By carrying out virtual mutations involving the *in silico* permutation of the order of the genes in the bacteriophage genome, for example the position of the gene coding for the RNA polymerase, researchers observed a slowing down of the replication of the bacteriophage that is practically equivalent to that measured *in vivo*. They came to the conclusion that **the arrangement of genes in a natural virus is optimal for its proliferation and thus for its survival, in accordance with the concept of Darwinian selection.**

Thanks to the creation of increasingly powerful software, the aim of virtual biology is to simulate signaling and metabolic pathways. In the longer term, the aim is to understand the molecular and cellular processes that direct embryo development, or to test the effects of drugs of known target on the metabolic behavior of the cell. **Metabolic engineering** (which is already well developed) comprises two types of models, **stoichiometric models** and **kinetic models**. Stoichiometric models describe metabolic networks in the stationary state, based on analytical data. Kinetic models combine stoichiometric information and that concerning the catalytic capabilities of the enzymes in a metabolic network.

In Canada, the Cyber-cell project plans to model the overall functioning of the machinery which, in the bacterium, includes its **metabolism** and its **proliferation**. The aim of the AfCS (Alliance for Cellular Signaling), which was launched in the USA, is to understand how **signaling** occurs in **cells such as the B lymphocyte, the macrophage or the cardiac cell** in response to different types of stress. The techniques that are used range from identification of all signaling network proteins to the evaluation of the flow of circulating information and to the integration of the data acquired into theoretical models. The European **nerve synapse** project makes use of similar procedures, with its long-term hope of linking the functioning of nerve cells with the **cognitive and behavioral functions of living beings**. This is a sizable challenge. In fact, there is far from being a real consensus concerning the principle of a demarcation between, on the one hand, cognitive functions such as language or memory, which are located in precise zones of the brain, and which could be reduced to physicochemical processes, and, on the other hand, forms reflective thought that are expressed through the creative imagination or by judgements concerning ethics or esthetics, the notion of personal responsibility, or even pictorial, architectural or musical beauty. Should we see the human soul as the programmer of a superb computer that never ceases to develop from the embryonic state onwards, like John ECCLES (1903 - 1997) and others, or should we admit, like Jean-Pierre CHANGEUX (b. 1936), Stanislas DEHAENE (b. 1965), Daniel DENNETT (b. 1942) and others, that thought is not transcendent, and that it is intrinsically dependent on the brain, which is considered to be a neurochemical system, and thus look for the secret of the individuation of the human being in brain information storage mechanisms with retrocontrol loops associated with subtle neuron architectures, or, in short, refer to a sort of TURING machine? Whatever the case, in this domain, as in others, simple animal models are used in order to identify elementary processes that are able to explain easily-tested functions such as the memory in anatomical and physiological terms. This is the case for the sea slug or sea hare (Chapter III-6.1) which, despite its rudimentary cognitive capabilities, provides information that can be used to reconstruct higher cognitive functions, present in the brains of mammals. It is clear that the cognitive sciences have reached a stage in which they are emerging from their infancy (Chapter IV-3.2). Now, ingenious computing methods and a basis for reflection that has spread beyond the confines of psychology and philosophy, are available to them. They have set themselves the goal of producing an artificial intelligence, using ultrarapid computers as well as software that is able to model the operation of neural networks and to come close to the performance of human intelligence in terms of the power of their reactivity and their memorization. At present, many other biological systems are being subjected to multiparametric exploration, with the aim of producing models. This is the case, for example, with the program of the differentiation of certain white blood cells, the neutrophils (Chapter III-2.2.4), from precursors located in the bone marrow, a differentiation that leads to the emergence of functions such as phagocytosis that are implemented in the fight against microbial infections [60].

In a domain that is closer to mechanical science, hydrodynamics, the digital simulation of the cardiovascular system has already made it possible to represent the physical phenomena associated with the propagation of a wave in deformable arteries during a cardiac contraction, in the form of equations, with a good approximation [61].

In short, from a monoparametric approach that often began as being essentially and necessarily reductionist, the experimental method, applied to living beings, has become a **"globalized", or synthetic, multiparametric approach**, the aim of

---

[60] K. THEILGARD-MÖNCH, B.T. PORSE and N. BORREGAARD (2006) *"Systems biology of neutrophil differenciation and immune response"*. Current opinion in immunology, vol. 18, pp. 54-60.

[61] J.F. GERBEAU and D. CHAPELLE (2005) *"Simulation numérique du système cardiovasculaire"*. Medicine/Sciences, vol. 21, pp. 530-534.

which is to understand the dynamics of molecular interactions in defined biological systems. Making use of data obtained, the hope is to use mathematical processing to simulate the overall functioning of a cell, organ or organism. This new paradigm of the experimental method ("systems biology" [62]) is not limited to a simple accumulation of observations concerning a given biological system and their abstraction in mathematical form. The originality of this approach is that it formulates **predictions** of changes in the behavior of a system as a function of the manipulation of parameters such as substrate concentration, the presence of inhibitors, and so on. The mathematical processing of experimental data, with a view to learning about the functioning of complex systems by modeling, is supported by the **technosciences**, particularly biocomputing. It is linked not only to the enormous sum of accumulated knowledge concerning the structures and functions of living beings in the post-genomic era, and to the notion that the life of a cell depends on multiple networks of molecular interactions and thousands of enzyme reactions located in its different organelles, but also to the information that comes to it from its environment. A first type of modeling is based on observations made or experiments carried out on an easy-to-study model system. Laws are drawn up from this. This **so-called "bottom-up" (or synthetic) procedure**, which proceeds from the simple to the complex, makes it necessary to have a set of very precise biochemical data. This great precision is all the more imperative in that any deviation, even a minimal one, in the integration of an experimental result can generate a mathematical model that is apparently plausible but which is unconnected to the living reality. The **reverse, "top-down" (analytic) procedure** proceeds from the overall operation of an organ and its theoretical analysis towards the specific mechanisms of its components. It takes into account the functioning of complex integrated systems such as the nerve and endocrine systems, immune and reproduction systems and the system controlling homeostasis, descending in stages towards the cellular, molecular and genetic levels.

In the end, an understanding of living beings involves the management of an amazing capital of experimental data. This makes it necessary to consider all of the genes (genome), all of the transcripts coding for the proteins (transcriptome), all non-coding RNAs (RNAome), all proteins expressed in a particular cell type (proteome) and all metabolites (metabolome) as a function of the enzyme catalyzers expressed and the energetics that underlie the catalyzed reactions, and, finally, to connect upstream events (mutations of genes and interference of messenger RNAs, chemical modifications to amino acid residues in the proteins) to phenotypical modifications on the scale of the whole organism (phenome) (Figure IV.19). The goal of integrated or integrative biology ("systems biology") is therefore to put **living beings into equations**, that is, to represent them in virtual systems for which the behavior, accessible by means of calculation, can be predicted as a function of

---

62 H.V. WESTERHOFF (2005) *"Systems biology in action"*. Current Opinion in Biotechnology, vol. 16, pp. 326-328.

modifying parameters. In addition to the possibilities that are opened up in terms of a deeper understanding of physiological mechanisms, such virtual systems could be used for the design of new drugs or for the manufacture of economically valuable biomolecules.

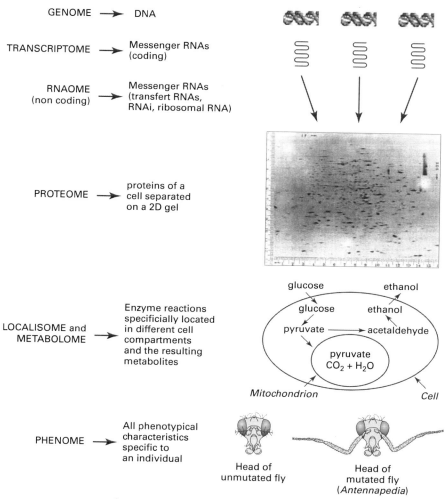

A genome mutation affect the phenome.

The diagram illustrates the different levels of complexity in the pathway that goes from all the genes together (genome) to all of the expressed characteristics (phenome) in the living being, passing via coding RNAs (transcriptome) and non-coding RNAs (non-coding RNAome), all the proteins (proteome), all addressing systems in the cell compartments (localisome) and all of the metabolic pathways (metabolome).

Figure IV.19 - The post-genomic era: from the genome to the phenome

At a scientific meeting held in Sheffield, England, in January of 2005 [63], with the theme *Systems biology: will it work?*, an argumentative discussion of the advantages as well as the disadvantages of an integrated, mathematized biology was useful in that it included a reminder that most of the parameters used in "systems biology" come from studies that are carried out *in vitro* on purified enzymes, and that it is not sufficient to know the value of the Michaelian parameters (V_{max} and K_m) in order to reach biological reality. In fact, *in vivo*, many enzymes record variations in activity that are hard to control due to allosteric type regulation or interenzyme contact; several enzymes of a metabolic pathway being able to interact to form a **metabolon**. However, by compacting several enzymes that catalyze contiguous reactions in a metabolic pathway, a metabolon considerably increases the catalytic efficiency of this pathway. Another element of uncertainty arises from the protein density of the cell medium, and also from the fact that covalent modifications of enzymes can introduce a change in endocellular location (nucleus, organelles of the cytoplasm...). Nevertheless, an approximative approach could limit itself to dealing with biological systems in modular terms, i.e., to considering them as being made up of a number of black boxes, each black box containing a series of reactions being processed mathematically together, with an input and an output. There is still a long way to go if we place ourselves on the cellular scale, but the end of the pathway seems even further away if we envisage the organism as a whole, taking into account the remote interactions between organs involving the interplay of chemical mediators.

The brain plays a critical role in the dialogue between different organs, and in the regulation of the energy equilibrium in higher animals. This equilibrium can be disturbed by fasting or intense, prolonged muscular activity, or by an overabundant diet. The corrective response comes from a deep region of the brain, the hypothalamus, via the secretion of different types of peptides, some of which stimulate the appetite and others of which suppress it [64].

While taking into account the multitude of parameters that affect the complexity of living beings on an individual level, the theoretical approach to the study of cell function by modeling has the advantage that it produces predictions and provides information about the validity of conclusions and of theories based on experiments that are old and accepted in the absence of contradictory elements. This was the case for the theory that stated that the state of activation of a gene is determined only by the presence in its environment of transcription factors. Recent studies concerning the level of gene transcription in isolated cells have shown that there are **probabilistic-type factors** which mean that a given gene in a given cell can be

---

63 M.P. WILLIAMSON (2005) *"Systems Biology: will it work?"*. Biochemical Society Transactions, vol. 33, pp. 503-506.
64 P. DOWELL, Z. HU and M.D. LANE (2005) *"Monitoring energy balance: metabolites of fatty acid synthesis as hypothalamic sensors"*. Annual Review of Biochemistry, vol. 74, pp. 515-534.

activated at any moment. A review which came out in 2005 [65] sums up this subject. In this review, the authors use modeling to analyze the behavior of cells in the process of differentiation during embryogenesis. Their Darwinian model, which associates contingency and selectivity, competes advantageously with the determinist (or instructive) model, based on an all-or-nothing logic, that has been implicitly accepted up until now. The Darwinian model takes into account the occurrence of stochastic events at gene expression level, events that are partially linked to the structural modifications to the chromatin that depend on covalent modifications of an epigenetic nature (phosphorylation, methylation...). By basing itself on the existence of mutational fluctuations that arise by chance, associated with a self-regulation of gene expression, the model that is obtained shows that during embryogenesis a cell has a choice either to differentiate into another cell type or to remain in its initial state. Differentiated cells stabilize their own phenotype and, in their surroundings, stimulate the proliferation of foreign cell phenotypes. A harmonious equilibrium between these two processes is the necessary condition for the setting up of the steps that lead to the arrangement of different cell types during organogenesis, which take place in an apparently inescapable order, in the absence of disturbances. A break in this equilibrium leads to an anarchical cell proliferation. Generally speaking, from the point of view of experimental science, the lesson that can be drawn from current modeling experiments is that the Bernardian determinism that has prevailed as the essential foundation stone of the methodology applied to the study of living beings may find itself being requalified by the taking into account of **stochastic phenomena**. This is the case when the number of reacting molecules is low and the probability of stochastic events is non negligible. The modeling of such systems necessitates having recourse to a complex mathematical formalism. It remains true that determinist models for simulation of the dynamics of living beings, represented by classical differential equations, are more-or-less valid when the number of reacting molecules involved is high and the reactions supposedly take place in a homogeneous medium.

Should "systems biology" be regarded as a resurgence of a physiology that has been somewhat neglected over the last few decades, but has been reinvigorated by a salutary hybridization of biologists and model-makers? In any case, this is the intention of the "physiome" project [66] which has recently been launched on an international scale. It is also doubtless due to this state of mind that a trend which had gone out of fashion, involving the simulation of the performance of living beings by very elaborate concrete models, robots, is being reborn.

---

[65] B. LAFORGE, D. GUEZ, M. MARTINEZ and J.J. KUPIEC (2005) *"Modeling embryogenesis and cancer: an approach based on an equilibrium between the autostabilization of stochastic gene expression and the interdependence of cells for proliferation"*. Progress in Biophysics and Molecular Biology, vol. 89, pp. 93-120.

[66] P.J. HUNTER (2004) *"The IUPS physiome project: a framework for computational physiology"*. Progress in Biophysics and Molecular Biology, vol. 85, pp. 551-569.

4.3. BIOROBOTS AND HYBRID ROBOTS

An immense distance has been covered in just over two centuries, since the time when VAUCANSON presented automata in the forms of human figures, moved by ingenious springs and cogs, and giving the illusion that their movements were controlled by an intelligence, to a marveling public (Chapter II-6.4).

In the last decades of the 20th century, considerable progress was made in the understanding of the operation of the nervous system and in the development of technologies in which miniaturized electronics have come to the aid of already high-performance micromechanics. The brain being considered as an information processing machine, the aim is to understand the logic of this information machine by means of simulations on computers and, based on the results obtained, to construct robots whose electrical circuits take their inspiration from the operation of animal neurons. These robots are called **biorobots** or **animats**. Insects have been chosen as a reference for the construction of such creatures because of the relative simplicity of their nervous systems: several hundred thousand neurons, in comparison with the billions of neurons present in mammals (100 billion in Man). The fly's **system of vision** has been favored as a subject of study because of the possibility it offers of being able to record the electrical response of neurons that can be identified one by one. In the middle of the 1980s, in France, this inspired the pioneering work in biorobotics carried out by Nicolas FRANCESCHINI (b. 1942) and his team [67,68] (Figure IV.20). Their objective was to study how an animal can avoid obstacles by means of its ocular perception and its movement-detecting neurons, the operation of which the team just analyzed using microelectrodes and a microscope-telescope specially built for the purpose. The fly's composite eye has 3 000 elementary units or ommatidia, each carrying eight light receptor neurons. The electrical signals emitted by these neurons in response to captured light (at most a few dozen millivolts) are sent to subjacent neurons that are organized into three levels that correspond to the optical ganglions called the "lamina", "medulla" and "lobula". The lobula is a strategic decoding center which, because of the small number of neurons contained in it (sixty), has been the subject of in-depth electrophysiological investigation. Each of the sixty neurons of the lobula operates as a signal integrator. The neurons of the lobula send their messages to motor neurons involved in the contraction of small muscles that control the guidance and stabilization of the fly's flight.

Based on an exhaustive study of the neuron wiring of the fly's eye, FRANCESCHINI and his colleagues were able to reconstruct a facetted artificial eye that can retranscribe the light signals received optoelectronically. This artificial eye, the electronic

---

[67] N. FRANCESCHINI, J.M. PICHON and C.V. BLANES (1992) *"From insect vision to robot vision"*. Philosophical Transactions of the Royal Society, London, vol. B337, pp. 283-294.

[68] N. FRANCESCHINI (1999) *"De la mouche au biorobot: reconstruire pour mieux comprendre"*, in Cerveaux et Machines, V. BLOCH Ed., Hermès, Paris, pp. 247-270.

components of which correspond to around one hundred movement detectors in the fly, was incorporated into the head of a robot. The recorded light signals were transmitted to the moving components of the robot.

A *Photo N. FRANCESCHINI*

B *Photo F. RUFFIER et N. FRANCESCHINI*

C *Photo N. FRANCESCHINI et C. BLANES*

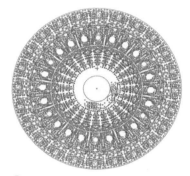

D *Photo N. FRANCESCHINI et C. BLANES*

A - Head of the blowfly, *Calliphora*, seen from the front, showing the two compound eyes with their multifacetted array. Each eye hides 40,000 photoreceptors that drive various image processors based on a few hundred thousand neurons.

B - "Elementary Motion Detector" (EMD) neuron and its evolution over fifteen years: on the left, first generation (1989), using Surface Mounted Device (SMD) technology, compared to a one franc coin from that period; on the right, the 2003 version of the highly-miniaturized hybrid (analog + digital) EMD circuit (mass 0.8 grams), compared with a one euro coin.

C - Autonomous vehicle (12 kg) able to move around in a field of obstacles that it does not know about in advance. Its vision is based on a genuine compound eye, whose circuits are inspired by those of the fly. It includes a network of 114 "motion detecting neurons", transcribed electronically according to the principle analyzed in the fly's eye by means of microelectrodes and a specially-constructed microscope-telescope. This network is arranged around a ring that is about thirty centimeters in diameter. The recently-constructed roboflies, Oscar and Octave, only weigh around one hundred grams.

D - Routing of the electronic components (resistances, condensers, diodes and amplifiers that operate in their thousands) soldered onto the six-layer printed circuit-board that provides the connection between the sensors and the steering motor on board the autonomous mobile robot shown in (**C**).

Figure IV.20 - Neuromimetic robot based on the operation of fly's eye
(construction and photographs by N. FRANCESCHINI *et al.*, Biorobotics Laboratory, Institute of Movement Science [UMR 6233, CNRS & University of the Mediterranean, Marseille], reproduced with permission)

Figure IV.20 illustrates the neuromimetic biorobofly constructed according to this principle. Completely autonomous because of its on-board power supply, this robot was able to move around at high speed (50 cm/s) in a cluttered area, avoiding the obstacles. This first "terrestrial" robofly, which was completed in 1991, was followed by several much lighter brothers and sisters: Fania, Oscar and Octave are **aerial roboflys** [69].

Constructed in 1999, Oscar is a captive robot that weighs around one hundred grams. It is equipped with an eye that reproduces the retinal microscanning of the fly's eye discovered by FRANCESCHINI, Oscar is able to rotate around a vertical axis because of its two diametrically opposed helices and can thus orient its view towards an object. If this object moves, Oscar follows it with its eye, up to an angular speed comparable to the tracking speed of the human eye.

Produced in 2003, Octave is another aerial robofly that is able not only to take off and distinguish a relief, but also to land automatically and to react sensibly to a contrary wind in a turbulent atmosphere. On board, it has an electronic visuomotive self-regulation system, the operation of which is based on the signal processing operations that, in the insect, carry out the automatic pilot functions [70]. The age of biorobotics, in which robots take their inspiration from animals, has only just begun [71]. If specimens are still so rare, this is because behaviors for which we have a good understanding of the underlying neuronal bases are also rare. At the time of writing, a robot rat named Psikharpax, with artificial muscles and a vision system that enables it to perceive objects in three-dimensional space, is being developed at the University of Paris VI.

Almost in the realm of science fiction, we find hybrid robots obtained by **hybridization of the living and the non-living**. This is the case for the hybrid robot produced by Japanese researchers, based on the silkworm moth. Control of the nervous system of this insect is spread throughout its body. If its head is cut off, it continues to fly, which gave rise to the idea of replacing the head with an electronic transistor system [72]. Using a remote measurement device, it was possible to explore certain behavioral aspects of the insect. Although the construction of hybrid robots may raise ethical objections, such technology is capable of giving rise to spectacular applications in the domain of prostheses. The neurological prostheses of the future will nevertheless require that a contact be made between living neurons and the electronic chips that are able to improve the inadequate processing of the physio-

---

69 N. FRANCESCHINI (2003) *"From insect vivion to robot vision. Re-construction as a mode of discovery"*, in Sensors and Sensing in Biology and Engineering, F. BARTH, J.A. HUMPHREY and T. SECOMB Eds, Springer, Berlin, pp. 223-235.
70 The 2003 version of the electronic EMD (Elementary Movement Detector) only weighs 0.8 grams. Its miniaturization was made possible by the use of a micro-controller.
71 N. CONSTANS (2006) *"L'essor des robots insectes"*. La Recherche, vol. 394, pp. 66-73.
72 R. TRIENDL (1999) *"Les biorobots: des insectes à puces"*. Biofutur, vol. 189, pp. 34-37.

IV - CHALLENGES OF EXPERIMENTATION ON LIVING BEINGS AT THE DAWN OF THE 21ST CENTURY   315

logical signal. Such a contact was produced recently in a German laboratory directed by Peter FROMHERZ [73] (b. 1942). A small network of snail neurons, chosen because of their large size, was cultured on the surface of a silicon chip. A signal emitted at one location of the chip was able to be transmitted to another location *via* the synapse connection between two neurons [73] (Figure IV.21).

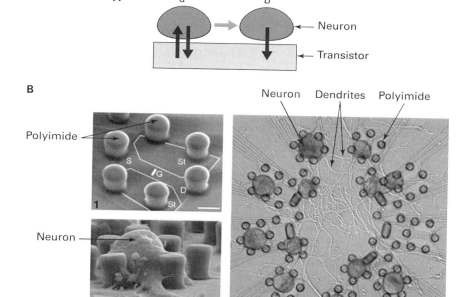

A - Principle of the assembly of a neuroelectric system with transistor/neuron coupling. Affected by stimulation of the transistor, the neuron **a** of the snail brain generates an electrical current that spreads to the neuron **b**, which interacts with the transistor.

B - Experimental production (**1**) and (**2**): electromicrograph showing the immobilization of a neuron cell blocked inside a network of six molecules of polyimide (20 μm white bar). (**3**): electromicrograph showing the dendrite expansions of neurons outside the polyimide barriers after a couple of days of growth (100 μm white bar).

Figure IV.21 - Production of an electrical interaction between snail brain neurons and a transistor
(reproduced from G. ZECK and P. FROMHERZ (2001) "Non invasive neuroelectronic interfacing with synaptically connected snail neurons immobilized on a semiconductor chip", *Proceedings of the National Academy of Sciences, USA*, vol. 98, pp. 10457-10462, National Academy of Sciences, USA, with the permission of P. FROMHERZ and of PNAS ; photograph by Max PLANCK Institut of Biochemistry/Peter FROMHERZ)

---

[73] G. ZECK and P. FROMHERZ (2001) *"Noninvasive neuroelectronic interfacing with synaptically connected snail neurons immobilized on a semiconductor chip"*. *Proceedings of the National Academy of Sciences, USA*, vol. 98, pp. 10457-10462.

On the molecular scale, **mitochondrial ATPase** or **ATP synthase**, with a size of around ten nanometers (Chapter III-6.2.1) was used recently for the manufacture of a biorobot that made its mark in the media as the **smallest known rotating molecular motor**. The membrane-type enzyme catalyzes the reversible reaction ATP + H_2O $\rightleftharpoons$ ADP + Pi. This enzyme therefore has a double function; hydrolysis and synthesis. For this reason it is called ATPase or ATP synthase depending on the physiological context in which it is involved.

In the mitochondria that oxidize metabolites, the enzyme operates like ATP synthase. It catalyzes the synthesis of ATP coupled to oxidation reactions. In the absence of respiration or of oxidizable substrates, the enzyme operates like ATPase; it catalyzes the hydrolysis of ATP. For ease of language, the enzyme will be designated here by the term ATPase. It should be remembered that mitochondrial ATPase includes two sectors, a hydrophobic sector, Fo, characterized as a proton channel located inside the mitochondrial membrane, and a hydrophilic sector, F1, carrying catalytic subunits that are arranged as if they were on a turret (see Figure III.18C). Fo contains two master parts of the **ATPase motor**, i.e., a **rotor** comprising an assembly of around ten so-called "c" subunits and a **stator** that corresponds to the "a" subunit. The "c" subunit assembly is attached to the "γ" subunit of the catalytic sector F1, which thus functions as a **rotor**.

In 1961, the British biochemist Peter MITCHELL (1920 - 1992) showed that phosphorylative oxidation in the mitochondria is associated with a transmembrane transfer of protons. The mechanism involved is said to be **chemiosmotic**. The most important experiment involved an almost serendipitous observation, carried out with a simple pH meter. When a current of oxygen was passed through a suspension of mitochondria in an unbuffered saline medium, in the absence of ADP and phosphate, an instantaneous acidification of the extramitochondrial medium occurred, shown by means of the pH meter electrode immersed in this medium. It was concluded that the sudden switch from anaerobiosis to aerobiosis, i.e., the start up of respiration, is correlated with an ejection of protons from the mitochondrial matrix to the extramitochondrial medium. Afterwards, this fact was linked with several others, the whole leading to the formulation of the chemiosmotic theory. Briefly, mitochondrial respiration generates a vectorial movement of protons from the interior to the exterior of the mitochondrion. Because of this, a proton concentration difference is established on either side of the mitochondrial membrane. The electrical potential that is created in this way is used by the mitochondrial ATPase in order to synthesize ATP from ADP and mineral phosphate. This process involves two correlated events:

- return movement of protons towards the inside of the mitochondrion across the Fo sector of the ATPase;
- rotation of the assembly of c subunits and the γ subunit that is interdependent with it.

We have therefore moved from electrical to mechanical energy. During its rotational movement, the γ subunit establishes contacts with the three catalytic subunits of the F1 sector, in succession. One after the other, each of the three catalytic subunits in contact with the γ subunit undergoes a change in the conformation of its active site, which is at the origin of the synthesis of ATP. In the absence of mitochondrial respiration, the reverse process occurs. The ATP is hydrolyzed into ADP and mineral phosphate, and the energy released at each of the three catalytic subunits is used to rotate the γ subunit in the reverse direction to that which accompanies the synthesis of ATP.

The existence of a rotational movement of mitochondrial ATPase, which had been suggested on the basis of biochemical arguments [74] and of structural data [75] was authenticated by Masasuke YOSHIDA and his co-workers in Japan in 1997, thanks to an imaging technique [76]. In a first step, the molecular system was simplified by being limited to the catalytic F1 sector of the enzyme. A methodological trick was employed: **genetic engineering** was used to modify the α and β subunits of this sector by fixing polyhistidine chains to them. Because of the strong affinity between polyhistidine and nickel ions, the F1 sector α and β subunits were immobilized on a medium covered with nickel ions (carried by an organic molecule). An actin filament labeled with a fluorescent ligand was attached to the end of the F1 sector γ subunit. This assembly made it possible, under a **fluorescence microscope**, to display a **rotational movement of the actin arm carried by the γ subunit affected by the addition of ATP and by its hydrolysis into ADP and mineral phosphate**. A similar rotational movement of the γ subunit carrying a metal microbar was observed by an American research group [77]. Remarkably, in 2004, after having fixed a metal microbead onto the γ subunit, the Japanese researchers [78] demonstrated synthesis of ATP from ADP and mineral phosphate by rotating the γ subunit by means of rotation of the magnetic bead, induced by magnets. Thus, the experimental coupling of a mechanical force and a chemical synthesis was demonstrated. In 2005, the Japanese research team [79] succeeded in photographing

---

74 P.D. BOYER (1997) *"The ATP-synthase – a splendid molecular machine"*. Annual Review of Biochemistry, vol. 66, pp. 707-749.

75 J.P. ABRAHAMS, A.G.W. LESLIE, R. LUTTER and J.E. WALKER (1994) *"Structure at 2.8 Å of F1-ATPase from beef heart mitochondria"*. Nature, vol. 370, pp. 621-628.

76 H. NOJI, R. YASUDA, M. YOSHIDA and K. KINOSITA (1997) *"Direct observation of the rotation of F1-ATPase"*. Nature, vol. 386, pp. 299-302.

77 R.K. SOONG, G.D. BACHAND, H.P. NEVES, A.G. OLKHOVETS, H.G. CRAIGHEAD and C.D. MONTEMAGNO (2000) *"Powering an inorganic nanodevice with a biomolecular motor"*. Science, vol. 290, pp. 1555-1558.

78 H. ITOH, A. TAKAHASHI, K. ADACHI, H. NOJI, R. YASUDA, M. YOSHIDA and K. KINOSITA Jr (2004) *"Mechanically driven ATP synthesis by F1-ATPase"*. Nature, vol. 427, pp. 465-468.

79 H. UENO, T. SUZUKI, K. KINOSITA and M. YOSHIDA (2005) *"ATP-driven stepwise rotation of Fo-F1 ATP synthase"*. Proceedings of the National Academy of Sciences, USA, vol. 102, pp. 1333-1338.

the rotational movement of the enzyme powered by ATP under the microscope, this time looking at the **entire ATPase complex, F1Fo**. After having attached a gold microbead onto the "c" subunits of the Fo sector, to act as a probe, the researchers were able to confirm that the rotational movement of these subunits depended on the hydrolysis of ATP into ADP and phosphate (Figure IV.22). **The whole of the mitochondrial ATPase (ATP synthase) does, in fact, function as a molecular rotational motor** powered by a proton flow, rather like an industrial rotational motor powered by a fossil fuel or electricity. The analogy is a striking one; the γ subunit of the enzyme corresponds to the motor driveshaft and the "c" subunits correspond to the motor itself. Because of its association with non-living structures, for example metal bars or gold beads, which are carried along in the rotational movement of the enzyme, it is possible to speak of **molecular biorobots**. This domain, in which nanomachines use macromolecules from the living world, has only just opened up, but its future is full of promise.

The story of scientific progress made with respect to the mechanisms of phosphorylative oxidation via the functioning of mitochondrial ATPase, from the time of MITCHELL's experiment with the pH meter until the time of the manufacture of YOSHIDA's biorobots, is an exemplary one. It is typical of the way in which a mode of thought evolves over time, from a primary discovery resulting from serendipity or an experiment "to see what happens", leading to the proposal of the existence of a mechanism, to a carefully programmed project which, because of its inventive technicity, shows the validity of the proposed mechanism, and, in addition, demonstrates its future utilitarian value.

Nowadays, certain biotechnologists dream of being able to "synthesize life"[80] in terms of cells that are able to imitate the performance of living cells. The concept of the "Lab-in-a-cell" is coming to the fore[81,82]. Nevertheless, it would be necessary to design an artificial cell that is an authentic replica of a living cell, and which benefits from all the attributes of a living cell. This is not achievable at the moment. Thus, the current aim of nanobiotechnology is limited to scheduling the construction of artificial cells that are relatively simple both in composition and in function, for example, a microvesicle edged with a lipid membrane, containing a system of protein synthesis expressed from a short sequence of DNA, as well as a system of ATP synthesis able to supply the energy necessary for this protein synthesis.

---

80 J.W. SZOSTAK, D.P. BARTEL and P.L. LUISI (2001) *"Synthesizing life"*. Nature, vol. 409, pp. 387-390.

81 A. POHORILLE and D. DEAMER (2002) *"Artificial cells: prospects for biotechnology"*. Trends in Biotechnology, vol 20, pp. 123-128.

82 H. ANDERSSON and A. VAN DEN BERG (2004) *"Microtechnologies and nanotechnologies for single-cell analysis"*. Current opinion in biotechnology, vol. 15, pp. 44-49.

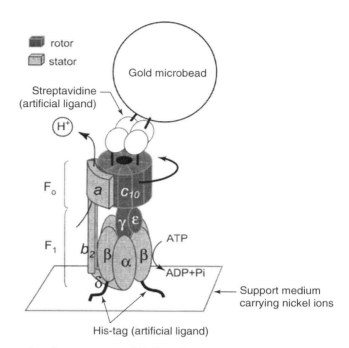

Demonstration of a rotational movement of F1Fo mitochondrial ATPase (ATP synthase), induced by ATP. ATPase or ATP synthase (reversible catalysis enzyme that hydrolyzes or synthesizes ATP) has two sectors (see Figure III.18). The membrane sector, Fo, comprises an assembly of a dozen so-called c (rotor) subunits and an a (stator) subunit. The other, extra-membrane sector, F1, is catalytic. It comprises three β catalytic subunits and three α non-catalytic subunits arranged in a ring, in alternating order. At the center of the ring is the γ subunit which is attached to the c subunits of the Fo sector. Subunits δ, ε and b stabilize the whole of the molecular complex. In the experiment illustrated in this figure, subunits α and β of the F1 sector of the enzyme have been genetically modified to include polyhistidine chains (His-tag, artificial ligand). Due to the interaction of these chains with nickel ions (linked to an organic molecule) covering a solid support medium, the α and β subunits are immobilized. In addition, a gold microbead is fixed onto the ring of the Fo sector c subunits by means of a chemical device (streptavidin molecule, artificial ligand). Following the addition of ATP, rotation of the microbead attached to the ring of Fo sector c subunits is observed by microscopy on a black background. This rotation is dependent on (and at the same time an indicator of) the rotation of the c subunits, itself led by the rotation of the γ subunit in contact with the catalytic β subunits. Note the ejection of protons. When the enzyme functions as ATP synthase, the proton movement takes place in the opposite direction.

Figure IV.22 - Mitochondrial ATPase, rotational nanomachine
(reproduced from H. UENO, T. SUZUKI, K. KINOSITA and M. YOSHIDA (2005) "ATP-driven stepwise rotation of Fo-F1 ATP synthase", *Proceedings of the National Academy of Sciences, USA*, vol. 102, pp. 1333-1338, National Academy of Sciences, USA,
with the permission of M. YOSHIDA and of PNAS)

5. THE DESIGN AND MEANING OF WORDS IN THE EXPERIMENTAL PROCESS

"Progress in biology is possibly mainly tributary to the drawing up of concepts or principles [...]. In the process of elaborating concepts, which marks scientific progress in biology, there is sometimes a crucial step, when we realise that a more-or-less technical term that we had previously considered to cover a given concept, in fact covers a mixture of two (or more) concepts."

Ernst MAYR
Translated from a French Translation entitled
History of Biology. Diversity, evolution and heredity - 1989

On the margins of the modeling and the difference in mathematized systems that comprise theoretical biology, particularly *in silico* biology, **concepts** are mental representations, often image-filled and idealized ones, of fundamental mechanisms that are deduced on the basis of experimental results. From the imaginary domain of the probable, they extrapolate constructions of the mind that are in phase with the facts and experimental data, within a reflective projection that gives them their meaning and makes it possible to make certain predictions.

There are **premonitory concepts**. This was the case for the concept of the **reflex arc** that associates movement with sensation. This concept was already present in the ideas of DESCARTES (Chapter II-3.4), but it took more than a century before the theory of the existence of a reflex arc was supported by BELL and MAGENDIE's demonstration of the existence of relay centers for sensory and motor nerves in the spinal chord (Chapter III-1). There have been premonitory concepts that, while they were demolished at the time they were first proposed, were shown to be completely accurate a few decades later. In the middle of the 19th century, the German pathologist Jacob HENLE (1809 - 1885) needed a healthy dose of imagination and audacity in order to oppose the theory of the "miasma", a theory that was taught as a dogma, with a new theory that not only explained the spread of contagious diseases by microscopic beings, but also formulated the criteria for validating this theory, i.e., isolation of the pathogenic agent and its development in culture away from the diseased organism, then reproduction of the original pathology after injection of the pathogenic agent, which has been isolated, characterized and multiplied in a culture, into a model animal. Thirty years would go by before the formulation of KOCH's postulates, based on experimental evidence (Chapter III-4). We may ask ourselves whether or not history is currently repeating itself in the case of spongiform encephalopathies that affect humans and animals, for which, according to the thesis of Stanley PRUSINER [83] (b. 1942), the **prion**, as an infectious protein, is responsible.

---

[83] S.B. PRUSINER (1998) *"Prions"*. Proceedings of the National Academy of Sciences, USA, vol. 95, pp. 13363-13383.

Other **evocative concepts** hold the keys that open doors to domains that are unknown, but are potentially rich in information. It is thus that the double helix DNA structure proposed by CRICK and WATSON, based on the complementarity of Adenine-Thymine and Cytosine-Guanine bases (Chapter IV-1.1.1), led to the concept of **DNA replication** with reconstruction of a double strand that is identical to the original double strand. The concept of DNA replication spurred on Matthew MESELSON (b. 1930) and Franklin STAHL (b. 1929) to develop an experimental protocol based on the labeling of the DNA nucleotide bases of the enterobacterium *E. coli* with a heavy isotope of Nitrogen, ^{15}N, and on the differentiation of monocatenary DNA strands in the process of synthesis by measurement of their density, as analyzed by centrifugation in cesium chloride gradients. In the same vein, JACOB and MONOD's discovery of regulatory genes (Chapter IV-1.1.1) gave rise to the concept of the **operon** which, in the bacterium, defines a genetic unit comprising structural genes and regulatory genes. The concept of the **regulation of gene expression**, extended to higher eukaryotes, makes it possible to explain the phenomenon of differentiation in cells with specific activities (muscle cells, nerve cells, epithelial cells…) by the silencing of certain genes and the activation of others. Within the framework of bioenergetics, the **chemiosmotic theory** put forward by MITCHELL, in order to explain the coupling of mitochondrial respiration with ATP synthesis (Chapter IV-4.3), gave rise to consideration of the concepts of **transmembrane transport of metabolites** and of **vectorial metabolism**.

Some **generalizing concepts** that carry a **unifying virtue** within them are known. One such is the concept of **compartmentation**. The cell is no longer considered to be a bag of enzymes, as used to be the case. It is now considered to be a compartmented structure in which each type of compartment corresponds to a type of organelle delimited by a membrane and characterized by specific functions. Thus, because of the genetic material that is present in it, the nucleus of the cell holds the information necessary for the manufacture of proteins. The mitochondria, which are called cell power plants, are in charge of oxidizing the products of cell catabolism and using the resulting energy for the synthesis of ATP from ADP and mineral phosphate. The lysosomes are the garbage collectors of the cell. Among the functions carried out by peroxisomes is the partial breakdown of very long chain fatty acids. The endoplasmic reticulum and the GOLGI apparatus are involved in the maturation and the secretion of proteins. The ribosomes represent the machinery upon which messenger RNAs are displayed in order to be decoded into proteins. A sign of the extreme sophistication of this setup is that the membranes of the endocellular compartments are not sealed common walls. They contain proteins that act as selective transporters of metabolites or highly specific ion channels, allowing the exchange of messages throughout the cell. Thus, each organelle, informed of the condition of the others, is able to adjust its own activity to ensure the greatest harmony of the whole. This **conditioned compartmentation** at cell level may be compared to the **socialization** of **human communities**. While endocellular organelles are compartments delimited by membranes, there are **non-**

membrane-bound compartments in the cell, such as protein complexes in which two, three or even more proteins are closely linked. Often, these are enzymes that catalyze reactions that are contiguous in a metabolic pathway. Being **compacted** into a complex known as a metabolon results in an increased **efficiency** of the flow of metabolites by facilitating the channeling of this flow.

Concepts evolve, often adjusting their representations according to accumulated knowledge. A good example of this is the **evolution of the concept of the gene** since its formulation at the beginning of the 20th century. The term "genetics" was created in 1906 by the English naturalist William BATESON (1861 - 1926). The term "gene" was introduced three years later by the Dane Wilhelm JOHANNSEN. This term designated a principle which, in the chromosomes of fertilized egg, and in an intentionally vague manner, was supposed to have an influence on the phenotype of the progeniture. During the same period, the term "locus" appeared out of the experiments carried out by the American Thomas Hunt MORGAN on the drosophila, a locus being defined as a region of a chromosome which, when altered by a mutation, leads to a modification of the phenotype of the living organism. Based on cross-breeding experiments carried out on hundreds of drosophila mutants, MORGAN and his co-workers drew up the first genetic maps. By chance, the salivary glands of the drosophila have a particular characteristic; the nuclei of their cells contain giant chromosomes called polytenes, which result from the association of a hundred replicate copies of chromosomes that, after staining, are visible under the optical microscope. On these chromosomes, it is possible to distinguish colored bands separated by clear bands. It was observed that specific mutations had specific effects on the arrangement and number of these bands. The material contained in the bands was therefore the site of mutations. In the middle of the 1930s, the listing of more than 3 500 bands made it possible to construct a **cytological map** that was already highly detailed. The **concept of the gene**, the material basis of inheritance, took root. The sporadic mutagenic effect of X-radiation in the drosophila, which was shown by the geneticist and biophysicist Hermann MÜLLER (1890 - 1967), led the Austrian physicist Erwin SCHRÖDINGER to question the **sporadic event** which, at the level of a target of a few dozen atoms, determines a mutation. He postulated that the target is located in the chromatin of the chromosomes, organized as an aperiodic crystal. The chemical nature of this target was identified with DNA, following bacterial transformation experiments (AVERY, MACLEOD and MCCARTHY, 1944) and experiments concerning bacterial infection by the bacteriophage (HERSHEY and CHASE, 1952) (Chapter IV-1.1.1). This is how the idea that **gene = DNA** was born.

The simple and reassuring idea that **one gene → one enzyme**, which was deduced from mutation experiments carried out by BEADLE and TATUM on the mold *Neurospora crassa* (Chapter III-6.1), had only a limited lifetime. A first stumbling block appeared when it was shown that the activity of a gene, and in consequence its contribution to the phenotype, depends on nucleic elements outside the gene. The definition of the term "gene" was then extended to include promoting and regulatory sequences. In the case of the *lac* operon of *Escherichia coli*, these sequences are located just upstream of the site where transcription begins. However, in eukaryotes, a regulatory sequence may be distant from the gene that must be transcribed and sometimes it may be involved in the regulation of several genes (Chapter IV-1.1.1). In the 1970s, the idea of the existence of the **mosaic gene** in eukaryotes appeared. A gene was now thought of as an assembly of several exons that originally in the chromosome are separated by introns. The **alternative splicing** of these pieces of genes gives rise to numerous possibilities for reconstitution, i.e., many messages coding for many different proteins. Thus, however useful the concept of the gene has been with respect to its ability to generate discussion and to provoke experimentation concerning the molecular machinery responsible for the transmission of the hereditary characteristics, we can see that the term itself has not ceased to be the subject of readjustments, since the time it was first formulated.

Certain concepts are matched with metaphors. While some **metaphorical concepts**, particularly those that make use of images designed to grab the imagination, and to be easy to understand, tend to take liberties with the realities of living beings, they can also shed light on unsuspected mechanisms in sectors that have been neglected. Metaphorical concepts are not a current fashion. It should be remembered that in his *Passions of the Soul* (1649), DESCARTES, when asked "how limbs can be moved by objects of the senses and by the mind without the help of the soul," responds that this takes place "in the same way as the movement of a watch is produced only by the force of its spring and the arrangement of its cogs." Later on, with LAVOISIER's clear vision of the vital role of oxygen, and his comparison of respiration with combustion, the concept of the **chemistry of life**, combined with that of **bioenergetics**, came to the fore, and was at the heart of studies on the metabolism. Chemical reactions that liberate and absorb heat were substituted for the cogs of Cartesian mechanics. The second half of the 20th century saw the birth and development of the concept of the **program**, a concept with computer technology connotations, which was destined to explain the phenomena of inheritance. This concept began to fill out from the moment when it became certain that, in its nucleotide sequence, DNA contains the necessary information for the construction of the protein material of cells. For a certain period of time, the passion for molecular genetics eclipsed the interest that had previously been given to metabolic chemistry. The powerfulness of the metaphorical concept may be measured according to the effect it has in pushing scientific research in particular directions, with the results this has on society. Thus, during the 17th and 18th centuries, the study of human pathology was impregnated with a strong **iatromechanical current**. **Physiological chemistry** and its corollary, **pathological chemistry**, which emerged as disciplines in their own right in the 19th century and achieved full expansion in the 20th century, are our inheritance from LAVOISIER and the concept of the **chemistry of life**. At the turn of the 21st century, taking the concept of the **program** as a basis, the fantastic advances made in the deciphering of genomes, the comparison of their structures and the listing of their anomalies, have opened up new horizons in domains as varied as the developmental and evolutionary sciences and the pathology of hereditary diseases.

Discussion about concepts *necessarily* leads to a brief discussion of **scientific semantics**, as shown by the few examples given in the previous pages. As we have just seen, the word **gene** that was put forward by JOHANNSEN around one century ago did not have the same meaning at that time as it has now, a meaning that still remains fluid. The **GMO**, an acronym meaning Genetically Modified Organism, which has been the subject of vehement diatribes over the last few years, becomes much less of an object of passion if it is considered within the context of evolution. After all, for the last two to three billion years, living organisms have been genetically modified constantly by spontaneous mutations, which is why the human beings that we are today are able to discuss them!

The term **cloning** is another example of a semantic misunderstanding that leads to inaccurate interpretation and arouses the passions. The primary meaning of the term cloning is the multiplication and the identical reproduction of a living cell. The simplest and most unambiguous example is that of bacterial cloning, a bacterial cell producing millions of cells that are identical to the original cell by its multiplication in a nutritive medium. The term **animal reproductive cloning** does not carry exactly the same semantic weight. It should be remembered that, in eukaryotes, the preliminary act of the cloning procedure involves the injection of the nucleus (with 2n chromosomes) from a somatic cell into an enucleated oocyte (Chapter IV-2.3.2; see also Chapter IV-6.1). The somatic cell nucleus, by providing its genetic equipment, gives the being that will develop in the uterus a phenotype that is practically identical to that of the somatic cell donor, but nevertheless not completely identical, as the cytoplasm of the enucleated ovum, with its mitochondria, provides a small but non-negligible fraction of genes, the mitochondrial genes. As for therapeutic cloning (in the absence of uterine implantation), this is used for the manufacture of differentiated cells that may be grafted into the individual who has donated the original somatic cell, with no immune-related rejection occurring. This is **non-reproductive cloning**. The passionate argument that has arisen because the term cloning is bandied about in an ill-considered fashion illustrates the confusion that can result from a lack of precision in the use of certain terms with a high level of media impact.

6. *THE EXPERIMENTAL METHOD, UNDERSTANDING OF LIVING BEINGS AND SOCIETY*

"All the major problems of the relations between society and science lie in the same area. When the scientist is told that he must be more responsible for his effects on society, it is the applications of science that are referred to [...]. No government has the right to decide on

the truth of scientific principles, nor to prescribe in any way the character of the questions investigated."

Richard FEYNMAN
The meaning of it all - 1963

The progress of science is linked to that of **civilization**. It is in keeping with the state of mind, the beliefs, the lifestyle and the thought patterns of societies. In Ancient Greece, where manual work was considered to be servile, science remained essentially theoretical, confined to logic and dialectics, and strongly attached to questions of philosophy. The birth of experimental science in the 16th and 17th centuries went hand-in-hand with the rehabilitation of **manual work**.

The technical side dominates in modern biology, which seeks to solve problems concerning the "how", rather than to address philosophical problems concerning the "why". As Ian HACKING (b. 1936) says in *Representing and Intervening* (1983), nowadays engineering, and not theorizing, is the greatest proof of **scientific realism**, which leads to the minimization of philosophical thought. In *A Skeptical Biochemist* (1992), the Polish-born American biochemist Joseph FRUTON (1912 - 2007) emphasizes the contrast between the 19th century and the first half of the 20th century, when eminent scientists were still interested in the ideas of the professional philosophers of the history of the sciences concerning the progress of experimental research and, in contrast, the end of the 20th century, when philosophy and the experimental sciences pretended to ignore one another. This is doubtless partly because the history of biology has become the history of **biotechnologies** to such an extent that, according to some, the objects being explored are so familiar that they are now part of the life of society. In *Pandora's Hope* (1999), Bruno LATOUR (b. 1947) considers that the current confrontation between subject and object, in which the researcher-subject explores the structure and function of the object, is being transformed into a human-nonhuman dialogue, in which the non-human-object becomes "socialized". Taking yeast as an example, LATOUR writes that it has been "working for millenia in the brewing industry, but now it works in a network of thirty laboratories where its genome is mapped, humanized and socialized like a code, a book, or a program of action that is compatible with our ways of coding, counting and reading [...]. Non-humans have become automatons, admittedly without rights, but much more complex than material entities." LATOUR visualizes the human-nonhuman associations in the form of collectives that are organized into strata that implement the technical, the political, the social, the ethical, the ecological... The technosciences correspond to one of these strata, the sociotechnical stratum that is directly linked to the stratum of political ecology. In the same spirit, the Belgian philosopher Gilbert HOTTOIS (b. 1946), in his *Philosophies of the Sciences, Philosophies of techniques* (2004) remarks that "laboratories produce things that go off to live their lives in society and in Nature." Thus, bacteria, yeasts or genetically modified plants are able to produce drugs such as insulin, growth hormone and vaccines for human medicine. These drugs become part of and indispensable to life in society. They are evaluated according to their market value by the companies that patent, manufacture and sell them, and according to the comfort they bring to the patients to whom they are administered. The financial management that results from their consumption becomes a worry for those responsible for public health, while their manufacture by specialized companies generates industrial activity and economic growth which may be measured according to how fashionable they are and how they sell.

For a long time, **society**, while benefiting from scientific progress, remained indifferent to the experimental method, that is to say, the way in which knowledge progresses. In the last decades of the 20th century, society became aware, via information concerning the occasionally demonized exploits of genetic engineering, that science can "take liberties" with the human being. Populations were well informed about the effects that genetic engineering could have on the mortality rates of pathologies such as cancer and diabetes or on degenerative illnesses of the nervous system, and about the closeness of possible solutions. However, they were also warned about the risks to which science was exposing humankind. Remembering certain tragic episodes concerning HIV-contaminated blood transfusions, growth hormone and mad cow disease, and certain Cassandra-like predictions, such as a catastrophic epidemic of spongiform encephalopathy that has happily yet to appear, society shows reservations when the media inform its members of new feats of modern technology. Political authorities, for their part, afraid of potential problems, tend to follow the **principle of precaution, which in fact hides a fear of risk**. However, evaluating risk involves not being afraid of it but understanding it in a lucid and courageous fashion. Informed by the media, which often use sensationalism, the citizen is increasingly calling into question whether certain practices involving the biosciences, such as cloning, or certain mercantile transactions such as the taking out of patents concerning gene sequences, or even experimentation on live animals, are well-founded.

6.1. HUMAN CLONING CENSURED BY CODES OF BIOETHICS

"The problem of experimentation on Man is no longer a simple problem of technique. It is a problem of value. From the moment that biology concerns Man no longer simply as a problem, but as instrumental to the search for solutions concerning him, the question arises of deciding whether the price of knowledge is such that the subject of the knowledge is able to consent to become the object of his or her own knowledge. We have no difficulty here in recognizing the still open debate concerning Man as a means or an end; an object or a person. This is to say that Human Biology does not contain in and of itself the answer to questions concerning its nature and its significance."

Georges CANGUILHEM
Knowledge of Life - 1965

Written at a time when people were far from imagining how molecular biology was going to expand, the prophetic words of Georges CANGUILHEM (1904 - 1995) have maintained their philosophical validity. Manipulation of the human embryo, whether this involves its creation by cloning or the modification of its genetic inheritance, obviously leads to the need to consider the societal, religious and political points that arise from the domain of bioethics and are a reflection of the period in which we are living. Until recently, advances made in biology left moralists indifferent. This ceased to be the case when scientific experimentation began to look at the human embryo with a view to utilitarian ends in the health domain. The specter of cloning was brandished without any clear distinction being made between reproductive cloning and therapeutic cloning. Biology became demonized. However, as biologist Pierre CHAMBON (b. 1931) said in an interview in the French journal *Biofutur*: "In absolute terms, biology is unable to tell us whether the cloning of a human being is moral or immoral, it simply tells us whether it is biologically possible."

The birth Dolly the sheep in 1997 (Chapter IV-2.3.2) triggered a virulent debate because now that the cloning of an animal had become possible, that of a human being became envisageable. The media sensationalized this debate all the more in that it was exacerbated by debate concerning GMOs (Chapter IV-2.1). The Dolly affair became a **problem of society**. Up until then, the biosciences had been happy just to try and understand the mechanisms that explained the functions of living beings, but now, with the advent of GMOs and cloning, it became obvious that a forbidden barrier had been crossed and that Man had the power not only to transform but also invent himself. Faced with this **desacralisation of Nature**, the need arose for some philosophical reflection. This was given the name of **bioethics**, which is the title of the book, *Bioethics, A bridge to the future*, which was written by the American biologist Van Rensselaer POTTER (1911 - 2001) in 1971.

The term bioethics covers philosophical considerations that range from the biosphere to the human person. Bioethics tries to give a wider meaning to the moral codes which, in human societies, depend on ancestral traditions. It aims to prescribe that which is desirable according to the Kantian maxim of the categorical imperative. In his *What is Bioethics?* (2004), the Belgian historian Gilbert HOTTOIS reminds us that bioethics are above traditional morals, the latter being a set of norms that are most often spontaneously respected as good habits, without any critical reflection being involved, while bioethics, on the other hand, arises out of critical thought, analysis, discussion and the evaluation of established mores. Over the last few years, the problems that are targeted by bioethics have moved towards today's burning issues. Human cloning is an example.

While allegations of the transcendence of Man in Nature may lead to **human reproductive cloning** being considered as a crime, strictly scientific considerations lead to an emphasis on the lack of responsibility shown by a few zealots, given the hazards involved in cloning in animals, such as the need to use a large number of oocytes in order to achieve success in cloning, the very low viability of the cloned embryos and the development of serious functional anomalies in the clones that survive. Even supposing that scientific progress will one day overcome these difficulties, human reproductive cloning will come up against an insurmountable obstacle, the cloned subject's fear of finding that he or she is identical to the relative from whom his or her genetic inheritance comes. After all, the notion of manipulation of the human ovule with the aim of serial reproduction has often haunted science fiction stories. In *Brave New World* (1932), Aldous HUXLEY (1894 - 1963) gives an apocalyptic vision of the budding of human eggs that produce hundreds of identical twins which are conditioned into classes and subclasses while being raised in jars, depending on the quality of the nutritive substances they are given. In *The Artificial Uterus* (2005), Henri ATLAN (b. 1931) predicts that the raising of human fetuses in jars could well become an alternative to uterine gestation in a distant future. Let it be understood that human reproductive cloning, which is no longer part of the domain of science fiction, as it has become feasible, must be considered as being reprehensible because it goes beyond the limits of reason, and is a denial of human transcendence. Man as subject cannot be considered as an object.

The problem of **therapeutic cloning** is quite different, although it leads to reticence and prohibition because the demarcation between therapeutic and reproductive cloning depends mainly on whether a cloned embryo is implanted in a uterus. While, at the time of writing, therapeutic cloning has been prohibited in France, Germany and other countries, it is tolerated in Great Britain. In the USA, the prohibition only applies to publicly-financed researched, while each state has its own legislation, which is relatively flexible.

The objective of therapeutic cloning is to provide patients with tissues that arise from their own selves, and are therefore **immunocompatible** and able to be grafted without there being any risk of rejection (Chapter IV-2.3.2). It is based on the removal of somatic cells from the subject to receive the graft and the transfer of the nuclei of these cells into enucleated oocytes. The stem cells that are obtained after the first divisions are stimulated using appropriate growth factors. Depending on the factor used, the stem cells differentiate to form a type of tissue (hepatic, muscular, nerve...) that can be used as a graft. Such a procedure may be envisaged for patients who have suffered a serious, invalidating trauma, for example section of the spinal chord. A graft of immunocompatible nerve cells might make it possible to re-establish nerve continuity. A similar type of therapy has been considered for PARKINSON's disease, the cause of which is a degenerescence of certain cells of the encephalon (Chapter IV-2.3.1). Given the hopes that are raised by the possibility of such therapies, and the fact that, after all, such therapeutic cloning is the equivalent to an autograft, even if the ways in which the graft is obtained are slightly tortuous, the demonization and rejection of such practices should be reconsidered, calmly and coolly.

Another option for therapeutic cloning is the correction of mutations identified in the **mitochondrial genome** of a woman wishing to have children. It is, in fact, the mother's ovum that provides the fertilized egg with its complement of mitochondria that are indispensable for its viability. The manipulation involves inserting the nucleus of a fertilized ovum from the mother, obtained by artificial insemination, into an enucleated oocyte taken from a woman who is not suffering from the mitochondrial defect. The cytoplasm of the enucleated oocyte provides the stock of functional mitochondria that are indispensable to normal cell function in the future embryo.

In a domain of the bioethics, in which rational objectivity comes up against deliberately technophobic religious and cultural considerations, it is useful to remember certain legal and legislative paradoxes. Thus, in France, after having been considered to be a criminal act that was subjected to severe repression by the law up until 1975, the right to have an abortion before the end of the third month of pregnancy became not only authorized but also protected by law. It is interesting to note that in the 13th century, THOMAS AQUINAS, The Father of the Church, had acknowledged that a fetus only becomes "animated" by the implantation of the soul by Holy Will in the third month after fertilization. Another subject to be considered is **pre-implantation genetic diagnosis (PGD)**, in which human embryos that have been fertilized *in vitro* are sorted in order to find those that are without defects, a practice which is on the verge of being a deviation in the direction of eugenics. Nevertheless, PGD is the basis of a practice that is either already legalized or is in the process of being so in several European countries, the creation of so-called **designer babies**. A typical example is that of a designer baby arising from an embryo whose immune profile to that of an older sibling who is suffering from leukemia. In this case, there is good reason to hope that a graft of immunocompatible blood cells from the designer baby into the sibling who is suffering from leukemia will save the latter from death.

Out of the disharmony of opinions that arising from cultural tradition, religious conviction or simply scientific pragmatism, the American biologist and philosopher H. Tristram ENGELHARDT (b. 1941), in *The Foundations of Bioethics* (1996) proposes a **lay bioethics** that is based upon the **principle of permission**. Lay bioethics advocates tolerance while admitting that this tolerance in no way prevents anyone from taking up a personal position; it means that each human being has a moral sensitivity as well as the ability to reason and to choose within a defined limit of non-harmfulness and of justice. The individual is free to modify his or her destiny, or to manipulate his or her nature by genetic interventions because, adds ENGELHARDT, "there is no lay moral foundation to prohibit such an intervention."

6.2. THE PATENTABILITY OF LIVING BEINGS

When a researcher or the research organization that the researcher belongs to files for a patent for an invention with a patent office, it is necessary to demonstrate the novel and utilitarian nature of this invention. If a patent is accepted, this gives the person or body that filed it the exclusive right to make use of the invention over a pre-determined period of time, generally 20 years, which is a means of protection, or, if desired, to allow others to make use of the invention by issuing a license to do so. In the domain of living beings, there has sometimes been confusion between invention and discovery. In 1991, Craig VENTER, known for his participation in the sequencing of the human genome, filed a demand for a patent covering the sequences of 2 700 fragments of recombinant DNA (cDNA) called EST (Expressed Sequence Tags) that are obtained by reverse transcription from human brain messenger RNAs, in the name of the NIH (National Institutes of Health) at Bethesda (USA). The patent specified that ESTs could be used as probes to characterize genes that are potentially involved in neurological ailments. The resulting outcry led the NIH to withdraw its patent demand.

In fact, the patenting of living beings has a long history that goes back to the patent that was filed in 1895 in France by Louis PASTEUR, and then in 1873 in the USA, for the use, in brewing, of a yeast culture that was free from pathogenic bacteria. From this historical perspective, the case of Ananda CHAKRABARTY (b. 1938) set a legal precedent. In 1994, CHAKRABARTY filed a demand with the US patent office for a patent relating to a *Pseudomonas* type bacterium which, by genetic modification, had acquired the ability to digest crude oil. His demand was refused. After an appeal and many legal battles, the United States Supreme Court overturned the patent demand refusal, the basis of the judgement being that any modified micro-organism is a product of human ingenuity and has a specific name, characteristics and use.

Thus, from 1980 onwards, the arrival of an era of patents derived from genetic engineering was indicative of how this discipline was growing. In December of that year, Stanley COHEN and Herbert BOYER, acting on behalf of the University of Stanford, patented a nucleic chimera comprising a recombinant DNA carried by a vector. In 1982, a patent concerning the growth hormone gene was awarded to the University of San Francisco. In 1984, the University of California at Berkeley obtained a patent for the human insulin gene. In 1985, the American company *Pioneer Hi-Bred* succeeded in patenting a variety of corn in which genetic modification has led to an increased synthesis of tryptophan, an amino acid that is indispensable for animal feed. In 1988, the Genentech company acquired a patent for the gene coding for human gamma interferon. This was followed in Japan by a patent for the gene coding for beta interferon. In the same year Harvard University patented the OncoMouse, a transgenic mouse whose susceptibility to cancer is greatly increased. After this, several species of transgenic animals were patented for utilitarian purposes, such as the production of human alpha-1-antitrypsin taken from the milk of transgenic goats and used for the treatment of cystic fibrosis. The frenetic patenting of living beings has reached the domain of natural products arising from the plant world in tropical regions, the immensely varied essences arising from these plants being full of pharmacological potential. The potential for producing drugs of a considerable commercial value from such plants is very high. Here we return to the problem of the patenting of genetically modified, cultivatable plants (GMPs) (Chapter IV-2.1). Thus, the experimental method, the principle of which is to acquire pure knowledge, finds itself led astray in its applications. Whatever the motives that are given, particularly for manipulations that give rise to the manufacture of marketable products, the patenting of genomes for mercantile ends shows the regrettable, but unfortunately inevitable, direction in which the very spirit of a science, molecular biology, which half a century ago wished to be at the heart of an understanding of living beings, has drifted.

6.3. ANIMAL EXPERIMENTATION VERSUS THE FIGHT FOR ANIMAL RIGHTS

The suffering of animals that are being experimented upon gives rise to a moral problem. The end of the 19th century saw large-scale demonstrations against **vivisection** and repeated demands for it to be abolished. Today, there is renewed vigor in the call for the **abolition of vivisection**, without any real coherent basis. This desire to stop experimentation on animals ignores the imperatives of contemporary medicine, which must meet the challenge of pathologies whose increasing incidence is worrying, such as cardiovascular diseases, diabetes, cancer, and the degenerative illnesses that are linked with aging or are of genetic origin. It is true that animal experimentation inevitably leads to questions. Are the stakes involved in a particular experiment, in terms of the acquisition of new knowledge, worth the suffering of an animal used in that experiment? Is it not necessary to ensure that the experimental protocol is well-documented, that it is not redundant, or even that it has been the subject of previous studies carried out on cells in culture? It is easy to see the size of the methodological chasm that separates contemporary physiology from that of the time of Claude BERNARD, when cell culture techniques were not yet being used, when the main instrument used was the scalpel and the researcher, using his or her imagination and creativity, had to develop specific protocols that were able to validate or refute a working hypothesis. **Each period in history operates in its own way according to its moral laws and its technical capabilities.** The bloody operations carried out by MAGENDIE and by Claude BERNARD in the 19th century, which were tolerated at this time despite criticisms from antivivisectionists, would not be permitted today. Nevertheless, it is true that the physiologists of the 19th century, by means of the results of their experiments, wove a tapestry of new knowledge on which contemporary biologists were going to work and without which the level of understanding the modern science would be much lower than it is.

Animal experimentation remains indispensable in many areas of physiological investigation, in genomics, in toxicology and in pharmacology. It is a **precondition for clinical trials of any new drug**, being used to test for the drug's efficacy, its metabolism and any toxicity. However, not all data arising from animal experimentation can be extrapolated to man. The margin of uncertainty can be reduced by means of comparative trials on several animal species. Because of their phylogenetic proximity to man, primates may seem to be the solution for experimentation prior to the application of a drug in Man. This was the case for the development of a vaccine against hepatitis B. It has been proposed the grafting of stem cells in Man should be preceded by experimentation in apes, in order to ensure the absence of tumorization over the long term. However, the researcher is confronted with a dilemma: should he or she ensure the safety of Man with respect to possible deleterious effects or respond to ethical demands that recognize the very great genomic similarities between Man and the chimpanzee.

A consideration of cloning, patenting and animal experimentation practices illustrates the excesses of the experimental method in domains where political authorities consider themselves able to legislate. Administrative decisions, often made in the absence of any dialogue with scientific authorities, can have serious consequences. Thus, given the pretext of strict obedience to the principles of bioethics, which are a matter of tradition, and while certainly respectable, are nevertheless arguable, and also given the pretext of a sickly and unconsidered fear of the risk involved in certain experimental practices, and the absence of an intelligent evaluation of this risk, research, which until recently took place in a motivating atmosphere of liberty, may, over the long term, be weighed down with a highly prejudicial handicap and a limitless sense of discouragement.

7. *THE PLACE OF THE SCIENTIFIC RESEARCHER IN THE CHANGING ROLE OF BIOTECHNOLOGY*

In the 17th and 18th centuries, experimental research, which was still in an emergent phase, was mainly artisanal, and in the hands of rare scholars. It took form during the 19th century in the West, particularly actively in Germany, and became operational in the 20th century, under the aegis of governmental authorities, with the creation of Institutes, the programmed recruitment of researchers and the allocation of renewable budgets. Modern science, based on the principles of the experimental method, came to the fore much later in the East than in the West. The globalization of knowledge has meant that at present experimental science, in all domains, including that of the life sciences, has spread throughout the world, with even those countries that had become relatively backward in these domains because of their isolation catching up rapidly. Nevertheless, it is true that the progress of the experimental sciences in the USA and in the United Kingdom has been distinguished by pragmatic management of these countries' science policies, based on the excellence and the high degree of autonomy of their Universities and Research Institutes with respect to recruitment and choice of subjects of study. The efficacy of this policy in the life sciences may be judged by the number of researchers who have won Nobel prizes since the second world war (at the time of writing, more than 80 in the USA and twenty or so in Great Britain as opposed to only 4 in France).

In France, research on living beings is carried out in the laboratories of Universities, in Institutes connected with Higher Education and in laboratories that are run by large organizations such as the National Scientific Research Center (CNRS), the National Institute of Health and Medical Research (INSERM), the National Institute of Agronomic Research (INRA), the Atomic Energy Commission (CEA), the National Institute of Research in Computer Processing and Automation (INRIA), the National Center for Space Studies (CNES) and the French Institute of Research on the Seas and Oceans (IFREMER). Equivalent bodies exist in countries other than France, some of them being institutes that are dedicated solely to research, and some being university laboratories that associate research and teaching. At the beginning of the 20th century, the function of researcher was most often associated with that of a professor occupying a chair at a University, surrounded by a few assistants, the professor directing the research work in his area of specialization. Now, within a period of a few decades, the status of researcher has been modified greatly. Today we talk of **research careers** classified according to level of expertise and technicality. Management, or the supervision of career paths and the control of financing, is carried out by an administration that is itself highly hierarchical. The **scientific process** has undergone a **metamorphosis**, shown by changes in the behavior of researchers not only within the institutions in which they work but also in their relationships with the media, the political sphere and society. The teaching of the life sciences needs to take this into account.

7.1. FUNDAMENTAL RESEARCH FACED WITH THE METAMORPHOSIS IN THE EXPERIMENTAL METHOD

"Long ago, there was a time when scientists recounted the exact circumstances of their discoveries, without shame, even when their recital showed up the fragility of their forecasts or an indecent collaboration on the part of every bit of luck. Such times are past, and the researchers of today often like to make us believe that they only find what they are looking for. The thousands of pages PASTEUR's lab books provide an opportune reminder to us (and to program-makers or impatient users) that it is just as difficult to ask a question as to answer it, that a scientific discovery often occurs after a long, winding path, that rather than following the fashion, it is preferable to follow one's ideas, particularly if they are good ones, and are in advance of the fashion."

Jean JACQUES
Molecular Dissymmetry, *in "PASTEUR, Workbooks of a Scholar"* - 1995

Current technological progress, the accumulation of the scientific knowledge, the institutionalization of the public research and many other factors are disrupting a ritual of the experimental process that had survived until the middle of the

20th century, and even beyond. The experimental life sciences of the 21st century will necessarily see themselves remodeled with respect to their objectives and procedures.

7.1.1. A new strategy in the organization of research

Faced as it is by an increasingly tough international competition, the scientific community is also subject to restrictions in terms of operation and prospectives. An organization into small teams of a few researchers gathered around a boss, working in **friendly interaction**, is increasingly giving way to **large groupings** that sometimes seem like consortiums. Focused on research subjects that are deemed to be "cost-effective", these **superstructures** are encouraged, or even imposed, in the sadly illusive hope that the will lead to greater efficacy. The person in charge of such large groups is taken up with everyday management tasks and by maintaining good relations with the administrative bodies on which his or her organization's survival depends. He or she may become distanced from the experimentation and forget the intellectual motivations that in the past caused his or her competence to be recognized. It should be emphasized that the secret of future successes lies in situations where young researchers are in direct contact with their bosses, and where friendly interaction with a **known master** teaches the apprentice researcher how to learn, how to think and how to experiment in a critical fashion.

Preoccupied by the rapid expansion of the scientific population, accompanied by the creation of laboratories whose operation necessarily requires financing, often on a large scale, political authorities, giving way to the requirements of media-fed public opinion, are interfering more and more, via administrative relays, in the control of the objectives of experimental research. **Short-term objectives**, considered to be "visible", are favored. *A priori*, the viability of a project is judged according to the scientific context of the period and its **impact on society**, insofar as the project looks at health problems with a high degree of media coverage (cancer, degenerative illnesses, viral infections…) and often in agreement with a consensus that avoids going against the orthodoxy of the moment. This leads to a rigid management of projects that are financed and controlled according to objectives that have been fixed in advance, and that are all the more easily accepted by state authorities when they are somewhat fantastic in character. However, fundamental research proceeds from a **playful activity**, and for this reason, its efficacy is dependent on the passion of the researcher for the problem that he or she is studying. In contrast to what is believed by the narrow-minded, the effectiveness of a researcher in terms of discoveries depends upon the liberty that is given to this researcher, assuming, of course, that this liberty is underpinned by criteria of confidence such as the researcher's scientific past, his or her motivation, and judgements made concerning the researcher by impartial peers. It should not be forgotten that the determination of the three-dimensional structure of hemoglobin by Max PERUTZ (Chapter III-6.2.1) took around twenty years of solitary, uninterrupted and untiring labor. The theoretical and technical tricks that led to this success helped to open up the domain of the structures of giant macromolecules, several dozen kilodaltons in size, which no-one had dared study before.

Anyone who uses the experimental method realizes that while fundamental research must be organized, **it cannot be scheduled**. Such a person knows that the pathways to discovery are convoluted, and that an inexplicable observation that appears unexpectedly during an experiment can sometimes, if the researcher is sufficiently perspicacious, be the beginning of an adventure that leads to a discovery. It was to just such a convoluted path that the Belgian biologist Christian DE DUVE (b. 1917) alluded in his speech when he received the NOBEL prize for Medicine and Physiology in 1974. After working at the University of Saint Louis in the USA, DE DUVE, who had taken up a post at the University of Louvain, Belgium, decided to look at a research theme that had received a great deal of media coverage, diabetes and insulin. It was while operating on one of the subcellular fractions obtained from ground rat's liver, and analyzing certain of the enzyme activities of these fractions, that he was surprised to find, in one of them, enriched with mitochondria, a phosphatase activity that, paradoxically, increased with time, while the enzyme activities specific to the mitochondria declined. This was an activity belonging to organelles that were contaminating the mitochondria. Dropping all research on diabetes, DE DUVE set out to identify and characterize these unknown organelles. He discovered that they were involved in the breakdown (lysis) of molecules that are undesired by the cell and, for this reason, he called them **lysosomes**. The discovery of lysosomes helped to open a new chapter in cell biology and to attribute a molecular cause to diseases with serious prognoses whose etiology had remained a mystery up until then. These diseases were given the label **lysosomal diseases**. These diseases result from the absence of a lysosomal enzyme that is responsible for the breakdown of a given metabolite. The accumulation of this non-broken-down metabolite in the lysosomes leads to cell malfunction, which causes the lysosomal disease. As DE DUVE said jokingly, if he had carefully followed the experimental process laid down in his diabetes research project, and if he had not given way to the temptation of **"playing hooky"** or **"playing truant"** he would never have mounted the podium in Stockholm. In the same way, Henri-Gèry HERS [84] (1923 - 2008), a cell pathologist at the internationally renowned Louvain school, remarked in an article published in the review *Médecine/Sciences*: "I believe we would obtain maximum value for the money devoted to research if we were willing to distribute it to those who have been shown to be productive, according to their needs, and without asking them for a program." HERS concluded, in a tone that was deliberately playful, but thought-provoking, "such a simple system would lead to unemployment for a large number of administrators, which is why I suspect that it will never be adopted."

---

84 H.G. HERS (1991) *"Réflexion sur la recherche médicale"*. *Médecine/Sciences*, vol. 7, pp. 169-171.

Research has its own set of ethics, driven by **anticonformity** and the **creative imagination**, capable of shaking up firmly-anchored ways of thinking and established hierarchies, and of leaving the researcher the freedom to express him or herself and to experiment off the beaten paths. As ECCLES says in *Evolution of the brain and creation of the conscience*, it is important to distinguish between **intelligence** and **imagination**. Intelligence is measured according to the rapidity and depth of understanding and clearness of expression. It may be measured and even given a numerical value. The same is not true for the imagination, a more subtle, unmeasurable phenomenon that cannot be learned. The imagination is one of the levers that is able to lift the boulder that hides scientific truth. The imagination is the ultimate weapon of research, which shakes up the knowledge acquired by the intelligence. Nevertheless, the imagination must be tempered by a good critical sense that is able to perceive potential sources of artifacts, both in sophisticated instruments that act as so many black boxes from which already manufactured information emerges and in genetic or chemical cell exploration methods whose specificity must be carefully checked.

The benefits that can sometimes be gained from prospective research that is far from dogma that is rooted in sterilizing tradition, the way in which knowledge progresses, most often by moving away from any orthodoxy, the way discoveries appear unexpectedly on the fringes of carefully put together projects, all of these points are **matters for reflection** for those in power in the worlds of politics, economics and industry.

7.1.2. A new way of circulating knowledge

Publication is an essential tool for communicating scientific knowledge, and is the judgement criterion for committees in charge of evaluating the creativity of a researcher. In order to have meaning, a publication must provide information that is sufficiently innovative with respect to parallel work carried out in other laboratories. Here again, **media coverage** has quietly infiltrated the scene. Its role is all the more perverse in that the rating of a publication is estimated according to its **impact index**, or, roughly speaking, the renown of the scientific journal in which it is published. Curiously, it has happened that articles that would later be considered to be of primary importance have been rejected by highly prestigious journals, simply because the facts mentioned in the article and the conclusions made have not coincided with the orthodox opinions of the period and the traditionalist spirit of the journal's editorial committee. This was the case for an article which the biochemist Hans KREBS (1900 - 1980) submitted to the British journal *Nature* in 1937. In this article KREBS described a series of experiments showing that an endocellular metabolite, pyruvate, product of glycolysis, is completely degraded during a cycle of enzyme reactions. This degradation cycle would later be recognized as the central pivot of the intermediate metabolism. Called upon to judge revolutionary scientific considerations, and unable to perceive their importance, *Nature*'s editorial committee rejected the article. KREBS then sent his article to a journal with a relatively restricted audience, *Enzymologia*. It was accepted and published in the two months that followed. The importance of the concept that was put forward in the article ensured that its author gained international recognition, leading to his winning the NOBEL prize for Physiology and Medicine in 1953.

For the researcher, publication is a way of making his or her work known. It is also the way in which the researcher learns about the work of others. While the rhythm at which publications in the life sciences appeared increased slightly in the first half of the 20th century, the second half of that century saw a great acceleration in this rhythm, leading to a difficult-to-manage proliferation of reviews and books. It has been estimated that in the last thirty years the volume of publications in the biological domain has increased five-fold; in the preceding twenty years it had already doubled.

This accumulation of publications makes it harder for the researcher to judge the quality of the huge mass of published articles, even in the highly targeted domains that are within his or her area of expertise. The researcher, therefore, will deliberately choose a particular article according to the prestige of the journal in which it is published, which is not an inviolable criterion of quality. In addition, any judgement concerning the pertinence of a scientific article necessitates a dissection of the subtleties of the methodology, the well-groundedness of the experimental protocol and the validity of the results, by means of a careful examination of tables of results and graphs, and, finally, the logic of the discussion. This restrictive yet absolutely necessary requirement limits the number of articles that are likely to be screened. However, this is not the worse fault of publication today. There is another problem that is much more worrying. Many documentation centers have reacted to this inflation in the scientific press by equipping themselves with computing facilities that are able to find, in data banks, articles that have been selected on the basis of a key word index, and to display them on screens. While acknowledging that this constitutes an inescapable change in the transmission of scientific know-how, it should be recognized that in browsing through the pages of a high-quality scientific review, it is possible to come across an article containing an innovative idea or a useful technique, an advantage that is less available when using the on-line system of scientific publication that is most prevalent nowadays.

Mention should also be made of the requirement to publish frequently and within short time frames, for reasons of competitivity, when aspiring to obtain jobs or promotions, or even just to obtain recognition, this requirement being another factor that is prejudicial to fundamental research. It is the cause of worrying excesses, such as experiments that are hastily published and non-reproducible, or even the falsification of experimental results, occasionally within a context of considerable media coverage. Although such practices, which are the exception rather than the rule, are rapidly detected and condemned in a scientific culture where information circulates freely, the publicity that they incite, which reaches society at large via the media, leads to an overall discrediting of experimental research.

At present, one of the most noticeable trends in scientific publication is that of **collectivism**. While, in the 19th century, scientific articles were usually published in the name of a single author, occasionally two authors, and very rarely more than two, nowadays publications are often co-authored by several people, and when the work involves the analysis of structures, or the sequencing of genomes, several dozen researchers may be co-authors. From being the work of individuals, research has become collective. In domains whose complexity requires a wide selection of techniques that may range from physics to genetics, the hybridization of specific areas of expertise is certainly indispensable, and this requires the collaboration on a particular project of researchers who are sometimes physically remote from one another. The downside for the researcher, particularly one who is young, is that this requires him or her to abandon **individuality** and **creativity**. Both collectivism and inflation in scientific publication are facts that are an integral part of contemporary science, facts which reflect an irreversible trend that it would be difficult to obviate.

Over the last few years, scientific publication has been subject to a type of restraint, in that certain "sensitive" data in the domain of molecular biology might be used for the manufacture of biological weapons in a form of terrorism known as bioterrorism. Thus, the means of synthesizing *de novo* viruses (influenza virus, poliomyelitis) and the possibility of modifying their tropism by "directed molecular evolution" (change from a sexual tropism to a respiratory tropism for the AIDS virus) have been the subject of publications in prestigious journals. Given sufficient means, terrorist pharmacists could well make use of such data in order to carry out malicious actions with catastrophic consequences [85].

7.1.3. A new horizon for cross-disciplinarity

In order to please a public that is eager for progress and the sensational, politicians favor, by means of targeted financing, the types of organization that appeal to their sensibilities, such as the **technological platforms**. While recognizing that such platforms are now an integral part of the landscape of research on living beings, and that they must therefore be taken into account, and while acknowledging that projects which implement the latest technologies in different domains need to be federated, it is nonetheless vital not to underestimate the potential creativity of small groups of researchers, a point that was expressed by one of the greatest of contemporary biologists, Arthur KORNBERG (1918 - 2007), winner of the NOBEL prize for Physiology and of Medicine, in a speech given in 1997: "As I view the steady growth of collective science and big science, the greatest danger I see is a dampen-

---

[85] P. BERCHE (2006) *"Progrès scientifiques et nouvelles armes biologiques"*. Médecine/Sciences, vol. 22, pp. 206-211.

ing of individual creativity and reversion to the old politics – the inevitable local politics that infects every group and institution."

However, conscious of the metamorphosis that is occurring in the experimental method, and faced with a particularly inventive and all-conquering technology, **fundamental research in the life sciences must come to terms**. A century ago, fundamental research and technological research interacted all the more directly because they were both in their infancy. This is no longer the case. Management of the ever-increasing amount of knowledge in the life sciences, and the degree of sophistication achieved by bioengineering techniques and instruments, is widening a gap that makes **dialogue increasingly laborious**. However, dialogue appears to be a guarantee of future progress. The solution can only come from an increase in **cross-disciplinarity**, which should begin with university teaching and the establishment of a recruitment policy that advocates the cohabitation of talents from different educational backgrounds in the same laboratory. Fortified by such **hybrid expertise**, while maintaining its share of originality and liberty in the choice of problems to be studied, fundamental research on living beings can only be enriched by a marriage of reason with biotechnology. Convinced of the necessity for such a marriage, Stanley FIELDS, the inventor of the double hybrid method (Chapter IV-4.1), in an article entitled "The interplay of Biology and Technology" (*Proceedings of the National Academy of Sciences, USA*, 2001, vol. 98, pp. 10051-10054), concludes,: "It is at the interfaces of biology and other sciences that many of the future discoveries will be made, at the interfaces of biology and engineering that these discoveries will come to be exploited, and at the interfaces of biology and ethics and law that their consequences for society will be decided."

The desired dialogue between biology and technology also implies the breaking down of barriers that too often isolate fundamental research and so-called applied research, and the facilitating of consistent interaction between the discoveries made in the academic institutions and their application for utilitarian ends in private companies. This is where the twin demons of money and power raise their heads. Already, at the turn of the 1980s, A. Bartlett GIAMATTI [86] (1938 - 1989), who was then president of Yale University in the USA, commenting on American university policies, spoke of a "ballet of antagonisms" between, on the one hand, commercial companies that are interested in the rapid cost-effectiveness of any new therapeutic advance and, on the other hand, non-profit university laboratories. Recently, James J. DUDERSTADT [87] (b. 1973), emeritus President of the University of the Michigan, argued that the University is a "counter-hierarchical" organism. In fact, its members are free to carry out the research that pleases them and to think

---

[86] A.B. GIAMATTI (1981) *The University and the public interest*. New York, Athenum.
[87] J.J. DUDERSTADT (2000) *A University for the 21ᵗʰ century*. Ann Arbor University of Michigan Press.

in the ways that they wish to think, in any case within an academic norm that considers itself as being free from the constraints dictated by private interest groups. Until recently, such behavior was considered as a sort of ethic which arose out of the University conscience and dignity. The crumbling away of this ethic in the final decades of the 20th century coincided with the rise of biotechnologies and the large-scale filing of patents relating to molecular genetics techniques that could be applied to the manipulation of living beings, by researchers in the public sector. The intrusion of the American private sector into public research laboratories, in the form of collaborations with transfer of "sensitive" information from the public to the private, has become such a worrying problem that drastic control measures have had to be taken. Within this context, the American federal government, in February 2005, issued a certain number of prohibitions targeting the National Institutes of Health (NIH) of Bethesda, particularly with respect to the retribution of researchers for services rendered to industry [88]. These stands call for thought concerning the **place that is currently held in Universities with respect to fundamental research**. Without arguing against the efficacy of major research institutes, it is nevertheless necessary to remember the part played by the University in this domain. The University is not only the place where knowledge, both as it is now, in its current state of advancement, and as it has been, it is also the place where knowledge must be created by fundamental research.

For the last few decades, under pressure from state policies, and also as a function of an improvement in social status, the world of the university has opened up to a wider public, leading to an influx of students that is sometimes so enormous that the task of teaching them has become overwhelming. Because of this, the share of their time that university researchers can, in practice, devote to their research tasks has shrunk. This situation is highly prejudicial to the mission to innovate, which should be a priority. It is, in fact, during their University studies that the thought patterns of young students are forged by contact with teachers who not only instruct them, but also educate them by inspiring in them a motivation and an enthusiasm that gives rise to hope. How could this be true if the teaching faculty did not itself participate in scientific creation?

7.2. THE EXPERIMENTAL METHOD TAUGHT AND DISCUSSED

"What can teaching, ex cathedra, do to guide the researcher? Nothing, obviously. The researcher is trained in the laboratory. And the first stroke of genius on the part of a future researcher is to find a good boss. Such a find will open up the royal road to success. The road will be opened – but the researcher must travel along it. A researcher may be taught many things. He or she can become familiar with techniques and with equipment. She or he

---

[88] R. STEINBROOK (2005) *"Standards of Ethics at the National Institutes of Health"*. New England Journal of Medicine, vol. 352, pp. 1290-1292.

can be assigned a problem to resolve. However, what is essential for the researcher is to know how to understand relationships between phenomena that seem unrelated, and to be able to progress from the particular to the general. A boss may develop such qualities in a gifted young researcher, but intuition is a gift; it cannot be taught."

André Lwoff
Games and Combats - 1981

While the Bernardian style experimental method, based on a working hypothesis aroused by an observation, followed by implementation of an experimental protocol, is still extant in the life sciences, and while "serendipity" is still the origin of great discoveries, "big science", underpinned by sophisticated biocomputing or bioinformatics procedures, is intruding more and more, while genomics and proteomics are not far behind. The methods and instruments developed by the biotechnosciences have led to profound modifications in the ways that the structures and functions of living beings are investigated. For example, by varying multiple parameters in DNA chips or protein chips, at the same time, the experimenter is able to ask questions that lead to grouped all-or-nothing answers (Chapter IV-1.1.3). In combinatory chemistry, screening makes it possible to detect a molecule that is active for a given pathology from among a multitude of molecules (Chapter IV-3.4). The mathematical simulation of metabolic networks or of signaling chains is already well under way (Chapter IV-4.2). Given this new technological outlook and the hope that it can provide rapid solutions to health problems subject to considerable media coverage, the **teaching of biology in universities** must not be limited to a description of current advances, no matter how brilliant and promising they may be. This teaching should return to its origins, be a **reminder of history**, and should not hesitate to use examples to illustrate how a major discovery can arise from a long period of wandering in the wilderness. In practical terms, while being conscious of the extraordinary complexity of living nature, and carefully avoiding the dangers of simplification, it is important to remember that the reductionist method was a necessary path to an understanding of the integrated, modelized biology that is emerging nowadays. At present, certain people call reductionism **naive**, but this is only the case insofar as we have faith in recent advances in integrated biology [89]. With this in mind, it should be noted that the deciphering of the protein synthesis mechanism in prokaryotic microorganisms (Chapter IV-1.1.1) was, along with the discovery of the genetic code, a jumping-off point for an inventory of similar, but noticeably more sophisticated, mechanisms in eukaryotic organisms. The reductionist "one gene, one enzyme" dogma, formulated on the basis of BEADLE and TATUM's experiments on the mold *Neurospora crassa* (Chapter III-6.1) was a necessary prerequisite to a considerably more elaborate understanding of the relationship between the genotype and the

---

[89] K. STRANGE (2005) *"The end of "naïve reductionism": rise of systems biology or renaissance of physiology". American Journal of Physiology. Cell physiology,* vol. 288, pp. C968-C974.

phenotype. The way in which the nucleic acid and protein units in the tobacco mosaic virus spontaneously organize themselves (Chapter III-7.3) acted as a basis for thought concerning the self-organization of macromolecular complexes in the cell. These few examples underline the fact that it is difficult to comprehend the scientific research process if we only refer to experiments carried out in the present, and if we do not have a clear idea not only of the way in which hypotheses, even false ones, were once formulated, but also of the way in which experimental work, which may have led to failures, was once carried out, or, in brief, if we do not look back at the past. Let us add that it is occasionally good for us to show some **humility** when we take the trouble to examine the past. Thus, the processes involved in the phagocytosis of bacteria by innate immune cells (neutrophils, macrophages), which are today studied in the greatest detail with particularly refined technical facilities, had already been perceived more than a century ago by METCHNIKOFF, and even analyzed, admittedly with the clumsy means at his disposal, but with such accuracy that none of the conclusions formulated at that time have yet been disproved (Chapters III-2.2.4 and III-6.2.5).

The experimental method applied to the Life Sciences, the history of its birth and of its development, the way in which it is regarded by political and societal authorities, and, finally, the dependencies that are developing at present between the technosciences, human medicine and the different branches of the economic sector, all of these aspects should be covered by **university teaching that includes not only the pure sciences, but also the human, political and economic sciences, as well as philosophy**.

The student should not be saturated with book-learning, but he or she should be taught to reason, to imagine and to criticize, not to accumulate knowledge in an indigestible catalogue, but to ask questions about the way in which certain, carefully chosen, items of knowledge have been acquired, and not to deliberately accept science in its current state without knowing what it was like in the past. He or she should understand what pathways of thought led to dogmas that were established and taught as truths being refuted, and favor experimentation, with its risks and questions, rather than well-smoothed, abstract theoretical presentations without rough edges. These should be the principles of teaching that is designed to **open up young minds to creativity**.

In Anglo-Saxon countries, the worlds of industry and research that welcome the graduate manage to communicate with one another, but these worlds ignore one another in France, or at least remain reserved, a situation which is prejudicial from the economic point of view. If we look at the pharmaceutical industry in particular, we see that only half a century ago the pharmacopeia was limited to plant extracts or active agents isolated from these plants, with antibiotics quietly beginning to make their appearance. In the last decades of the 20th century, a great technological leap forward was made, with completely new methods in bioengineering, combinatory chemistry, and the finding of therapeutic targets in macromolecules, and this created a hiatus that severely handicapped countries that were unprepared for it. France, with its biological fundamental research training that is out of phase with that of the Anglo-Saxon countries, fell behind, and continues to be behind, a situation that is prejudicial for its economy. The remedy for this does not lie in incantatory speeches. It requires a volontarist policy for the management of experimental research. Generally speaking, the fact that the major engineering schools in France, which recruit the scientific intellectual elite, students being chosen by competitive exams that select for intelligence rather than imagination, are unable to impose upon their students an end-of-course thesis that would authenticate their engineering degree, should not be tolerated. In contrast to other countries, in France only a small percentage of engineers have received doctoral training or had to present a thesis before entering their careers. The French dual system of major engineering schools and universities, which, a century ago, made sense for the economy of that period, has become completely obsolete, and deserves a courageous revision.

8. CONCLUSION
LOOKING AT THE PRESENT IN THE LIGHT OF THE PAST

"There is a question, much older than modern science, which has never ceased haunting certain men of science: that of the conclusions that the existence of science and the contents of scientific theories can lead to concerning the relationships that humankind has with the natural world. Such conclusions cannot be imposed by science as is, but they are an integral part of the metamorphosis of this science."

I. PRIGOGINE and I. STENGERS
The New Alliance. Metamorphose of Science - 1986 (2nd edition)

In the 1950s – 1960s, the hybridization of the techniques of genetics, biochemistry and biophysics gave birth to molecular biology. With the resolution of the double helix structure of DNA, the demonstration of its replication, the elucidation of the mode of expression of its nucleotide sequence as a sequence of amino acids in proteins and finally the deciphering of the genetic code, biology underwent a revolution of an amplitude similar to that which, at the end of the 19th century, saw a blossoming of the seeds of cell biology.

The last decades of the 20th century represented the utilitarian era of molecular biology. The introduction of **genetic engineering** into biological experimentation dates to the beginning of the 1970s. It was at this time that techniques were developed that made it possible to transfer a fragment of genomic DNA from one species into the genome of another species. Genetic engineering now fills a predominant position in the Life Sciences, supported by increasingly effective biocomputing or bioinformatics techniques. It is easy to understand that expertise and a high degree of knowledge about fundamental research is necessary in order to be able to master or even invent the genetic engineering techniques that are indispensable if we are going to produce biomolecules with a therapeutic impact, such as those that are currently being used in the pharmaceutical domain: insulin, growth hormone, blood coagulation factors, vaccines, etc. The **engineering sciences** that make up the greater part of contemporary biotechnology have now come to the fore in many domains of the Life Sciences. It is thus that a modernistic and original way of investigating Nature has come into being. **A multiparametric model, in which biocomputing or bioinformatics and high-throughput screening reign, is added to, or even substituted for, the Bernardian model for the experimental method, based on observation, an** *a priori* **hypothesis, and experimentation to verify this hypothesis by varying a single parameter at a time.** The aim of this globalized approach is to integrate the multiple reactions that take place almost simultaneously in different locations of a cell into a coherent whole, to rationalize the interpretation of the dialogue that operates between the different endocellular organelles, and finally to discover how the exchanges of information between cells in an organ and between organs in multicellular organisms are set up. We are therefore witness to the emergence of an integrated biology that has been labeled "systems biology". Its long-term objective is to model the functioning of living beings and to theorize them. Its development is encouraged by the perspective of consequences that could revolutionize certain sectors of the human economy and of public health. Today, concrete, mechanical models, in the form of biorobots and hybrid robots, and, very recently, molecular motors are added to abstract models that are based on the logic of mathematics and algorithms, ushering in the era of nanobiomachines. Becoming more utilitarian, the life sciences are imperceptibly detaching themselves from traditional philosophical concepts that try to explain the modes of reasoning of the researcher, or even to impose a framework for thought that is likely to orient his or her way of doing research.

Looking at genetic inheritance, contemporary experimentation has shown that at all levels of the tree of Nature, including Man, this inheritance can be modified. Aware of his or her ability to influence the functioning and the destiny of living beings, **the researcher is confronted with the dilemma of a desire for knowledge versus a questioning of the use to which discoveries may be put.** There has never been such a real divorce between the world of phenomena that are understood by the experimenter and the world of noumena whose intelligibility is foreign to our senses. There has never been such a wide gap between the biotechnosciences, whose possibilities are coming to be seen as limitless, and a reflective analysis of thought, which wanders between freedom of action and prohibition.

As society becomes aware of the potential applications of discoveries made concerning living beings, **problems of bioethics**, particularly those involving reproduction, have become problems of public interest. Cloning and the production of stem cells are subjects that give rise to diatribes and passions. In the near future, genotyping, which is the result of progress in pharmacogenetics, could usher in a new form of customized medicine. Elsewhere, the cognitive sciences that are bringing together philosophy and psychology in the domains of computer technology and artificial intelligence, and which are tackling the processes of thought, the creative imagination and memory, will no doubt be the subject of the considerable questioning concerning research on living beings with which the experimental method will be confronted in the 21st century.

When faced with the way in which biotechnologies have erupted into the life of society, the mind travels back to the allegorical illustration that embellishes Francis BACON's *Novum Organum* (see Figure II.19), showing vessels returning from unknown lands, loaded with precious cargoes and returning to port having sailed past the pillars of Hercules. At present, the challenge has been partially met, but a great deal remains to be done. Innumerable cargoes have already reached port, but what will be the destiny of this precious merchandise? After all, the seeds of the idea of **technoscience** were already in place in the 17th century, in the philosophy of Francis BACON and Robert BOYLE (Chapter II-6). BACON recommended that the governments of the time promote experimental science by the creation of laboratories equipped with high-performance instruments and libraries, by the organization of researchers into teams and by appropriate financing. The utilitarian ends of scientific research were underlined. BOYLE imagined a situation in which laboratories were open to society and researchers were able to accept criticism. Given innovations that upset tradition, protestations arose. The pneumatic machine or vacuum pump was the subject of the fameuse diatribe between BOYLE and the philosopher HOBBES (Chapter II-6.2). HOBBES criticized the validity of BOYLE's conclusions, drawn from experiments that he qualified as doubtful. Following his words, he came to see in the discoveries of experimental science a possible threat to the power of governments and the hierarchical layout of society. Such overcautious opposition to the pursuit of knowledge is in no way anecdotal, it is still a reality, with the uprooting of genetically modified plants and the veto that has been placed in certain areas on stem cell research. This type of opposition is also shown when pressures or even vetoes are in operation that take into account more the opportunism of the moment than an in-depth understanding of science and of its history and that forget that freedom of the mind is a guarantee of its creativity, because, just as in the world of Arts and Letters, the world of Scientific Research is situated outside those norms that can be modulated by state decrees. The creativity of the researcher cannot be manufactured on demand. Where it exists, it still needs to be detected and encouraged.

Chapter V

EPILOGUE

"There are ancient cathedrals which, apart from their consecrated purpose, inspire solemnity and awe [...]. The labors of generations of architects and artisans have been forgotten, the scaffolding erected for their toil has long since been removed, their mistakes have been erased, or have become hidden by the dust of centuries. Seeing only the perfection of the completed whole, we are impressed as by some superhuman agency [...]. Science has its cathedrals, built by the efforts of a few architects and of many workers."

G.N. LEWIS and M. RANDALL
Thermodynamics and the Free Energy of Chemical Substances - 1923

For today's young researcher, carrying out an experiment according to a carefully established protocol, gathering results, discussing these results and writing a report of these results with a view to publication; all of these activities take place within the framework of a standard routine. As a consequence, a researcher does not really pause to enquire **when** wise men and women began to adopt the intellectual and manual gymnastics that have been called the experimental method or **who** was at the origin of this innovation. Today's scientific researcher would perhaps be surprised to learn that the scientific method that he or she practices daily was invented only four centuries ago, and that it blossomed in an atmosphere of curiosity, revolt and creativity. In the life sciences, the start of the scientific method in the 17th century was signalled by HARVEY's demonstration that the blood flowing to all organs makes a complete circuit of the body, during which it leaves the heart and returns to it. During the same period, experimental science as applied to the physical world was also undergoing unprecedented growth. This sudden explosion in the process by which knowledge is acquired was a fruit that was long in ripening, with a tortuous history going all the way back to Ancient Greece, where a logical system of reasoning was created for the first time. It was in Ancient Greece, where there was a succession of philosophers who were skilled in the art of dialectics, and a political context that was open to debate, that the idea that the imaginary should not be disconnected from the rational, and that certitudes which are based on tradition can be revised, took shape.

To learn about the sequence of discoveries, and to understand how supposedly untouchable dogmas were swept away by the results of a single experiment, is to recognize the role played by the scientific method in the construction of our knowledge. Let us take the case of a very simple experiment that aims to show that contraction of a muscle is not accompanied by a modification in the muscle mass. An isolated frog's leg is immersed in a jar that is partially filled with water. The muscle mass is made to contract by means of an electrical discharge or another device. It is possible to verify that the muscle contraction does not lead to any variation in the level of water in the receptacle. The conclusion is reached that there is no change in the volume of the muscles of the frog's leg. This raw fact, while interesting, is only an outline, it lacks epistemological clothing, that is to say, it needs to be situated within its historical and scientific context. When this experiment was carried out for the first time by the Dutch naturalist SWAMMERDAM in the 17th century, GALEN's spirit theory was still in vogue. It was assumed that the nerves, which were considered to be hollow tubes, carried a fluid containing the "spirit" and that this fluid, entering or leaving a muscle, led to the volume modifications associated with contraction or relaxation of the muscle. By showing that a muscle maintains the same volume when at rest and when it is contracted, SWAMMERDAM's experiment cast doubt, or even discredit, on the "spirit" theory. Thus, when epistemological considerations are taken into account, the experimental fact takes on a greater meaning. The resulting debate leads to discussion, in the field, concerning established dogmas and their sterilizing power as obstacles to the progress of scientific thought.

Applying the scientific sethod to the world of living beings, in order to extract their secrets, involves the use of appropriate tricks and devices in order to modify the parameters that determine whether a function is fulfilled. This involves evaluating the influence of these parameters on the accomplishment of the function being explored, so that, in the end, an attempt can be made to explain its mechanism. Technological engineering has played a major role in the growth of experimental science, to the point where, at present, it has become the inescapable basis of those life sciences given the flattering name of biotechnosciences. The Greek name τέχνη, from which the word "technique" is derived, may be translated just as well as "art" or "industry" as it can be as "ruse" or "trick". To sum up, the technique is a deliberate ruse designed to penetrate the complexities of living beings. The risk is the creation of artefacts that divert away from the truth. The existence of such a risk did not escape the philosophers of Ancient Greece, as we are reminded by Mirko GRMEK in his book *Medea's Cauldron*. In the Greek legend of MÉDÉA, an old king, PELIAS, advised by his daughters, decides to try and become young again, and calls upon a famous magician, MÉDÉA. In order to convince the king, MÉDÉA carries out youth-restoring experiment on an animal. She takes an old ram, has it put to death, cuts it up and boils the remains in a cauldron. A huge cloud of steam rises up. When the steam has blown away a young and handsome ram is seen. Unaware of the trick that has been played upon him, PÉLIAS hands his body over to MÉDÉA. Obviously, the experiment fails. The legend of MÉDÉA has a symbolic value, and it is not without reason that the wise men of Ancient Greece, with a few exceptions such as the famous physician GALEN, hesitated about manipulating living beings in order to better understand how they functioned. While most medieval scholastics, inheritors of Greek philosophy, suffered the same hesitations, some rebellious minds such as Robert GROSSETESTE and Roger BACON dared to challenge established dogma and extol the virtues of experimentation rather than dream up nebulous theories based on speculation. The **why** of philosophers began to be replaced by the **how** of experimentors.

During the Renaissance, in a sociopolitical climate that was troubled but also bore audacious ideas that challenged preferences and beliefs, the Academies and Universities, particulalry in Italy, brought together scholars who were driven by a curiosity to know and the desire to experiment. They asked questions about "how" and invented methods for answering these questions. From an intellectual process that, at the end of the Middle Ages, was happy to observe Nature, wonder at it, consider it sacred, make comparisons between living species and draw up classifications, scholars moved on to experimental practice. Experimentation based on the use of ingenious instrumentation took over from observation and from an approach to the phenomena of Nature that simply makes use of the senses. Experimental results were quantified. The degree of error was evaluated. Scientific knowledge, made instrumental and objective, became separate from the knowledge acquired from the senses alone, with their share of subjectivity. Thus began a scientific revolution of a scale equal to that known in Ancient Greece when the philosophers laid down the bases of logic and reasoning. In the same way, as states began to distance themselves from ecclesiastical powers, science, in becoming experimental, gained its autonomy with respect to metaphysical concepts. In a Europe that today is searching for coherence in the memory of its roots, it is not unuseful to remember that it was in the West that the scientific method was discovered, a method that was the origin of the amazing way in which human knowledge took off, along with its applications in health and economics. The 17th and 18th centuries saw the birth and strengthening of the scientific method, as they were also witness to its theorization. Fundamental principles were laid down. During this period, there was a scientific eclecticism that led to dialogue between physicists, chemists and naturalists. The accumulation of knowledge and the increasing complexity of science gradually led to a compartmentalization of knowledge and an isolation of scientific disciplines.

In the 19th century, biology became a discipline in its own right and underwent a dazzling growth that benefitted from technical innovations in the techniques of physics and chemistry. The Bernardian doctrine of experimental determinism took over as the biologist's bible. From the 19th century, modern biology was born, with the formulation of fundamental concepts such as cell theory, transformation by the selection of species and the transmission of hereditary characteristics according to quantifiable laws. The theory of evolution had the effect of renewing scientific philosophy. A century later, the implementation of protein and DNA sequencing techniques in a wide range of animal, plant and microbe species exemplified the similarities of sequences across the evolutionary scale and provided irrefutable proof of the concept of transformism. It was also the 19th century that saw the unprecedented development of heavy industry, mass movement of populations from the countryside to towns and a reorganisation of the social order that led to politico-religious crises. These upheavals are not unlike those that were seen in the states in the West during the Renaissance, which were a prelude to the birth of modern science.

In the 20th century, faced with the complexity of the living world, a reductionist approach was, for a time, the only way out of the impossibility of understanding the living world in its entirety. We went from the exploration of the organ to that of the cell and then to that of the endocellular organelle. Thought was given to anatomy and the function of the macromolecules that inhabit the different compartments of the cell. The steps that led from the coded message of the genome to protein structures were untangled. Molecular interactions, metabolic networks and signalling systems were explored. Breaking down the complexity in order to resolve its enigmas was not just an act of naive reductionism. It was a necessary, Cartesian process, moving towards an attempt to understand living beings. It was not free from the researcher's understanding that the knowledge acquired thanks to these processes showed the **possible**, but not necessarily the **real**, and that the dynamics of living beings did not constitute a simple adding up of partial mechanisms, but instead the result of an interaction and an intrication of these mechanisms, with unexpected effects as a result, such as, for example, the smoothing out, or, in contrast, the exacerbation of certain reactions shown *in vitro*.

While the biology of the 19th century and the first decades of the 20th century can be distinguished by the large scale use of the methods of analytical chemistry and precise, high-performance measurement techniques inherited from physics, the biology of the end of the 20th century was marked by genetics. At the turn of the 21st century, with the deciphering of the human genome, hundreds of prokaryotic genomes and dozens of eukaryotic genomes, with access to increasingly numerous data arising from the inventory of proteins, stored and analyzed in computers, and also the application of atraumatic methods for the exploration of complex living structures, there is a new challenge to be met, which involves understanding the dialogues that come and go between the thousands of molecular components organized into networks within endocellular compartments. The reductionist approach that was prevalent in the 20th century is tending to give way to a holistic process that is accompanied by a mathematisation and a simulation of mechanisms at the center of the life of a cell. In this new approach to the life sciences, engineering and information technology have become essential agents.

With the emergence of the biotechnosciences, the biologist, not content to master Nature, has come to subjugate it. He or she ceases to consider it as sacred, with a view to making use of it to meet his or own needs. Thus, beyond an understanding of the role of DNA in the mechanisms of cell division and differentiation, the biologist, by invalidating genes or replacing them, is today able to modify genetic inheritance in the way that he or she wishes, and, as a consequence, to alter the phenotype of a living species. From a discoverer, the biologist has become an inventor and constructor of objects that imitate the creations of Nature. Big instruments, technical platforms and high-throughput systems, which are part and parcel of contemporary research, fascinate people with their performance. Mirages for young minds, their contribution to the discovery of new mechanisms needs to be evaluated and discussed with discernement and without passion. Although it is less aggressive than the scientificism of the end of the 19th century, modern "technologism" nevertheless involves research in a vision that is strongly based on the totipotentiality of "big science". Given the inescapable evolution of knowledge, and faced with often excessive media coverage of any discovery that is announced, and the ever-increasing demands of an ill-informed public, political authorities are tempted to become involved. In this context, it is obvious that conflicts are springing up between, on the one hand, biotechnosciences that are keen to transmit to man the practises that are being mastered in animals, and, on the other hand, bioethical imperatives that cling to the concept of the transcendancy of Man in Nature.

BIBLIOGRAPHY

REFERENCES SPECIFIC TO EACH CHAPTER

CHAPTER I

ARISTOTE, *Historia Animalium*. English translation: D'Arcy Wentworth THOMPSON. 1910. Clarendon Press. *Histoire des animaux d'ARISTOTE*. French translation: J. BARTHÉLEMY-SAINT HILAIRE. Vol. I, 1883. Hachette, Paris.

ARISTOTE, *De anima. De l'âme*. Text set by A. JANNONE. French translation and notes: E. BARBOTIN. 1989. Tel-Gallimard, Paris.

ARISTOTE, *Organon. De Sophisticis elenchis. VI - Les réfutations sophistiques*. French translation and notes: J. TRICOT. 1995. Vrin, Paris.

ARISTOTE, *Organon. Categoriae, de Interpretatione. I - Catégories, II- De l'interprétation*. French translation and notes: J. TRICOT. 1997. Vrin, Paris.

AVICENNE, *Dânèsh-Nâma*. 1331-1955. Téhéran. *Le livre de science. I - Logique, Métaphysique, II - Science naturelle, Mathématiques*. French translation: Mohammad ACHENA, Henri MASSÉ. 2nd ed. revised by Mohammad ACHENA 1986. Les Belles Lettres/UNESCO, Paris.

BAUDET Jean, *De l'outil à la machine. Histoire des techniques jusqu'en 1800*. 2003. Vuibert, Paris.

BODÉÜS Richard, *ARISTOTE. Une philosophie en quête de savoir*. 2002. Vrin, Paris.

BOURGEY Louis, *Observation et expérience chez ARISTOTE*. 1955. Vrin, Paris.

BOYDE T.R. Caine, *Foundation Stones of Biochemistry*. 1980. Voile et Aviron, Hong Kong.

CARON M. et HUTIN S., *Les alchimistes*. 1959. Seuil, Paris.

CHEVREUL Eugène, *Résumé d'une histoire de la matière depuis les philosophes grecs jusqu'à LAVOISIER*. 1875. Mémoires de l'Académie des Sciences de l'Institut de France, Vol. 39. Gauthier-Villars, Paris.

CROMBIE Alistair Cameron, *AUGUSTINE to GALILEO: The History of Science AD 400-1650*. 1952. The Falcon Press, London. Ed. revised 1969. Penguin. *Histoire des Sciences de SAINT AUGUSTIN à GALILÉE (400 - 1650)*. French translation: Jacques D'HERMIES. Vol. I, 1959. Presses Universitaires de France, Paris.

DJEBBAR Ahmed, *Une histoire de la science arabe. Introduction à la connaissance du patrimoine scientifique des pays d'Islam. Entretiens avec Jean ROSMORDUC*. 2001. Seuil, Paris.

GALIEN, *Œuvres médicales choisies*. French translation: Charles DAREMBERG, text and notes: André PICHOT. Vol. I *De l'utilité des parties du corps humain*. 1994. Gallimard, Paris.

GANZENMÜLLER Wilhelm, *Die Alchemie im Mittelalter*. 1938. Paderboorn. *L'alchimie au Moyen Âge*. French translation: G. PETIT-DUTAILLIS. 1940. Aubier-Montaigne, Paris.

GÉRARDIN Lucien, *L'alchimie. Tradition et actualité*. 1972. Bibliothèque de l'irrationnel, Paris.

GRIMAUX Edouard, *LAVOISIER 1743 - 1794, d'après sa correspondance, ses manuscrits, ses papiers de famille et d'autres documents inédits*. 2nd ed. 1996. Félix Alcan, Paris.

LAFONT Olivier, *D'Aristote à Lavoisier. Les étapes de la naissance d'une science*. 1994. Ellipses, Paris.

LASZLO Pierre, *Qu'est-ce que l'alchimie*. 1996. Hachette, Paris.

LE BLOND J. M., *Logique et méthode chez ARISTOTE. Etude sur la recherche des principes dans la physique aristotélicienne*. 4th ed. 1996. Vrin, Paris.

LECLÈRE Albert, *La philosophie grecque avant SOCRATE*. 2nd ed. 1908. Bloud & Cie, Paris.

LLOYD Geoffrey Ernest Richard, *Magic, reason and experience. Studies in the origins and development of Greek science*. 1979. Cambridge University Press, Cambridge, GB.

LLOYD Geoffrey Ernest Richard, *Early greek science, THALES to ARISTOTLE*. 1970. Chatto and Windus, London, and *Greek science after ARISTOTLE*. 1973. Chatto and Windus, London. *Une histoire de la science grecque*. French translation: Jacques BRUNSCHWIG. 1990. La Découverte, Paris.

LOUIS Pierre, *La découverte de la vie. ARISTOTE*. 1975. Hermann, Paris.

PETIT Paul, *Précis d'histoire ancienne*. 6th ed. corrected by André LARONDE 1986. Presses Universitaires de France, Paris.

PICHOT André, *La naissance de la science. 1. Mésopotamie, Egypte*. 1991. Gallimard, Paris.

PICHOT André, *La naissance de la science. 2. Grèce présocratique*. 1991. Gallimard, Paris.

ROBIN Léon, *La pensée grecque et les origines de l'esprit scientifique*. 1973. Albin Michel, Paris.

RODIER Georges, *Etudes de philosophie grecque*. 1926. Vrin, Paris.

SCHRÖDINGER Erwin, *Nature and the Greeks*. 1954. Cambridge University Press, Cambridge. UK. *La nature et les Grecs*. French translation and notes: Michel BITBOL, Annie BITBOL-HESPÉRIÈS, preceded by *La clôture de la représentation* by Michel BITBOL. 1992. Seuil, Paris.

SOUTIF Michel, *L'Asie, source de sciences et de techniques. Histoire comparée des idées scientifiques et techniques de l'Asie*. 1995.
Grenoble Sciences-EDP Sciences, Les Ulis, France.

SOUTIF Michel, *Naissance de la physique. De la Sicile à la Chine*. 2002.
Grenoble Sciences-EDP Sciences, Les Ulis, France.

TATON René, *La science antique et médiévale. Des origines à 1450*. 1994.
Quadrige-Presses Universitaires de France, Paris.

CHAPTER II

ADAM Charles, TANNERY Paul (Eds), *Œuvres de Descartes. Le Monde. Description du corps humain. Passions de l'Ame. Anatomica. Varia*. 1986. Vrin, Paris.

BAUDET Jean, *De l'outil à la machine. Histoire des techniques jusqu'en 1800*. 2003.
Vuibert, Paris.

BENSAUDE-VINCENT Bernadette, STENGERS Isabelle, *Histoire de la chimie*. 2001.
La Découverte & Syros, Paris.

BERNHARDT Jean, *HOBBES*. 2nd ed. 1994. Presses Universitaires de France, Paris.

BLAMONT Jacques, *Le chiffre et le songe. Histoire politique de la découverte*. 1993.
Odile Jacob, Paris.

BLAY Michel, *La science du mouvement - De GALILEE à LAGRANGE*. 2002. Belin, Paris.

BLAY Michel, HALLEUX Robert (Eds), *La science classique XVIe-XVIIIe siècle. Dictionnaire critique*. 1998. Flammarion, Paris.

BOUTIBONNES P., *VAN LEEUWENHOEK. L'exercice du regard*. 1994. Belin, Paris.

BOYLE Robert, *New experiments Physico-Mechanicall, Touching the Spring of the Air, and its Effects (Made, for the Most Part, in a New Pneumatical Engine) Written by Way of Letter to the Right Honorabgle Charles Lord Vicount of Dungarvan, Eldest Son to the Earl of Corke*.1660. H. Hall, Oxford.

BOYLE Robert, *The Sceptical Chymist, or, Chymico-Physical Doubts & Paradoxes*. 1661.
J. Cadwell, London.

BOYLE Robert, *New experiments Physico-Mechanical, Touching the Air: Whereunto is Added A Defence of the Authors Explication of the Experiments. Against the Objections of Franciscus LINUS and Thomas HOBBES*. 1662. H. Hall, Oxford.

CALLOT Emile, *La renaissance des sciences de la vie au XVIe siècle*. 1951.
Presses Universitaires de France, Paris.

CANGUILHEM Georges, *La formation du concept de réflexe aux XVII et XVIIIe siècles*.
2nd ed. 1977. Vrin, Paris.

COMTE Hubert, *Le Microscope. La traversée des apparences*. 1993. Casterman, Paris.

CROMBIE Alistair Cameron, *AUGUSTINE to GALILEO: The History of Science A.D 400 - 1650*. 1952. The Falcon Press, London. Revised ed. 1969. Penguin, London. *Histoire des Sciences de SAINT AUGUSTIN à GALILÉE (400 - 1650)*. French translation: Jacques D'HERMIES. Vol. 2, 1959. Presses Universitaires de France, Paris.

DAGOGNET François, *Les grands philosophes et leur philosophie. Une histoire mouvementée et belliqueuse.* 2002. Seuil, Paris.

DAUMAS Maurice, *LAVOISIER, théoricien et expérimentateur.* 1955. Presses Universitaires de France, Paris.

DEBUS Allen G., *Man and Nature in the Renaissance.* 1994. Cambridge University Press, New York.

DELAVAULT Robert, *Les précurseurs de la biologie. De l'anatomie à la biologie expérimentale.* 1998. Corsaire Editions, Orléans, France.

DESCARTES René, *Traité de l'homme.* 1664.

DESCARTES René, *Discours de la méthode.* Text and commentaries: Etienne GILSON. 5th ed. 1976. Vrin, Paris.

DUCHESNEAU François, *Les modèles du vivant de DESCARTES à LEIBNIZ.* 1998. Vrin, Paris.

FLOURENS Pierre, *Histoire de la découverte de la circulation du sang.* 2nd ed. 1857. Garnier Frères, Paris.

FOISNEAU Luc (Ed.), *La découverte du principe de raison. DESCARTES, HOBBES, SPINOZA, LEIBNIZ.* 2001. Presses Universitaires de France, Paris.

GRIMAUX Edouard, *LAVOISIER 1743 - 1794, d'après sa correspondance, ses manuscrits, ses papiers de famille et d'autres documents inédits.* 1896. Félix Alcan, Paris.

GUYÉNOT Émile, *Les sciences de la vie aux XVIIe et XVIIIe siècles. L'idée d'évolution.* 1941. Albin Michel, Paris.

HARVEY William, *Exercitatio anatomica de motu cordis et sanguinis in animalibus.* 1628. Francfort. *Dissertation anatomique sur le mouvement du cœur et du sang chez les animaux.* French translation: Charles RICHET (1869). 1990. Christian Bourgeois, Paris.

HOLMES Frederic Lawrence, *LAVOISIER and the Chemistry of Life. An Exploration of Scientific Creativity.* 1985. The University of Wisconsin Press, Madison, WIS, USA.

HUME David, *An Enquiry concerning Human Understanding.* 1748. *Enquête sur l'entendement humain.* French translation and notes: André LEROY. 1977. Aubier-Montaigne, Paris.

INGENHOUSZ Jan, *Expériences sur les végétaux.* 1780. Didot, Paris.

JACOMY Bruno, *Une histoire des techniques.* 1990. Seuil, Paris.

KEYNES Geoffrey, *A bibliography of the writings of Dr William HARVEY 1573 - 1657.* 2nd ed. revised 1953. Cambridge University Press, Cambridge, UK.

KLEMM Frederick, *Technik, eine Geschichte ihrer Probleme.* Karl Alber, Fribourg-Munich. *Kurze Geschichte der Technik.* 1954. Verlag Herder, Freiburg im Breisgau.

Histoire des techniques. French translation: A.M. LAHR-DEGOUT. 1966. Payot, Paris.

LAFONT Olivier, *D'ARISTOTE à LAVOISIER. Les étapes de la naissance d'une science*. 1994. Ellipses, Paris.

LA METTRIE Julien Offroy DE, *L'Homme-Machine*. Preceded by *Lire LA METTRIE* by Paul-Laurent ASSOUN. 1981. Denoël/Gonthier, Paris.

LAVOISIER Antoine-Laurent DE, *Traité élémentaire de chimie*. 1789. Paris.

LAVOISIER Antoine-Laurent DE, LAPLACE Pierre-Simon DE, *Mémoire sur la chaleur*. 1920. Gauthier-Villars, Paris.

LICOPPE Christian, *La formation de la pratique scientifique. Le discours de l'expérience en France et en Angleterre (1630 - 1820)*. 1996. La Découverte, Paris.

MADDISON R.E.W., *The life of the honourable ROBERT BOYLE F.R.S.* 1969. Taylor & Francis, London.

MAIRET Gérard, *Léviathan HOBBES*. 2000. Ellipses, Paris.

MAURY Jean-Pierre, *A l'origine de la recherche scientifique : MERSENNE*. 2003. Vuibert, Paris.

MAYR Ernst, *This is biology: the science of the living world*. 1997. Belknap Press of Harvard University Press, Cambridge, MA, USA. *Qu'est ce que la biologie ?* French translation: Marcel BLANC. 1998. Fayard, Paris.

MERSENNE Marin, *Les nouvelles pensées de GALILÉE*. French translation: R.P. Marin MERSENNE. Critical edition by Pierre COSTABEL, Michel-Pierre LERNER. Vol. I and II, 1973. Vrin, Paris.

NEEDHAM Joseph, *Science and civilisation in China*. 1954. *La science chinoise et l'Occident*. French translation: Eugène SIMION, R. DESSUREAULT, J.-M. REY. 1973. Seuil, Paris.

POUSSEUR Jean-Marie, *BACON 1561 - 1626. Inventer la science*. 1988. Belin, Paris.

QUIRICO Tiziana (Ed.), *Les siècles d'or de la médecine. Padoue XVe - XVIIIe siècles*. 1989. Electa, Milano, Italy.

RÉTHORÉ F., *CONDILLAC ou l'empirisme et le rationalisme*. 1864. Auguste Durand, Paris.

ROGER Jacques, *Les sciences de la vie dans la pensée française du XVIIIe siècle*. 1971. Armand Colin, Paris. New ed. 1993. Albin Michel, Paris.

ROSSI Paolo, *La nascita della Scienza moderna in Europa. La naissance de la science moderne en Europe*. French translation: Patrick VIGHETTI. 1999. Seuil, Paris.

SHAPIN Steven et SCHAFFER Simon, *Leviathan and the Air-Pump, HOBBES, BOYLE, and the Experimental Life*. 1985. Princeton University Press, Princeton, NJ, USA. *Leviathan et la pompe à air. HOBBES et BOYLE entre science et politique*. French translation: Thierry PIÉLAT, Sylvie BARJANSKY. 1993. La Découverte, Paris.

SHAPIN Steven, *The scientific revolution*. 1996. The University of Chicago Press. Chicago, ILL, USA. *La révolution scientifique*.
French translation: Claire LARSONNEUR. 1998. Flammarion, Paris.

WEST John B., *"Robert BOYLE's landmark book of 1660 with the first experiments on rarefied air"*. 2005. *Journal of Applied Physiology*, Vol. 98, pp. 31-39.

CHAPTER III

BACHELARD Gaston, *La formation de l'esprit scientifique. Contribution à une psychanalyse de la connaissance objective*. 13th ed. 1986. Vrin, Paris.

BACHELARD Gaston, *Le nouvel esprit scientifique*. 1938.
Presses Universitaires de France, Paris.

BALIBAR Françoise, PRÉVOST Marie-Laure (Eds), *PASTEUR, Cahiers d'un savant*. 1995. CNRS Editions, Paris.

BARBEROUSSE Anouk, KISTLER Max, LUDWIG Pascal, *La philosophie des sciences au XXe siècle*. 2000. Flammarion, Paris.

BERNARD Claude, *Introduction à l'étude de la médecine expérimentale*. 1865.
J.-B. Baillière, Paris. Chronology and preface by François DAGOGNET. 1966. Garnier-Flammarion, Paris.

BERNARD Claude, *Leçons sur les phénomènes de la vie communs aux animaux et aux végétaux*. 1878. J.-B. Baillière, Paris.

BERNARD Claude, *Recherches sur une nouvelle fonction du foie considéré comme organe producteur de matière sucrée chez l'homme et les animaux*.
Doctoral thesis 1853, Paris.

BERNARD Claude, *Principes de médecine expérimentale*. 1947.
Presses Universitaires de France, Paris.

BICHAT François-Xavier, *Recherches physiologiques sur la vie et la mort*. 1852. Masson, Paris.

CASTIGLIONI Arturo, *Storia della Medicina*. 1927. Unitas, Milano, Italy. *Histoire de la médecine*. French translation: J. BERTRAND, F. GIDON. 1931. Payot, Paris.

CÉLINE Louis Ferdinand, *SEMMELWEIS*. 1999. Gallimard, Paris. From CÉLINE's doctoral thesis: *La vie et l'œuvre de Philippe Ignace SEMMELWEIS*, 1924.

COLEMAN William, *Biology in the nineteenth century. Problems of Form, Function, and Transformation*. 1977. Reprinted 1990.
Cambridge University Press, Cambridge, UK.

COMTE Auguste, *Cours de philosophie positive*. Vol. III contenant: *La philosophie chimique et la philosophie biologique* 1838. 1908. Schleicher Frères, Paris.

COMTE Auguste, *Philosophie des Sciences*. Texts selected by Jean LAUBIER. 1974.
Presses Universitaires de France, Paris.

COMTE Auguste, *Discours sur l'ensemble du positivisme*. Introduction, notes, chronology by Annie PETIT. 1998. Flammarion, Paris.

COURNOT A.A., *Considérations sur la marche des idées et des évènements dans les temps modernes*. Vol. I, 1872. Hachette, Paris. *Œuvres complètes* Vol. IV, André ROBINET (Ed.) 1973. Vrin, Paris.

DAREMBERG Charles, *La Médecine. Histoire et doctrines*. 2nd ed. 1865. Didier/J.-B. Baillière, Paris.

DEBRU Claude (Ed.), *Qu'est-ce que la physiologie ? Achèvement et renaissance. Histoire des Sciences Physiologiques*. 1997. Vrin, Paris.

DELACAMPAGNE Christian, *Histoire de la philosophie du XXe siècle*. 2000. Seuil, Paris.

DUMAS Jean-Baptiste, *Leçons sur la philosophie chimique*. 1836. Impression 1878. Gauthier-Villars, Paris.

DUTROCHET M. Henri, *Mémoires pour servir à l'histoire anatomique et physiologique des végétaux et des animaux*. Vol. 1, 2, 3, 1837. J.-B. Baillière, Paris.

FLECK Ludwik, *Entstehung und Entwicklung einer wissenschaftlichen Tatsache*. 1935. Benno Schwabe & Co. *Genesis and Development of a Scientific Fact*. 1979. T.J. TRENN and R.K. MERTON (Eds), foreword by Thomas KUHN. The University of Chicago Press, Chicago, ILL, USA. *Genèse et développement d'un fait scientifique*. French translation: Nathalie JAS. 2005. Les Belles Lettres, Paris.

FRUTON Joseph Stewart, *Proteins, Enzymes, Genes. The interplay of chemistry and biology*. 1999. Yale University Press, New Haven, London.

GENDRON Pierre, *Claude Bernard. Rationalité d'une méthode*. 1992. Vrin, Paris.

GILLE Bertrand, *Les ingénieurs de la Renaissance*. 1964. Hermann, Paris.

GRMEK Mirko, *Claude BERNARD et la méthode expérimentale*. 1973. Librairie Droz, Genève.

HAYAT M.A., *Principles and Techniques of Scanning Electron Microscopy. Biological Applications*. Vol. 6, 1978. Van Nostrand Reinhold Company, New York.

HOLMES Frederic Lawrence, *Claude BERNARD and animal chemistry*. 1974. Harvard University Press, Cambridge, MA, USA.

KUHN Thomas Samuel, *The Structure of Scientific Revolutions*. 1962. The University of Chicago Press, Chicago, ILL, USA. *La structure des révolutions scientifiques*. French translation: Laure MEYER 1983. Champs-Flammarion, Paris.

LATOUR Bruno, *PASTEUR, une science, un style, un siècle*. 1994. Librairie académique Perrin, Paris.

LEDUC Stéphane, *Théorie physico-chimique de la vie et générations spontanées*. 1910. A. Poinat, Paris.

LEDUC Stéphane, *La biologie synthétique, étude de biophysique*. 1912. A. Poinat, Paris.

LICHTENTHAELER Charles, *Geschichte der Medizin*. 1975. Deutscher Ärzte-Verlag GmbH, Köln-Lövenich. *Histoire de la médecine.* French translation: Denise MEUNIER. 1978. Fayard, Paris.

MAGENDIE François, *Précis élémentaire de physiologie*. 1825. Méquignon-Marvis, Paris.

METCHNIKOFF Elie, *L'immunité dans les maladies infectieuses*. 1901. Masson, Paris.

METCHNIKOFF Elie, *Trois fondateurs de la médecine moderne PASTEUR - LISTER - KOCH*. 1933. Félix Alcan, Paris.

NEEDHAM Joseph, *The Sceptical Biologist*. 1928. Chatto, Windus, London.

NICOLLE Charles, *Destin des maladies infectieuses*. 3rd ed. 1937. Félix Alcan, Paris.

OREL Vítězslav, *MENDEL*, English translation: Stephen FINN. 1984. Oxford University Press. *MENDEL 1822 - 1884. Un inconnu célèbre*. French translation: Françoise ROBERT. Correspondence with Carl NÄGELI translated from german by J.-R. ARMOGATHE. Postface by J.-R. ARMOGATHE. 1985. Belin, Paris.

OREL Vítězslav, *Gregor MENDEL. The first geneticist*. English translation: Stephen FINN. 1996. Oxford University Press, Oxford, New York, Tokyo.

PANZA Marco, PONT Jean-Claude, *Les savants et l'épistémologie vers la fin du XIXe siècle*. 1995. Albert Blanchard, Paris.

PRIESTLEY Joseph, *Experiments and Observations on Different Kinds of Air (1774-1786)*. *Expériences et observations sur les différentes branches de la physique avec une continuation des observations sur l'air*. French translation: M. GIBELIN. Vol. 1 and 2, 1782; Vol. 3, 1783. Nyon l'aîné, Paris.

RIEDMAN Sarah R., *Antoine LAVOISIER. Scientist and Citizen*. 1967. Abelard-Schuman, London, New York, Toronto.

SAMUEL Edmund, *Order in Life*. 1972. Prentice Hall, Englewood Cliffs, NJ, USA.

UMBREIT W.W., BURRIS R.H., STAUFFER J.F., *Manometric techniques. A manual describing methods applicable to the study of tissue metabolism*. 4th ed. 1964. Burgess Publishing Company, Minneapolis, Minnesota, USA.

WURTZ A.D., *Introduction à l'étude de la chimie*. 1885. Masson, Paris.

CHAPTERS IV & V

ATLAN Henri, *Entre le cristal et la fumée. Essai sur l'organisation du vivant*. 1979. Seuil, Paris.

ATLAN Henri, *L'utérus artificiel*. 2005. Seuil, Paris.

BACHELARD Gaston, *La philosophie du non. Essai d'une philosophie du nouvel esprit scientifique*. 1940. 5th ed. 2002. Quadrige-Presses Universitaires de France, Paris.

BACHELARD Gaston, *Epistémologie*. Text selected by Dominique LECOURT. 6th ed. 1995. Presses Universitaires de France, Paris.

BERNARD Jean, *La Bioéthique*. 1994. Dominos-Flammarion, Paris.

CHANGEUX Jean-Pierre, *L'homme neuronal*. 1983. Fayard, Paris.

CHANGEUX Jean-Pierre, *Molécule et Mémoire*. 1988.
Dominique Bedou, Gourdon, France.

CHANGEUX Jean-Pierre, DEHAENE Stanislas, "*Modèles nouveaux des fonctions cognitives*" in *Philosophie de l'Esprit et Sciences du cerveau*. 1991. Vrin, Paris.

CHEVEIGNÉ Suzanne de, BOY Daniel, GALLOUX Jean-Christophe, *Les biotechnologies en débat*. 2002. Balland, Paris.

CHNEIWEISS Hervé, NAU Jean-Yves, *Bioéthique : avis de tempêtes. Les nouveaux enjeux de la maîtrise du vivant*. 2003. Alvik Editions, Paris.

COHEN John, *Human robots in myth and science*. 1966. George Allen and Unwin Ltd., London. *Les robots humains dans le mythe et dans la science*. French translation: Marinette DAMBUYANT, 1968. Vrin, Paris.

CORNISH-BOWDEN Athel, JAMIN Marc, SAKS Valdur, *Cinétique enzymatique*. 2005. Grenoble Sciences-EDP Sciences, Les Ulis, France.

DEBRU Claude, *Le possible et les biotechnologies. Essai de philosophie dans les sciences*, avec la collaboration de P. NOUVEL. 2003.
Presses Universitaires de France, Paris.

DENNETT Daniel, *Consciousness explained*. 1992. Back Bay Books.
La conscience expliquée. 1993. Odile Jacob, Paris.
See also Christian DELACAMPAGNE "*Daniel DENNET : L'âme et le corps ? No problem !*". 1999, *La Recherche*, vol. 323, pp. 102-104.

DOUCE Roland (Ed.), *Les plantes génétiquement modifiées*. Rapport sur la science et la technologie n° 13 de l'Académie des Sciences. 2002. TEC & DOC Lavoisier, Paris.

ECCLES John Carew, *The mind-brain problem revisited : the microsite hypothesis*, in J.C ECCLES, O.D. CREUTZFELDT (Eds) *The principles of design and operation of the brain*. Pontificae Academiae Scientiarum Scripta Varia, Vatican City. *Evolution du cerveau et création de la conscience. A la recherche de la vraie nature de l'homme*. French translation: Jean-Mathieu LUCCIONI, Elhanan MOTZKIN. 1992. Fayard, Paris.

FERENCZI Thomas (Ed.), *Les défis de la technoscience*. 2001.
Editions Complexe, Bruxelles.

FEYERABEND Paul, *Against method*. 1975. New Left Books, London. *Contre la méthode. Esquisse d'une théorie anarchiste de la connaissance*. French translation: Baudouin JURDANT, Agnès SCHLUMBERGER. 1979. Seuil, Paris.

FEYERABEND Paul, *Wissenschaft als kunst. La science en tant qu'art*.
French translation: Françoise PÉRIGAUT. 2003. Albin Michel, Paris.

FRANCESCHINI Nicolas, "*De la mouche au robot : reconstruire pour mieux comprendre*" in V. BLOCH (Ed.) *Cerveaux et Machines*. 1999, pp. 247 - 270. Hermès, Paris.

FRUTON Joseph S., *A skeptical biochemist*. 1992.
Harvard University Press, Cambridge, MA, USA.

GROS François, *Les secrets du gène*. 1986. Odile Jacob, Paris.

GROS François (Ed.), *Les Sciences du vivant. Ethique et société*. 2001. Odile Jacob, Paris.

HACKING Ian, *Representing and intervening*. 1983. Cambridge University Press, Cambridge, UK. *Concevoir et expérimenter. Thèmes introductifs à la philosophie des sciences expérimentales*. 1989. Christian Bourgeois, Paris.

HACKING Ian, *The social construction of what?* 1999. Harvard University Press, Cambridge, MA, USA. *Entre science et réalité. La construction sociale de quoi ?* French translation: Baudouin JURDANT. 2001. La Découverte, Paris.

HOTTOIS Gilbert, *Philosophie des sciences, philosophie des techniques*. 2004.
Odile Jacob, Paris.

HOTTOIS Gilbert, *Qu'est-ce que la bioéthique ?* 2004. Vrin, Paris.

HOUDÉ Olivier, KAYSER Daniel, KOENIG Olivier, PROUST Joëlle, RASTIER François (Eds), *Vocabulaire des sciences cognitives. Neuroscience, psychologie, intelligence artificielle, linguistique et philosophie*. 2003.
Quadrige-Presses Universitaires de France, Paris.

HUXLEY Aldous, *Brave New World*. 1932. *Le meilleur des mondes*.
French translation: Jules CASTIER. 1953. Plon, Paris.

JORDAN Bertrand, *Les marchands de clones*. 2003. Seuil, Paris.

KAHN Axel, PAPILLON Fabrice, *Copies conformes. Le clonage en question*. 1998.
Nil, Paris.

KELLER Evelyn FOX, *The century of the gene*. 2000. Harvard University Press, Cambridge, MA, USA. *Le siècle du gène*. French translation: Stéphane SCHMITT. 2003. Gallimard, Paris.

KELLER Evelyn FOX, *Making sense of life: explaining biological development with models, metaphors, and machines*. 2002. Harvard University Press. Cambridge, MA, USA. *Expliquer la vie. Modèles, métaphores et machines en biologie du développement*. French translation: Stéphane SCHMITT. 2004. Gallimard, Paris.

KITANO H., *Foundations of systems biology*. 2001. MIT Press, Cambridge MA, USA.

KRIMSKY Sheldon, *Science in the private interest: Has the lure of profits corrupted biomedical research?* 2003. Rowman & Littlefield, Lanham, Maryland, USA. *La recherche face aux intérêts privés*. French translation: Léna ROZENBERG. 2004.
Seuil, Paris.

LATOUR Bruno, *Pandora's hope. Essays on the reality of science studies*. 1999. Harvard University Press, Cambridge, MA, USA. *L'espoir de Pandore. Pour une vision réaliste de l'activité scientifique*. French translation: Didier GILLE. 2001.
La Découverte, Paris.

LECOURT Dominique, *La philosophie des sciences*. 2nd ed. 2002. Presses Universitaires de France, Paris.

LE DOUARIN Nicole, *Des chimères, des clones et des gènes*. 2000. Odile Jacob, Paris.

LEMIRE Laurent, *Alan TURING. L'homme qui a croqué la pomme*. 2004. Littératures-Hachette, Paris.

LENOIR Noëlle, MATHIEU Bertrand, *Les normes internationales de la bioéthique*. 2nd ed. 2004. Presses Universitaires de France, Paris.

LEWIS G.N., RANDALL M., *Thermodynamics and the Free Energy of Chemical Substances*. 1923. McGraw-Hill, New York.

MACH Ernst, *Analyse der Empfindungen*. 1922. *L'analyse des sensations. Le Rapport du physique au psychique*. French translation: F. EGGERS, J.-M. MONNOYER. 1996. Jacqueline Chambon, Nîmes, France.

MEYER Philippe, TRIADOU Patrick, *Leçons d'histoire de la pensée médicale. Sciences humaines et sociales en médecine*. 1996. Odile Jacob, Paris.

MORANGE Michel, *Histoire de la biologie moléculaire*. 1994. La Découverte, Paris.

MORANGE Michel, *La part des gènes*. 1998. Odile Jacob, Paris.

PAUTRAT Jean-Louis, *Demain le nanomonde. Voyage au cœur du minuscule*. 2002. Fayard, Paris.

POPPER Karl, *Objective knowledge*. 1979. Oxford University Press, Oxford. *La connaissance objective*. French translation: Jean-Jacques ROSAT. 1998. Champs-Flammarion, Paris.

POPPER Karl, *The Postscript to the logic of scientific discovery I. Realism and the aim of science*. 1982. *Le réalisme et la science. Post-scriptum à La Logique de la découverte scientifique I*. French translation: Alain BOYER, Daniel ANDLER. 1990. Hermann, Paris.

POPPER Karl, *The Postscript to the logic of scientific discovery II. The open Universe*. 1982. Hutchinson, London. *L'univers irrésolu. Plaidoyer pour l'indéterminisme. Post-scriptum à La Logique de la découverte scientifique II*. French translation: Renée BOUVERESSE. 1984. Hermann, Paris.

POPPER Karl, *La lezione di questo secolo*. 1992. *The lesson of this century* English translation 1997. *La leçon de ce siècle*. Interviewer: Giancarlo BOSETTI, followed by two essays from Karl POPPER on Liberty and the democratic State. French translation: Jacqueline HENRY, Claude ORSONI. 1993. Anatolia Editions, Paris.

PRIGOGINE Ilya, STENGERS Isabelle, *La nouvelle alliance. Métamorphose de la science*. 2000. Gallimard, Paris.

RIFKIN Jeremy, *The Biotech Century: Harnessing the Gene and Remaking the World*. 1998. Jeremy P. Tarcher/G.P. Putman's Sons, New York. *Le siècle biotech. Le commerce des gènes dans le meilleur des mondes*. French translation: Alain BORIES, Marc SAINT-UPÉRY. 1998. La Découverte & Syros, Paris.

RUFFIÉ Jacques, *Naissance de la médecine prédictive*. 1993. Odile Jacob, Paris.

SÉRALINI Gilles-Eric, *Ces OGM qui changent le monde*. 2004.
Champs-Flammarion, Paris.

SFEZ Lucien, *Le rêve biotechnologique*. 2001. Presses Universitaires de France, Paris.

STENGERS Isabelle, *Sciences et pouvoirs. La démocratie face à la technoscience*. 2002.
La Découverte, Paris.

STENGERS Isabelle et SCHLANGER Judith, *Les concepts scientifiques. Invention et pouvoir*. 1991. Gallimard, Paris.

TATON René, *La Société Moderne*. 1995.
Quadrige-Presses Universitaires de France, Paris.

TAYLOR Gordon Rattray, *Science of life. A picture of biology*. 1963. McGraw-Hill, New York. *Histoire illustrée de la biologie*. French translation: Colette VENDRELY. 1965. Hachette, Paris.

WATSON James, D., GILMAN Mickael, WITKOWSKI Jan, ZOLLER Mark, *Recombinant DNA*. 2nd ed. 1992. W.H. Freeman, New York. *ADN recombinant*. 1994. French translation: Olivier REVELANT. 2nd ed. De Boeck Université, Bruxelles.

WATSON James D., *A Passion for DNA: Genes, Genomes, and Society*. 2000.
Cold Spring Harbor Laboratory Press, USA. *Gènes, génomes et société*.
French translation: Jean MOUCHARD. 2003. Odile Jacob, Paris.

GENERAL WORKS CONCERNING BIOLOGY AND AND THE PHILOSOPHY OF SCIENCE

ALBERTS Bruce, JOHNSON Alexander, LEWIS Julian, RAFF Martin, ROBERTS Keith, WALTER Peter, *Molecular Biology of the Cell*. 5th ed. 2007.
Garland Science Textbooks-Taylor & Francis, London.

AMBOISE Georges, *Dix siècles de philosophie. Le rôle historique des idées générales*. 1946. Editions de Flore, Paris.

ANDLER Daniel, FAGOT-LARGEAULT Anne, SAINT-SERNIN Bertrand,
Philosophie des sciences I. 2002, Gallimard, Paris.

BENMAKHLOUF Ali, *RUSSELL*. 2004. Les Belles Lettres, Paris.

BERG Jeremy Mark, TYMOCZKO John L., STRYER Lubert, *Biochemistry*.
6th ed. revised 2006. W.H. Freeman and Company, New York.

BIEZUNSKI Michel (articles selected by), *La recherche en histoire des sciences*. 1983.
Seuil, Paris.

BLANC Marcel, *L'ère de la génétique*. 1986. La Découverte, Paris.

BOREL J.P., STERNBERG M., *Biochimie et Biologie Moléculaire illustrées*. 2000.
Frison-Roche, Paris.

BRAUDEL Fernand, *Grammaire des Civilisations*. 1987. Flammarion, Paris.

CANGUILHEM Georges, *La connaissance de la vie*. 1992. 2nd ed. revised 1998. Vrin, Paris.

CANGUILHEM Georges, *Le normal et le pathologique*. 1966. 5th ed. 1994. Quadrige-Presses Universitaires de France, Paris.

CANGUILHEM Georges, *Etudes d'histoire et de philosophie des sciences concernant les vivants et la vie*. 1968. 7th ed. 1994. Vrin, Paris.

CANGUILHEM Georges, *Idéologie et rationalité dans l'histoire des sciences de la vie. Nouvelles études d'histoire et de philosophie des sciences*. 2nd ed. revised 1993. Vrin, Paris.

CANGUILHEM Georges, *Ecrits sur la médecine*. 2002. Seuil, Paris.

CHANGEUX Jean-Pierre, *L'homme de vérité*. 2002. Odile Jacob, Paris.

CHAST François, *Histoire contemporaine des médicaments*. 1995. La Découverte, Paris.

COLEMAN William, LIMOGES Camille (Eds), *Studies in History of Biology*. 1981. The Johns Hopkins University Press, Baltimore, London.

COURNOT A.A., *Matérialisme, vitalisme, rationalisme. Etude sur l'emploi des données de la science en philosophie*. 1875. Hachette. *Œuvres complètes*, Vol. V, Claire SALOMON-BAYET (Ed.), 2nd ed. 1987. Vrin, Paris.

CREVIER Daniel, *AI: The tumultuous history of the search for artificial intelligence*. 1994. Basic Books, New York. *A la recherche de l'intelligence artificielle*. French translation: Nathalie BUCSEK. 1997. Champs-Flammarion, Paris.

DAGOGNET François, *Philosophie de l'image*. 1986. Vrin, Paris.

DAUMAS Maurice (Ed), *Histoire de la science*. 1957. La Pléiade-Gallimard, Paris.

DEBRU Claude, *L'esprit des protéines. Histoire et philosophie biochimiques*. 1983. Hermann, Paris.

DELIUS Christoph, GATZEMEIER Matthias, SERTCAN Deniz, WÜNSCHER Kathleen, *Geschichte der Philosophie: von der Antike bis Heute*. 2000. Könemann, Köln. *Histoire de la philosophie, de l'Antiquité à nos jours*. French translation: Christian MUGUET. 2000. Könemann Verlagsgesellschaft mbH, Köln.

DE WIT Hendrik C.D. *Ontwikkelingsgeschiedenis van de Biologie*. 1982. Pudoc, Wageningen, NL. *Histoire du développement de la biologie*. Vol. I. French edition 1992. Presses polytechniques et universitaires romandes, Lausanne, CH.

DUHEM Pierre, *L'aube du savoir. Epitomé du système du monde*. Text presented by Anastasios BRENNER. 1997. Hermann, Paris.

DUMAS Jean-Baptiste, *Mémoires de chimie*. 1843. Bechet, Paris.

DUMAS Marie-Noëlle, *La pensée de la vie chez LEIBNIZ*. 1976. Vrin, Paris.

DUPOUEY Patrick, *Epistémologie de la biologie. La connaissance du vivant*. 1997. Nathan, Paris.

DUPUY Jean-Pierre, *Aux origines des sciences cognitives*. 1999. La Découverte & Syros, Paris.

DURIS Pascal, GOHAU Gabriel, *Histoire des sciences de la vie*. 1997. Nathan, Paris.

EDELMAN Gerald M., *Bright air, brilliant fire: on the matter of mind*. 1992. Basic Books, New York. *Biologie de la conscience*. French translation: Ana GERSCHENFELD. 1992. Odile Jacob, Paris.

FELTZ Bernard, *La science et le vivant. Introduction à la philosophie des sciences de la vie*. 2003. De Boeck Université, Bruxelles.

FEYNMAN Richard P., *The meaning of it all*. 1999. Penguin, London, UK.

FORD B. J., *Images of Science*. 1992. The British Library.

FOUREZ Gérard, *La construction des sciences. Les logiques des inventions scientifiques*. 4th ed. 2002. De Boeck Université, Bruxelles.

GOHAU Gabriel, *Biologie et biologistes*. 1978. Magnard, Paris.

GORNY Philippe, *L'Aventure de la médecine*, 1991. Jean-Claude Lattès, Paris.

GRANGER Gilles-Gaston, *La science et les sciences*. 2nd ed. 1995. Presses Universitaires de France, Paris.

GRMEK Mirko D., *Le chaudron de Médée. L'expérimentation sur le vivant dans l'Antiquité*. 1997. Institut Synthélabo, Le Plessis-Robinson, France.

HACKING Ian, *An introduction to probability and inductive logic*. 2001. Cambridge University Press, Cambridge, UK. *L'ouverture au probable. Eléments de logique inductive*. HACKING Ian, DUFOUR Michel, 2004. Armand Colin, Paris.

HAMBURGER Jean (Ed), *La philosophie des sciences aujourd'hui*. 1986. Gauthier-Villars. Paris.

HÉDON E., *Précis de Physiologie*. 13th ed. revised, updated by Louis HÉDON 1943. Doin, Paris.

JACOB François, *La logique du vivant. Une histoire de l'hérédité*. 1970. Gallimard Paris.

JACOB François, *Le jeu des possibles. Essai sur la diversité du vivant*. 1981. Fayard, Paris.

JACOB François, *La souris, la mouche et l'homme*. 1997. Odile Jacob, Paris.

JACOB Pierre (Ed), *L'âge de la science. Lectures philosophiques. 2 - Epistémologie*. 1989. Odile Jacob, Paris.

JARROSON Bruno, *Invitation à la philosophie des sciences*. 1992. Seuil, Paris.

KAHN Axel, LECOURT Dominique, *Bioéthique et liberté*. 2004. Presses Universitaires de France, Paris.

KARP Gerald, *Cell and Molecular Biology: Concepts and Experiments*. 5th ed. revised 2007. John Wiley & Sons, New York.

KAY Lily E., *The Molecular Vision of Life. Caltech, The Rockefeller Foundation, and the Rise of the New Biology*. 1993. Oxford University Press, New York.

KLEIN Marc, *Regards d'un biologiste. Evolution de l'approche scientifique. L'enseignement médical strasbourgeois*. 1980. Hermann, Paris.

KOYRÉ Alexandre, *Etudes d'histoire de la pensée scientifique*. 1966. Presses Universitaires de France, Paris. 3rd ed. 1985. Gallimard, Paris.

LA COTARDIÈRE Philippe DE (Ed), *Histoire des Sciences de la préhistoire à nos jours*. 2004. Tallandier, Paris.

LALANDE André, *Vocabulaire, technique et critique de la philosophie*. 5th ed. 1947. Presses Universitaires de France, Paris.

LAMBRICHS Louise L., *La vérité médicale*. 1993. Robert Laffont, Paris.

LAUGIER Sandra, WAGNER Pierre (Eds), *Philosophie des sciences. Expériences, théories et méthodes*. 2004. Vrin, Paris.

LAUGIER Sandra, WAGNER Pierre (Eds), *Philosophie des sciences. Naturalismes et réalismes*. 2004. Vrin, Paris.

LECLERCQ René, *Traité de la méthode scientifique*. 1964. Dunod, Paris.

LECLERCQ René, *Histoire et avenir de la méthode expérimentale*. 1960. Masson, Paris.

LECOURT Dominique (Ed.), *Dictionnaire d'histoire et philosophie des sciences*. 1999. Presses Universitaires de France, Paris.

LE GUYADER Hervé, *Théories et histoire en biologie*. 1988. Vrin, Paris.

LEHNINGER A.L., NELSON D.L., COX M.M., *Principles of biochemistry*. 1993. Worth, New York.

LEWIN Benjamin, *Genes VIII*. 2008, Prentice Hall, Englewood Cliffs, NJ, USA.

LOCKE John, *An essay concerning human understanding*. 1690. 1975. Oxford University Press, Oxford, UK. *Essai sur l'entendement humain*. Vol. I and II. French translation: Jean-Michel VIENNE. 2001. Vrin, Paris.

LWOFF André, *Jeux et combats*. 1981. Fayard, Paris.

MAGNER Lois M., *A history of life science*. 2nd ed. 1994. Marcel Dekker, New York, Basel, Hong Kong.

MALHERBE MICHEL, POUSSEUR Jean-Marie (Eds), *Francis BACON. Science et Méthode*. 1985. Vrin, Paris.

MAUREL Marie-Christine, MIQUEL Paul-Antoine, *Programme génétique : concept biologique ou métaphore ?* 2001. Kimé, Paris.

MEULDERS Michel, CROMMELINCK Marc, FELTZ Bernard (Eds), *Pourquoi la science ? Impacts et limites de la recherche*. 1997. Champ Vallon, Seyssel, France.

MICHAUD Yves (Ed), *Les Nouvelles Thérapies*. 2004. Odile Jacob, Paris.

MICHEL Henri, *Les instruments des sciences dans l'art et l'histoire*. 1973. Albert De Visscher, Bruxelles.

MONOD Jacques, *Le hasard et la nécessité. Essai sur la philosophie naturelle de la biologie moderne*. 1970. Seuil, Paris.

NADEAU Robert, *Vocabulaire technique et analytique de l'épistémologie*. 1999. Presses Universitaires de France, Paris.

NEEDHAM Joseph (edited and introduced by), *The chemistry of life. Eight Lectures on the History of Biochemistry*. 1970. Cambridge University Press, Cambridge, UK.

NICOLLE Jean-Marie, *Histoire des méthodes scientifiques. Du théorème de* THALES *à la fécondation in vitro*. 1994. Bréal, Paris.

PANZA Marco, PONT Jean-Claude, *Les savants et l'épistémologie vers la fin du XIXe siècle*. 1995. Albert Blanchard, Paris.

PELMONT Jean, *Enzymes, catalyseurs du monde vivant*. 1997. Grenoble Sciences-EDP Sciences, Les Ulis, France.

PERUTZ Max F., *Is Science Necessary?* 1988. Dutton, New York. *La science est-elle nécessaire ? Essais sur la science et les scientifique*. French translation: Jacques GUIOD, Annette KELLY. 1991. Odile Jacob, Paris.

PENSO Giuseppe, *La conquête du monde invisible, parasites et microbes à travers les siècles*. 1981. Roger Dacosta, Paris.

PICHOT André, *Histoire de la notion de vie*. 1995. Gallimard, Paris.

POINCARÉ Henri, *La science et l'hypothèse*. 1902. Reprinted 1968. Champs-Flammarion, Paris.

POPELARD Marie-Dominique, VERNANT Denis, *Les grands courants de la philosophie des sciences*. 1997. Seuil, Paris.

PROCHIANTZ Alain, *Les anatomies de la pensée. A quoi pensent les calamars ?* 1997. Odile Jacob, Paris.

PROCHIANTZ Alain, *Machine - esprit*. 2001. Odile Jacob, Paris.

RENAN Ernest, *L'avenir de la science*. 1890. Reprinted with presentation, chronology and bibliography by Annie PETIT. 1995. Flammarion, Paris.

REY P., *Anatomie et Physiologie animales et végétales*. 1930. Vuibert, Paris.

ROGUE C., *Comprendre* PLATON. 2002. Armand Colin/VUEF, Paris.

RONAN Colin, *The Cambridge illustrated history of the world's science*. 1983. Newnes Vol. Twickenham, Middlesex, UK. *Histoire mondiale des sciences*. French translation: Claude BONNAFONT. 1988. Seuil, Paris.

ROSMORDUC Jean, *L'histoire des sciences*. 1996. Education-Hachette, Paris.

RUSSELL Bertrand, *The problems of philosophy*. 1912, Oxford University Press, London. *Problèmes de philosophie*. French translation: François RIVENC. 1989. Payot, Paris.

RUSSO François, *Nature et méthode de l'histoire des sciences*. 1983. Albert Blanchard, Paris.

SCHATZMAN Evry, *L'outil théorie*. 1992. Eshel, Paris.

SCHLEGEL Hans Günter, *Geschichte der Mikrobiologie*. 1999. Acta Historica Leopoldina, Deutsche Akademie der Naturforscher Leopoldina, Halle, Saale.

SCHRÖDINGER Erwin, *What is life ?* (1944), with *Mind and Matter* (1958, Cambridge University Press), and *Autobiographical sketches* (1958, Canto, Cambridge, UK). Reprinted 1995. Cambridge University Press, Cambridge, UK.***

SCHRÖDINGER Erwin, *Mind and Matter*. 1958. Cambridge University Press, Cambridge, UK. *L'esprit et la matière*, preceded by *L'Elision, essai sur la philosophie d'E. SCHRÖDINGER*. French translation: Michel BITBOL. 1990. Seuil, Paris.

SCHULZ Walter, *Der Gott der Neuzeitlichen Metaphysik*. 1957, 5th ed. 1974. Günther Neske, Pfullingen. *Le Dieu de la Métaphysique moderne*. French translation, presentation and notes: Jacques COLETTE, 1978. CNRS Editions, Paris.

SERRES Michel, (Ed.), *Eléments d'Histoire des Sciences*. 1989. Bordas, Paris.

SICARD Didier (Ed.), *Travaux du Comité Consultatif National d'Ethique*. 2003. Quadrige-Presses Universitaires de France, Paris.

SIMON G., DOGNON A., *Précis de Physique*. 1941. Masson, Paris.

SINGER Charles, *Histoire de la Biologie*. French translation: F. GIDON, 1934. Payot, Paris.

SINGER Maxine, BERG Paul, *Genes and Genomes*. 1991. University Science Books, Mill Valley, CA, USA. French translation: Jacques BÉCHET, Raymond CUNIN, Nicolas GLANSDORFF, André PIÉRARD. 1992. Vigot, Paris.

SOLER Léna, *Introduction à l'épistémologie*. 2000. Ellipses, Paris.

STENGERS Isabelle, *L'invention des sciences modernes*. 1995. Champs-Flammarion, Paris.

TATON René, *Causalités et accidents de la découverte scientifique*. 1955. Masson, Paris.

THE EUROPEAN DANA ALLIANCE FOR THE BRAIN, *Vision du cerveau : un rapport sur les progrès récents de la recherche sur le cerveau*. Update 2004. *Cerveau et Immunité*. 2004. Prilly, CH.

THUILLIER Pierre, *D'ARCHIMEDE à EINSTEIN. Les faces cachées de l'invention scientifique*. 1988. Fayard, Paris.

THUILLIER Pierre, *Science et société. Essais sur les dimensions culturelles de la science*. 1997. Librairie Générale Française, Paris.

VERGELY Bertrand, *Les philosophes du Moyen Âge et de la Renaissance*. 1998. Editions Milan, Toulouse, France.

VERGELY Bertrand, *Les grandes interrogations de la connaissance*. 2000. Editions Milan, Toulouse, France.

VERGNIOUX Alain, *L'explication dans les sciences*. 2003. De Boeck Université, Bruxelles.

VIENNOT Laurence, DEBRU Claude, *Enquête sur le concept de causalité*. 2003. Presses Universitaires de France, Paris.

VIGNAIS Pierre, *La biologie des origines à nos jours. Une histoire des idées et des hommes*. 2001. Grenoble Sciences-EDP Sciences, Les Ulis, France.

Virieux-Reymond A., *L'épistémologie*. 1966. Presses Universitaires de France, Paris.

Wagner Pierre (Ed.), *Les philosophes et la science*. 2002. Gallimard, Paris.

Watson James D., Baker Tania A., Bell Stephen P., Gann Alexander, Levine Michael, Losick Richard, *Molecular Biology of the Gene*. 6th Int. ed. 2008. Pearson Education-Benjamin Cummings, USA.

Wojtkowiak Bruno, *Histoire de la chimie. De l'alchimie à la chimie moderne*. 1998. TEC & DOC Lavoisier, Paris.

INDEX OF PERSONAL NAMES

A

ABBE Ernst (1840 - 1905) 176
ABÉLARD Pierre (1079 - 1142) 32,35
ACQUAPENDENTE (or AQUAPENDENTE) Gerolamo FABRIZZI, known by the name of FABRICIUS D'ACQUAPENDENTE (1533 - 1619) 50,54,55,61-63,74,130
ADAMS John (1735 - 1826) 128
ADANSON Michel (1727 - 1806) 67
ADDISON Thomas (1793 - 1860) 284
ADELARD of Bath (1070 - 1150) 32
AESCULAPUS (ASKLEPIOS) 14
AGRICOLA, Georg BAUER, known by the name of (1494 - 1555) 43,44
AIGNER Achim (b. 1965) 289
ALBERT the Great, Albert VON BOLLSTADT (around 1193 - 1280) 35,37,42
AL-BÎRÛNÎ (973 - 1048) 29
ALCMEON of Croton (6th c. BC) 14,15,24,63
ALCUIN, english name EALWHINE, latin name ALBINUS FLACCUS (735 - 804) 31
ALDROVANDI Ulisse (1522 - 1605) 82,197
ALEMBERT Jean LE ROND D' (1717-1783) 143
ALEXANDER III the Great (356 - 323 BC) 20,22,23
AL-KHUWÂRIZMÎ, or IBN MÛSÂ AL-KHUWÂRIZMÎ (783 - 850) 29
AL-MAMÛN (9th century) 30
AL-QURASHI, or IBN AL-NAFÎS (1210 - 1288) 30,60
AL-RÂZÎ, or RHAZES (865 - 923) 41
ALZHEIMER Alois (1814 - 1917) 168,287,292,294
AMBROS Victor Robert 262
AMICI Giovanni, Battista (1786 - 1863) 176
AMPÈRE André-Marie (1775 - 1836) 141
ANAXAGORUS of Clazomenae (500 - 428 BC) 13
ANAXIMANDER (610 - 545 BC) 9,10
ANAXIMENES (580 - 530 BC) 9,10,13
APOLLONIUS of Perga (262 - 190 BC) 26,27,31,143
ARBER Werner (b. 1929) 246
ARCHIMEDES (287 - 212 BC) 27,31
ARCY THOMPSON Wentworth D' (1860 - 1948) 225-227
ARISTARCHUS of Samos (310 - 230 BC) 26

ARISTOTLE (384 - 322 BC) 3,13,16,18-23,25, 26,29-31,34-38,45,46,50,53, 58,59,65,67,88-93,114,126,134
ARROUET François Marie, known by the name of VOLTAIRE (1694 - 1778) 143
ARYABHATA (6th century) 29
ASELLI Gasparo (1581 - 1626) 64
ASKLEPIOS (AESCULAPUS) 14
ASTRUC Jean (1684 - 1766) 78
ATLAN Henri (b. 1931) 328
AURELIUS AUGUSTINUS, saint AUGUSTINE (354 - 430) 28
AVERROES, IBN RUSHD, known by the name of (1126-1198) 29,61
AVERY Oswald Theodore (1877 - 1955) 243,322
AVICENNA, IBN SINĀ, known by the name of (980 - 1037) 25,29,61
AVOGADRO Amadeo (1778 - 1856) 140

B

BACHELARD Gaston (1884 - 1962) 235,236,241
BACON Francis (1561 - 1626) 113-120,123, 125,133,135,148,153,229,233,235,345
BACON Roger (1214 - 1294) 3, 30,35-37,42,98,349
BAER Karl Ernst VON (1792 - 1876) 65
BALARD Antoine Jérôme (1801 - 1879) 153
BALTIMORE David (b. 1938) 246,247,251
BARRESWIL Charles Louis (1817 - 1870) 155,156,164
BARTCH 151,152
BARTHEZ Paul Joseph (1734 - 1806) 122
BARTHOLIN Thomas (1616 - 1680) 64,77
BATAILLON Eugène (1864 - 1953) 231
BATESON William (1861 - 1926) 322
BAUER Georg, known by the name of AGRICOLA (1494 - 1555) 43,44
BAUHIN Jean (1541 - 1613) 67
BAUHIN Gaspard (1560 - 1624) 67
BEADLE George (1903 - 1989) 191,322,341
BECKER Joachim (1635 - 1682) 44
BECQUEREL Henri (1852 - 1908) 143
BEHRING Emil VON (1854 - 1917) 158,184
BELL Charles (1774 - 1842) 87,146,320
BELON Pierre (1517 - 1564) 67

BELOUSOV Boris P. (1893 - 1970) 228
BENABID Alim-Louis (b. 1942) 287
BERG Paul (b. 1936) 249
BERGSON Henri (1859 - 1941) 148,233
BERKELEY George (1685 - 1753) 120
BERNARD Claude (1813 - 1878) 1,77,102
127,146-150,155,156,164,166,
168,179,186,230,265,283-285,331
BERNOULLI Jacques (1654 - 1705) 128
BERNOULLI Jean (1667 - 1748) 128
BERT Paul (1833 - 1886) 146
BERTALANFFY Ludwig VON
(1901 - 1972) 180
BERTHELOT Marcellin (1827 - 1907) 232
BERTHOLLET Claude (1748 - 1822) 102,110
BERZELIUS Jöns (1779 - 1848) 140,232
BICHAT Xavier (1771 - 1802) 122
BIOT Jean-Baptiste
(1774 - 1862) 140,173,174
BI SHENG (11th century) 32
BLACK Joseph (1728 - 1799) 102,103
BLAMONT Jacques-Emile (b. 1926) 23
BOCK Jerome (1498 - 1554) 65
BOERHAAVE Hermann
(1668 - 1738) 78,79,122
BOHR Niels (1885 - 1962) 143
BOLLSTADT Albert VON, ALBERT the Great
(around 1193 - 1280) 35,37,42
BONNET Charles (1720 - 1793) 73,83,84,123
BORDEU Théophile DE (1722 - 1776) 122
BORELLI Giovanni Alfonso
(1608 - 1679) 76,78,79,81
BOUSSINGAULT Jean-Baptiste
(1802 - 1887) 140,155,179
BOVERI Theodor (1862 - 1924) 189
BOYER Herbert (b. 1936) 249,317,330
BOYLE Robert
(1626 - 1691) 1,44, 50,77, 88,96,97,99,
113,116-119,129,131,133,135,148,229,345
BRAHE Tycho (1546 - 1601) 89,90
BRAUDEL Fernand (1902 - 1985) 125,132
BRENNER Sidney (b. 1927) 195
BRETONNEAU Pierre (1787 - 1862) 284
BRIGGS Robert (1911 - 1983) 278
BRIGHT Richard (1789 - 1858) 284
BRINSTER Ralph (b. 1932) 251
BROCA Paul (1824 - 1880) 168,292
BROWN Patrick O. (b. 1954) 253,255
BRÜCKE Ernst Wilhelm VON
(1819 - 1892) 146,231
BRUNFELS Otto (1489 - 1534) 65
BRUNO Filippo, brother GIORDANO
(1548 - 1600) 89,91
BUCHNER Eduard (1860 - 1917) 180,221,230
BUDDHA (around 536 -480 BC, or according to historians 480 - 400 BC) 113

BUFFON Georges Louis LECLERC DE
(1707 - 1788) 35,67,83,143
BUNSEN Robert (1811 - 1899) 175
BURIDAN Jean (1295 - 1385) 38

C

CAESALPINUS, Andrea CESALPINO
(1519 - 1603) 61,62,67
CAGNIARD-LATOUR Charles
(1777 - 1859) 179
CALVIN, Jean CAUVIN, known by the name
of (1509 - 1564) 60,125
CAMPANELLA Tommaso (1568 - 1639) 116
CAMPBELL Keith H. (b. 1954) 278
CANGUILHEM Georges
(1904 - 1995) 146,161,326,327
CARLSSON Arvid (b. 1923) 288
CARNOT Sadi Nicolas (1796 - 1832) 141
CARREL Alexis (1873 - 1944) 220
CASSINI Jean Dominique (1625 - 1712) 128
CASTIGLIONI Arturo (1874 - 1953) 58
CATHERINE DE MEDICI (1519 - 1589) 127
CAVAZZANA-CALVO Marina 270
CAVENDISH Henry (1731 - 1810) 102,103
CAVENTOU Joseph Bienaimé
(1795 - 1877) 297
CECH Thomas (b. 1947) 248
CELSIUS Anders (1701 - 1744) 75
CESALPINO Andrea (CAESALPINUS)
(1519 - 1603) 61,62,67
CESI Federico (1583 - 1630) 128
CHAIN Ernst (1906 - 1979) 161-163,297
CHAKRABARTY Ananda Mohan
(b. 1938) 330
CHAMBERS William (1723 - 1796) 134
CHAMBON Pierre (b. 1931) 327
CHANGEUX Jean-Pierre (b. 1936) 305,307
CHARGAFF Erwin (1905 - 2002) 212,243
CHARLEMAGNE (742 - 814) 31
CHARLES I of England (1600 - 1649) 51
CHARLES II of England (1630 - 1685) 129
CHARLES QUINT or CHARLES V, Holy
Roman Emperor (1500 - 1558) 62
CHARLES V, CHARLES the Wise
(1338 - 1380) 38
CHASE Martha (1927 - 2003) 243,322
CHEVREUL Eugène (1788 - 1889) 139
CLAUSIUS Rodolf Julius Emmanuel
(1822 - 1888) 141
CLEMENT IV (pope) Gui FOULQUES or
le Gros (1200 - 1268) 36
CLEMENT VIII (pope) Ippolito
ALDOBRANDINI (1536 - 1605) 62
COHEN Stanley (b. 1937) 249,330
COHN Ferdinand (1828 - 1898) 84
COITER Volcher (1534 - 1600) 63

Index of Personal Names

COLBERT Jean-Baptiste (1619 - 1683)   128
COLLINS Francis (b. 1950)   248
COLOMBO Realdo (1516 - 1559)   61,62,130
COLOMBUS Christopher (1451 - 1506)   31,125
COMTE Auguste (1798 - 1857)   148,176, 186,229-231,237,238
CONDILLAC Etienne BONNOT DE (1715-1780)   111,123,143
CONDORCET Marie Jean Antoine (1743 - 1794)   116,143
CONFUCIUS (6th -5th centuries BC)   10
CONSTANTINE (emperor) (270 - 337)   28
COPERNICUS Nicolaus (1473 - 1543)   3,17, 89-91,99,135
CORRENS Carl (1864 - 1933)   187
COSTABEL Pierre (1912 - 1989)   130
COULOMB Charles-Augustin DE (1736 - 1806)   98
COURNOT Antoine Augustin (1801 - 1877)   7,73,91,129,139,157,161
CRICK Francis (1916 - 2004)   206,212,244,321
CROMBIE Alistair, Cameron (1915 - 1996)   8
CROMWELL Oliver (1599 - 1658)   51,126
CROOKES William (1832 - 1919)   142
CUENOT Lucien (1866 - 1951)   197
CURIE Marie (1867-1934)   143
CURIE Pierre (1859-1906)   143
CUVIER Georges (1769 - 1832)   20,67

D

DAGUERRE Louis Jacques (1787 - 1851)   142
DALTON John (1766 - 1844)   13,139,140
DANNA Katleen (b. 1945)   246
DARWIN Charles (1809 - 1882)   144,183,231
DAUBENTON Louis (1716 - 1800)   67
DA VINCI Leonardo (1452 - 1519)   93,127
DAVIS Ronald W.   253,255
DE DUVE Christian (b. 1917)   335
DE GRAAF Reinier (1641 - 1673)   65
DEBUS Allen G. (b. 1926)   100
DEHAENE Stanislas (b. 1965)   307
DEMOCRITUS (460 - 370 BC)   13,14,17,38,119
DENNETT Daniel C. (b. 1942)   307
DESCARTES René (1596 - 1650)   14,24,3957,76,78-80,86-88 90,93,94, 113,118,119,121,128,130, 133,135,143,147,148,229,292,320,323
DIDEROT Denis (1713-1784)   122,134,143
DIENERT Frédéric (1874 - 1948)   183
DIONIS Pierre (1643 - 1718)   57
DIOPHANTUS (3rd century)   27
DOLLOND John (1706 - 1761)   72,176
DÖPPLER Christian (1801 - 1853)   186
DRIESCH Hans (1867 - 1941)   184,233

DU BOIS-REYMOND Emil (1818 - 1896)   80, 80,141,146,148,171,172,231
DU FAY Charles François DE CISTERNAY (1698 - 1739)   98
DUCHENNE DE BOULOGNE Guillaume Benjamin (1806 - 1875)   141
DUDERSTADT James J. (b. 1973)   339
DUHEM Pierre (1861 - 1916)   7,235
DULONG Pierre-Louis (1785 - 1838)   140
DUMAS Jean-Baptiste (1800 - 1884)   100,140,155
DUNS SCOTUS John (1274 - 1308)   38,39
DUSSOIX Daisy   246
DUTROCHET Henri (1776 - 1847)   172,173,189

E

EAGLE Harry (1905 - 1992)   220
EALWHINE, latin name ALBINUS FLACCUS, known by the name of ALCUIN (735 - 804)   31
ECCLES John Carew (1903 - 1997)   307,336
EHRLICH Paul (1854 - 1915)   158,177,295
EIJKMAN Christian (1858 - 1930)   185
EINSTEIN Albert (1879 - 1955)   133,143,234
ELIZABETH I (England's queen) (1533 - 1603)   97,113,126
EMPEDOCLES of Agrigentum (490 - 438 BC)   12-14,19,112
ENGELHARDT Hugo Tristram (b. 1941)   329
ENGELMANN Theodor Wilhelm (1843 - 1909)   175,176
EPICURUS (341 - 270 BC)   13,14,119
ERASISTRATUS (310 - 250 BC)   24-26,34,59
ERASMUS of Rotterdam (1467 - 1536)   125
ERATOSTHENES (273 - 192 BC)   27
EUCLID (3rd century BC)   11,26,31
EUDOXUS of Cnidus (406 - 355 BC)   17
EUSTACHIUS, Bartolomeo EUSTACHI (1500 - 1574)   15,61,63

F

FABER Giovanni (1574 - 1629)   68
FABRI DE PEIRESC Nicolas Claude (1580 - 1637)   130
FABRICIUS D'ACQUAPENDENTE, Gerolamo FABRIZZI, known by the name of (1533 - 1619)   50,54,55,61-63,74,130
FAHRENHEIT Daniel (1686 - 1736)   75
FALLOPIUS, Gabrielle FALLOPIO (1523 - 1562)   61,63,130
FARADAY Michael (1791 - 1867)   141
FERCHAULT DE RÉAUMUR René Antoine (1683 - 1757)   75,78,79,84,128
FERDINAND II DE MEDICI (1621 - 1670)   128
FERMAT Pierre DE (1601 - 1665)   128,130

FERNEL Jean (1497 - 1558)	43
FEYERABEND Paul (1924 - 1994)	116,234,236
FEYNMAN Richard Phillips (1918 - 1988)	325
FIELDS Stanley (b. 1955)	300,339
FIRE Andrew (b. 1959)	262
FISCHER Alain (b. 1949)	270
FISCHER Emil (1852 - 1919)	141,179
FLECK Ludwig (1896 - 1961)	237
FLEMING Alexander (1880 - 1955)	161-163,297
FLEMMING Walther (1843 - 1905)	177
FLOREY Howard (1898 - 1968)	161-163,297
FLOURENS Pierre (1794 - 1867)	146,166
FODOR Stephen P.A.	255
FOURCROY Antoine DE (1755 - 1809)	102,110
FOUREZ Gérard (b. 1937)	124
FOURIER Joseph (1768 - 1830)	141
FRAENKEL-CONRAT Heinz (1910 - 1999)	224
FRANCESCHINI Nicolas (b. 1942)	312-314
FRANÇOIS I of France (1494 - 1547)	127
FRANKLIN Benjamin (1706 - 1790)	128
FRANKLIN Rosalind (1920 - 1958)	206,224
FRASCATORO Girolamo (1478 - 1553)	83
FRAUNHOFER Joseph VON (1787 - 1826)	174
FREDERIK II, King of Denmark (1534 - 1588)	89
FREIBERG Theodoric VON (1250 - 1310)	98
FROMHERZ Peter (b. 1942)	315
FRUTON Joseph S. (1912 - 2007)	325
FUCHS Leonard (1501 - 1566)	65
FUNK Casimir (1884 - 1967)	185

G

GALEN, Claudius GALENUS (131 - 201)	24-26,29-31,34,46,50, 56-63,77,86,127,134,135,348,349
GALILEO, Galileo GALILEI (1564 - 1642)	1,50,68,75,88,90-94, 99,119,122,130,135,164
GALVANI Luigi (1737 - 1798)	80
GAMA Vasco DA (1496 - 1524)	125
GARROD Archibald (1857 - 1936)	286,287
GÄRTNER Carl Friedrich (1772 - 1850)	186
GASSENDI Pierre (1592 - 1655)	14,119,130
GAY-LUSSAC Joseph (1778 - 1850)	110,140,179
GEBER, latin name for JĀBIR IBN HAYYĀN (721 - 815)	41
GEIGER Hans (1882 - 1945)	215
GEOFFROY SAINT-HILAIRE Etienne (1772 - 1844)	21
GEORGIAS of Leontium (487 - 430 BC)	13
GENSFLEISCH Johannes, known by the name of GUTENBERG (1400 - 1468)	33
GERARD of Cremona (1114 - 1187)	30
GERBERT of Aurillac, pope SYLVESTRE II from 999 to 1003 (938 - 1003)	32
GERHARDT Charles-Frederic (1816 - 1856)	140
GERLACH Joseph (1820 - 1896)	177
GESNER Konrad (1516 - 1563)	67
GIAMATTI Angelo Bartlett Bart (1938 - 1989)	339
GILBERT William (1540 - 1603)	97,98
GIORDANO (brother), Filippo BRUNO (1548 - 1600)	89,91
GLAUBER Johann Rudolf (1604 - 1668)	44
GLEY Emile (1857 - 1930)	167
GLISSON Francis (1597 - 1677)	64,79
GMELIN Leopold (1788 - 1853)	232
GOETHE Johan Wolfgang (1749 - 1832)	123
GOFFEAU André (b. 1935)	198
GOLGI Camillo (1844 - 1925)	177,213,214,227,321
GOMPERZ Theodor (1832 - 1912)	35
GRAVES Robert (1793 - 1853)	167
GRAY Stephen (1670 - 1736)	98
GREGORY IX (pope) Ugolino CONTI DE SEGNI (1145 - 1241)	33,35
GREW Nehemiah (1641 - 1712)	65,68,70
GRMEK Mirko Drazen (1924 - 2000)	1,200,348
GROSSETESTE Robert (1170 - 1253)	3,30,35-37,349
GUERICKE Otto VON (1602 - 1686)	95,96,98
GUTENBERG, Johannes GENSFLEISCH, known by the name of (1400 - 1468)	33
GUYÉNOT Emile (1885 - 1963)	102
GUYTON DE MORVEAU Louis Bernard (1737 - 1816)	102,110

H

HACKING Ian (b. 1936)	217,218,325
HAECKEL Ernst (1834 - 1919)	232
HALES Stephen (1677 - 1761)	77,172
HALL Marshall (1790 - 1857)	146
HALLER Albrecht VON (1708 - 1777)	79,80,122,236
HAMBURGER Jean (1909 - 1992)	121
HANSEN Emil Christian (1842 - 1909)	183
HARDEN Arthur (1865 - 1940)	221,222
HARRISON Ross (1870 - 1959)	220
HARVEY William (1578 - 1657)	1,25,50-59, 61,62,64,71,73,79,99,118, 119,127,130,131,135,347
HAÜY René Just (1743 - 1822)	160
HEGEL Georg Wilhelm Friedrich (1770 - 1831)	123
HEIDENHAIN Martin (1864 - 1949)	177

HEISENBERG Werner (1901 - 1976) 143
HELMHOLTZ Hermann Ludwig Ferdinand
  VON (1821 - 1894) 142,146,151,172,231
HENLE Jacob Friedrich Gustav
  (1809 - 1885) 320
HENRI IV (1553 - 1610) 57
HENRI Victor (1872 - 1940) 304,305
HENRY VIII (1491 - 1547) 126
HERACLITUS (550 - 480 BC) 10,11
HERO (or HERON) of Alexandria
  (1st century) 27,75
HEROPHILUS (330 - 260 BC) 24,34
HERS Henri-Géry (1923 - 2008) 335
HERSHEY Alfred (1908 - 1997) 243,322
HEVESY DE HEVES Georg (1885 - 1966) 215
HIPPARCHUS of Nicaea (2nd c. BC) 26,27
HIPPOCRATES (460 - 377 BC) 14,15,24,29,58
HOBBES Thomas
  (1588 - 1679) 97,116,117,126,130,345
HOMBERG Wilhelm (1652 - 1715) 44
HOOKE Robert (1635 - 1703) 68-70,96
HOPKINS Frederick Gowland
  (1861 - 1947) 219
HOPPE-SEYLER Felix (1825 - 1895) 142,176
HOTTOIS Gilbert (b. 1946) 325,327
HUME David (1711 - 1776) 120,233
HUNTINGTON George (1850 - 1916) 289,292
HUXLEY Aldous (1894 - 1963) 328
HUXLEY Thomas Henry (1825 - 1895) 232
HUYGENS Christian (1629 - 1695) 93,97,128
HWANG Woo Suk (b. 1953) 277
HYPATHIA (370 - 415) 27,28

I

IBN AL HAYTHAM (965 - 1039) 30
IBN AL-NAFÎS, or AL-QURASHI
  (1210 - 1288) 30,60
IBN ISHÂQ (809 - 877) 29
IBN MÛSÂ AL-KHUWÂRIZMÎ (783 - 850) 29
IBN RUSHD, known by the name of
  AVERROES (1126-1198) 29,61
IBN SINÂ, known by the name of
  AVICENNA (980 - 1037) 25,29,61
INGENHOUSZ Jan
  (1730 - 1799) 102,104,105,136,215
INGRAM Vernon Martin
  (1924 - 2006) 209,287

J

JÂBIR IBN HAYYÂN, known by the name
  of GEBER (721 - 815) 41
JACKSON John (1834 - 1911) 146
JACOB François
  (b. 1920) 152,183,233,243,244,321
JACQUES Jean 333
JANSEN Hans 68

JANSEN Zacharias (1580 - 1638) 68
JAMES I (1566 - 1625) 51
JOHANNSEN Wilhelm
  (1857 - 1927) 187,322,324
JOLIOT Frédéric (1900 - 1958) 143,215
JOLIOT-CURIE Irène (1897 - 1956) 143,215
JORGENSEN Richard (b. 1951) 263
JOULE James Prescott (1818-1889) 141
JUSSIEU Antoine DE (1686 - 1778) 123
JUSSIEU Antoine Laurent DE
  (1788 - 1834) 123
JUSSIEU Bernard DE (1699 - 1777) 67,84,123
JUSTINIAN (482 - 565) 28

K

KAMEN Martin (1913 - 2002) 104,215
KANT Immanuel (1724-1804) 120,121
KEILIN David (1887 - 1963) 176
KEKULÉ VON STRADONITZ Friedrich August
  (1829-1896) 140
KENDREW John (1917 - 1997) 206,207
KEPLER Johannes (1571 - 1630) 88-91
KING Thomas (1921 - 2000) 278
KIRCHOFF Gustav (1824 - 1887) 175
KIRWAN Richard (1733 - 1813) 102
KITASATO Shibasaburo (1852 - 1931) 158,184
KLIN 151,152
KLUG Aaron (b. 1926) 223,224
KNOOP Franz (1875 - 1946) 215
KOCH Robert (1843 - 1910) 152,158,
  181-183,194,237,285,320
KOCHER Theodor (1841 - 1917) 167
KOLBE Hermann (1818 - 1884) 230
KORNBERG Arthur (1918 - 2007) 338
KOYRÉ Alexandre (1902 - 1964) 82,88,92
KREBS Hans A. (1900 - 1980) 336,337
KROGH August (1874 - 1949) 199,215
KUHN Thomas (1922 - 1996) 235,236
KÜHNE Wilhelm (1837 - 1900) 165,166
KUNCKEL (or KUNKEL) VON LÖWENSTERN
  Johann (1638 - 1703) 44
KÜTZING Friedrich (1807 - 1897) 180

L

LA BOE Franz DE, known by the name of
  SYLVIUS (1614 - 1672) 64,78
LAENNEC René (1781 - 1826) 284
LAKATOS Imre (1922 - 1974) 236
LAMARCK Jean-Baptiste (1744 - 1829) 144
LA METTRIE Julien OFFROY DE
  (1709 - 1751) 121,229
LANGERHANS Paul (1847 - 1888) 167,276
LAPLACE Pierre Simon DE
  (1749 - 1827) 111,122,164
LATOUR Bruno (b. 1947) 325
LAURENT Auguste (1807 - 1853) 140

LAVOISIER Antoine Laurent DE
  (1743 - 1794)   97,98,102,103,**105-112**,
              123,136,140,164,236,237,239,323
LE BEL Joseph (1847 - 1930)   141,174
LECLERC DE BUFFON Georges Louis
  (1707 - 1788)   35,67,83,143
LEDUC Stéphane (1853 - 1939)   225-227
LEIBNIZ Gottfried Wilhelm
  (1646 - 1716)   119,120,128
LEONARDO DA VINCI (1452 - 1519)   93,127
LEOPOLD of Tuscany (1747 - 1792)   128
LEUCIPPUS of Abdera (5th century BC)   13
LEWIS Gilbert Newton (1875 - 1946) 143,347
LIEBIG Justus VON (1803 - 1873) 140,179,180
LINNAEUS Carl (1707 - 1778)   67,123
LISTER Joseph (1827 - 1912)   181
LOCKE John (1632 - 1704)   114,119
LŒB Jacques (1859 - 1924)   231
LOUIS Pierre (1787 - 1872)   284
LOUIS XIII (1601 - 1643)   57
LOUIS XIV (1638 - 1715)   57,126,127
LOUVOIS François-Michel LE TELLIER DE
  (1639 - 1691)   128
LUCRETIUS (97 - 55 BC)   14,21
LUDWIG Carl (1816 - 1895) 146,169,171,231
LULL Ramon (1232 - 1316)   42
LUTHER Martin (1483 - 1546)   125
LWOFF André (1902 - 1994) 134,152,155,341
LYSSENKO Trofim (1898 - 1976)   237

M

MACH Ernst (1838 - 1916)   232
MAGELLAN Ferdinand (1480 - 1521)   125
MAGENDIE François
  (1783 - 1855)   87,145-147,164,320,331
MAIMONIDES (1135 - 1204)   30
MALEBRANCHE Nicolas (1638 - 1715)   120
MALPIGHI Marcello
  (1628 - 1694)   55,56,68,70,71,76,136
MALUS Etienne Louis (1775 - 1812)   140
MARCUS AURELIUS (121 - 180)   25
MAREY Etienne Jules (1830 - 1904)   169
MARICOURT Pierre DE (13th century) 36,98
MARIOTTE Edme (1620 - 1684)   44,88
MARSHALL Barry (b. 1951)   146,152
MARTIN Archer (1910 - 2002)   177,210
MATHER Increase (1639 - 1723)   128
MAUPERTUIS Pierre Louis DE
  (1698 - 1759)   143
MAXWELL James Clerk (1831 - 1879)   141
MAYER Julius Robert VON
  (1814 - 1878)   142,231
MAYR Ernst (1904 - 2005)   122,320
MAZARIN Jules (1602 - 1661)   126
MCCARTY MacLyn (1911 - 2005)   243,322
MCCLINTOCK Barbara (1902 - 1992)   199

MCCULLOCH Ernest A. (b. 1926)   272
MCLEOD Colin (1909 - 1972)   243,322
MCMUNN Charles Alexander
  (1852 - 1911)   176
MEDICI Catherine DE (1519 - 1589)   127
MEDICI Ferdinand II DE (1621 - 1670) 128
MELLO Craig C. (b. 1960)   262
MENASCHÉ Philippe (b. 1950)   275
MENDEL Gregor
  (1822 - 1884) 144,186,187,189,197,287
MENDELEEV Dimitri (1834 - 1907)   140
MENTEN Maud (1879 - 1960)   305
MERING Joseph VON (1849 - 1908)   167
MERSENNE Marin (father)
  (1588 - 1648)   94,128,130
MESELSON Mathew (b. 1930)   321
METCHNIKOFF Elie
  (1845 - 1916)   157-160,184,217,342
MICHAËLIS Leonor (1875 - 1949)   188,305
MIESCHER Johann Friedrich
  (1844 - 1895)   243
MILL John Stuart (1808 - 1873)   233
MINKOWSKI Oscar (1858 - 1931)   167
MITCHELL Peter (1920 - 1992) 316,318,321
MITSCHERLICH Eilhard (1794 - 1863)   140
MIZUTANI S.   247
MOIVRE Abraham (1667 - 1754)   128
MONOD Jacques
  (1910 - 1976) 132,152,183,2444,305,321
MOORE Stanford (1913 - 1982)   213
MORGAGNI Giovanni Battista
  (1682 - 1771)   80
MORGAN Thomas Hunt
  (1866 - 1945)   187195,322
MO ZI (480 - 390 BC)   93
MÜLLER Hermann (1890 - 1967)   322
MÜLLER Johannes
  (1801 - 1858)   122,146,148,231
MULLIS Kary R. (b. 1944)   252

N

NAPIER John (1550 - 1617)   127
NATHANS Daniel (1928 - 1999)   246
NEEDHAM John Turbeville (1713-1781)   83
NEEDHAM Joseph (1900 - 1995)   132,233
NESTORIUS (380 - 451)   30
NEWTON Isaac (1642 - 1727)   44,88,90,93,
                          98,99,122,128,135
NICOLLE Charles (1866 - 1936)   15
NIÉPCE Joseph Nicéphore (1765 - 1833) 142
NOBEL Alfred (1833 - 1896)   15
NOMURA Masayasu (b. 1923)   224

O

OCKHAM William of (1280 - 1349)   36,38,39
O'FARREL Patrick (b. 1949)   210

OFFROY DE LA METTRIE Julien
(1709 - 1751) 121,229
OKEN Lorenz (1799-1851) 123
ORESME Nicolas (1320 - 1382) 38
ORIGEN Adamantius (185 - ca. 254) 28

P

PALADE Georges (b. 1912) 200
PALISSY Bernard (1510 - 1590) 44,127
PALMITER Richard (b. 1942) 251,252
PANDER Christian (1794 - 1865) 184
PAPPUS (4th century) 27
PARACELSUS (1493 - 1541) 41,43,44,77
PARÉ Ambroise (1509 - 1590) 43,127
PARISANO Emilio (1567 - 1643) 57
PARKINSON James (1755 - 1824) 274,276,
284,287,288,292,328
PARMENIDES (515-450 BC) 11,12
PASCAL Blaise
(1623 - 1662) 1,50,88,94,95,128,130,135
PASTEUR Louis (1822 - 1895) 83,135,140,141,
144,152,**153**,**154**,180,183,235,237,330,333
PATIN Gui (1601 - 1672) 57
PAULI Wolfgang (1900 - 1958) 143
PAULING Linus (1901 - 1994) 209
PAVLOV Ivan Petrovitch (1849 - 1936) 166
PAYEN Anselme (1795 - 1871) 165
PECQUET Jean (1622 - 1674) 64
PELLETIER Pierre Joseph(1788 - 1842) 297
PELOUZE Théophile (1807 - 1867) 156,164
PÉRIER Florin 95
PERKIN William (1838-1907) 177
PERRIN Jean (1870 - 1942) 143
PERSOZ Jean-François (1805 - 1868) 165
PERUTZ Max (1914 - 2002) 206,298,334
PETIT Alexis (1791 - 1820) 140
PETIT Pierre 94
PETRI Richard
(1852 - 1921) 161-163,181,182,193
PEURBACH Georg (1423 - 1461) 39
PEYER Hans Conrad (1653 - 1712) 64
PFLÜGER Max Eduard (1829 - 1910) 151,233
PHILIP II of Spain (1527 - 1598) 62
PHILO of Byzantium (3rd century BC) 27
PHILOLAOS of Croton (5th century BC) 11
PIAGET Jean (1896 - 1980) 228
PITTON DE TOURNEFORT Joseph
(1656 - 1708) 67
PLANCK Max (1858 - 1947) 143
PLATO (428 - 348 BC) 11,**16-18**,22,25,
29,31,34,45,46,226
POINCARÉ Henri (1854 - 1912) 153
POLLACK Pierre (b. 1950) 274
POPPER Karl Raimund
(1902 - 1994) 121,234,236
PORATH Jerker Olof (b. 1921) 213

PORTER Keith (1912 - 1997) 200
POTTER Van Rensselaer (1911 - 2001) 327
POUCHET Félix Archimède
(1800 - 1872) 153
PRIESTLEY Joseph
(1733 - 1804) 102-104,105,108,136,215
PRIGOGINE Ilya (1917 - 2003) 228,343
PRIMEROSE James (1592 - 1659) 57
PROCHASKA Georg (1749 - 1820) 87
PROTAGORAS of Abdera (485 - 410 BC) 13
PROUT William (1785 - 1850) 140
PRUSINER Stanley (b. 1942) 320
PTOLEMY Claudius (90 - 168) 26,31,36,88-90
PTOLOMY SÔTER (around 360 - 283 BC) 23
PUCÉAT Michel (b. 1961) 275
PURKINJE Johann (1787 - 1869) 147
PYTHAGORUS (570 - 480 BC) 10,16
PYTHEAS (4th century BC) 17

R

RANDALL M. 347
RAY John (1627 - 1705) 20,67
RÉAUMUR René Antoine FERCHAULT DE
(1683 - 1757) 75,78,79,84,128
REDI Franscisco
(1626 - 1697) 82,131,135,146,238
REGIOMONTANUS, Johannes MÜLLER VON
KÖNIGSBERG, known by the name of
(1436 - 1476) 39
REGNAULT Henri Victor
(1810 - 1878) 169,170
REICHERT Karl (1811 - 1883) 148
REMAK Robert
(1815 - 1865) 144,148,161,236
RENAN Ernest (1823 - 1892) 126
REVERDIN Jean-Louis (1842 - 1929) 167
REY Jean (1583 - 1645) 43,44,98,107
RHAZES, or AL-RÂZÎ (865 - 923) 41
RIOLAN Jean (1580 - 1657) 57,64
RITTENBERG David (1906 - 1970) 215
ROBERT of Chester (12th century) 30
ROBERVAL Gilles PERSONNE
or PERSONIER DE (1602 - 1675) 128
ROBIN Charles (1821 - 1885) 231
ROGER Jacques (1920 - 1990) 63
ROLANDO Luigi (1773 - 1831) 166
RONDELET Guillaume (1507 - 1566) 67
RÖNTGEN Wilhelm (1845 - 1923) 142,290
ROUX Wilhelm (1850 - 1924) 184
RUBEN Samuel (1913 - 1943) 104,215
RUBNER Max (1854 - 1932) 169
RUDBECK Olav (1630 - 1702) 64
RUDOLF II VON HABSBOURG
(1576 - 1602) 90
RUFFIÉ Jacques (1921 - 2004) 283,293
RUSKA Ernst (1906 - 1988) 200

RUSSELL Bertrand (1872 - 1970) 232
RUTHERFORD Ernest (1871 - 1937) 143
RUVKUN Gary (b. 1951) 262
RUYSCH Frederik (1638 - 1731) 65

S

SAGREDO Giovanni, Francesco
 (1571 - 1620) 90
SAINT VICTOR Hugues DE (1096 - 1141) 34
SALADIN (1138 - 1193) 30
SALVIATI filippo (1582 - 1614) 90
SANCTORIUS Santorio (1561 - 1636) 74
SANGER Frederick (b. 1918) 247
SCHEELE Carl Wilhelm
 (1742 - 1786) 102,**103**-105,108,112
SCHELLING Friedrich Wilhelm Joseph
 (1775 - 1854) 123
SCHLEIDEN Matthias Jacob
 (1804 - 1881) 144,160,231
SCHLICK Moritz (1882 - 1936) 232
SCHOENHEIMER Rudolf (1898 - 1941) 215
SCHRÖDINGER Erwin (1867 - 1961) 14,322
SCHROETER Joseph (1835 - 1894) 181
SCHWANN Theodor
 (1810 - 1882) 144,148,180,232
SEEMAN Nadrian (b. 1945) 259-261
SÉGUIN Armand (1767 - 1885) 107
SEMMELWEIS Ignac Fülöp
 (1818 - 1865) 151,152,185
SENEBIER Jean (1742 - 1809) 78,102
SERVETO (SERVET) Miguel
 (1509 - 1553) 30,60-62,135
SHERRINGTON Charles Scott
 (1857 - 1952) 166
SJÖSTRAND Fritiof (b. 1912) 200
SKODA Joseph (1805 - 1881) 284
SMITH Hamilton (b. 1931) 246
SOCRATES (470 - 399 BC) 8,13,16,17,18
SONG Ok-Kyu 300
SØRENSEN Søren Peter Lauritz
 (1868 - 1939) 188
SPALLANZANI Lazzaro (1729 - 1799) 73,
 78,79,83,84,135,164,238
SPINOZA Baruch (1632 - 1677) 120
STAHL Franklin (b. 1929) 321
STAHL Georg Ernst (1660 - 1734) 44,
 100-102,108,120,122,229,236
STAUDINGER Hermann (1881 - 1965) 200
STEIN William (1911 - 1980) 213
STELLATI Francesco (1577 - 1653) 68
STENGERS Isabelle (b. 1949) 343
STENON, STENSEN Niels (1638 - 1687) 64
STRATO (335 - 265 BC) 26
SVEDBERG Theodor (1884 - 1971) 213
SWAMMERDAM Jan
 (1637 - 1680) 65,66,86,348

SYLVESTER II (pope) GERBERT of Aurillac
 (938 - 1003) 32
SYLVIUS, Franz DE LA BOE, known by the
 name of (1614 - 1672) 64,78
SYNGE Richard (1914 - 1994) 210

T

TARTAGLIA, Niccolò FONTANA, known by
 the name of (1499 - 1557) 93,127
TATON René (1915 - 2004) 49
TATUM Edward (1909 - 1975) 191,322,341
TEMIN Howard M. (1934 - 1994) 247
TEMPIER Etienne (ca. 1210 - 1279) 37,38
THALES (625 - 550 BC) 8,9
THENARD Louis Jacques (1777 - 1857) 140
THEOPHRASTES (372 - 287 BC) 22,23,65,67
THOMAS AQUINAS (saint)
 (1225 - 1274) 18,22,35,37,329
THOMPSON Wentworth D'ARCY
 (1860 - 1948) 225,227
THUCYDIDES (460 - ca. 395 BC) 15
TIEDEMANN Friedrich (1781 - 1861) 155,232
TILL James Edgar (b. 1931) 272
TISELIUS Arne (1902 - 1971) 209
TORRICELLI Evangelista
 (1608 - 1647) 75,94,95,135
TOURNEFORT Joseph PITTON DE
 (1656 - 1708) 67
TREMBLEY (or TREMBLAY) Abraham
 (1700 - 1784) 82,84,85,135,272
TREVIRANUS Gottfried (1776 - 1837) 144
TROUSSEAU Armand (1801 - 1867) 284
TSCHERMAK-SEYSENEGG Erich VON
 (1871 - 1962) 187
TSVET (or TSWETT) Mikhaïl Semenovich
 (1872 - 1919) 210
TURING Alan (1912 - 1954) 227,307
TYNDALL John (1820 - 1893) 84

U

UNZER August (1727 - 1799) 87
URBAN VIII (pope) Maffeo BARBERINI
 (1568 - 1644) 91
UREY Harold (1893 - 1981) 215
USSING Hans (1911 - 2000) 215

V

VALENTIN Basil (15th century) 43
VAN BENEDEN Edouard (1846 - 1910) 189
VAN HELMONT Jan Baptist
 (1577 - 1644) 44,77,80-82,103
VAN LEEUWENHOEK Antonie
 (1632 - 1723) 56,69,72,73
VAN'T HOFF Jacobus Henricus
 (1852 - 1911) 141,174
VASCO DA GAMA (1496 - 1524) 125

INDEX OF PERSONAL NAMES

VAUCANSON Jacques DE
  (1709 - 1782)   27,122,312
VENTER Craig (b. 1946)   248,258,330
VESALIUS Andreas, latin name
  for Andries VAN WESEL
  (1514 - 1564)   3,61,62,127,130,135
VIETE François (1540 - 1603)   127
VIEUSSENS Raymond (1641 - 1715)   64
VIGNAIS Pierre, Vital (1926 - 2006)   202
VILLENEUVE Arnauld DE (1235 - 1311)   42
VINCI Leonardo DA (1452 - 1519)   93,127
VIRCHOW Rudolf (1821 - 1902)   144,148,236
VIVIANI Vincenzo (1622 - 1703)   94
VOGT Walter (18?? - 19??)   184
VOLTA Alessandro (1745 - 1827)   80,98
VOLTAIRE, François Marie ARROUET, known
  by the name of (1694 - 1778)   143
VRIES Hugo DE (1848 - 1935)   187

W

WAGNER Rudolph (1805 - 1864)   186
WALDEYER Wilhelm (1836 - 1926)   177
WALPOLE Horace (1717 - 1797)   150
WARBURG Otto (1883 - 1970)   169,170,188
WARREN Robin J. (b. 1937)   152
WATSON James
  (b. 1928)   206,207,212,244,271,321
WEAVER Warren (1894 - 1978)   191
WHARTON Thomas (1610 - 1673)   64
WHYTT Robert (1714 - 1766)   87
WIENER Norbert (1894 - 1964)   228
WILCOX Kent W.   246
WILKINS Maurice Hugh Frederick
  (1916 - 2004)   206
WILLIAMS Robley C. (1908 - 1995)   224
WILLIS Thomas (1621 - 1675)   64,86,166
WILMUT Ian (b. 1944)   278
WINOGRADSKY Sergei (1856 - 1953)   179
WITTGENSTEIN Ludwig (1889 - 1951)   232
WÖHLER Friedrich (1800 - 1882)   180,230
WOLF Kaspar Frederik (1733 - 1794)   73

X - Y - Z

XENOPHANES (570 - 480 BC)   11
XENOPHON (430 - 355 BC)   16
YOSHIDA Masasuke   317-319
YOUNG William (1878 - 1942)   221,222
ZENO of Citium (335 - 264 BC)   23
ZENO of Elea (490 - 420 BC)   11,13
ZERNICKE Fritz (1888 - 1966)   205
ZHABOTINSKI Anatol M. (b. 1938)   228
ZOSIMOS of Panapolis (saint)
  (4th century AD)   39

INDEX OF SUBJECTS

Numbers in **bold** type indicate reference to sections.
Numbers in *italic* type indicate reference to figures and portraits.

A

Academies	46,125,128,129
Agrobacterium tumefaciens	266
air	95
atmospheric	105,112
breathable	103,104,107,108
chalky	103,107
dephlogisticated	103,104,110
fire	103
fixed	103
mephitic or foul	103
vital	103,108,110
alchemists	40,46,136
Arab	110
dispensary	*42*
in the Middle Ages	*42*
medieval	4
alchemy	**39**
Arab	41
Chinese	41
Greek	41
medieval	17,39
Alexandria	26,29,134
allegory of the cave	17
ALZHEIMER's disease	292
analysis	
chemical	240
of diseases	239
of photosynthesis	*175*
of reality *via* the instrument	216
structural	142
analytical chemistry applied	
to physiological exploration	179
anatomy	
comparative	21,53,58,63,65,68
human	63
live	63
microscopic	63
microscopic tissular	68
pathological	80
anesthetics	167
animal	
cells	281
experimentation	331
heat	111
rights	331
spirits	87
transgenic	330
animalcules	72,83
animalculists	73
animal-machine	78
animism	122,123
anthrax bacillus	182
antibodies (idea of)	184
anticonformity	336
aphids (parthenogenesis of)	73
aplysia (sea hare or sea slug)	195
neurons	196
apoptosis	195
approach	
a priori	16
a posteriori	16
reductionist	221,222
Arab	
culture	30
influence	30
science	29
trigonometry	29
Arabidopsis thaliana	197,253
ARISTOTLE's philosophy	18,37
artificial intelligence	228
Arts	
Faculty of	33
liberal	34
mechanical	34
ATPase (ATP synthase)	
mitochondrial	203,316,*319*
motor	316
autoradiography	216

B

babies	
bubble	270
designer	329
BACON F.	*113*
doctrine	**113**
Novum Organum Scientiarum	*115*
principle	123

bacteria	72
pathogenic	198
photomicrographs	*182*
pure cultures	181
bacteriophage	193
DNA	243
T4	*190*
T7	193,306
Baghdad	30
balance	
high-precision	105
in physiology	*74*
BERNARD C.	*149*
determinist bible	148
Introduction to the Study	
of Experimental Medicine	*147*
Bible	22
big science	207,338,341,351
biochemistry	142
biochips	*256*
bioenergetics	111,112,142,323
birth of	**111**
bioethics	283,327
codes	326
lay	329
problems	344
bioinformatics	
(biocomputing)	252,304,341-344
biolistics	268
biological membranes	215
cryofracturing	201,*202*
biology	144
cell	181
developmental	181,184
integrated (or integrative)	298,300,308
molecular	6,181,191,193,243
structural	6,181
virtual	225,304,305
biorobot (animat)	312,314
biorobofly	314
BIOT's polarimeter	*174*
biotechnology	7,**242**,**290**,**332**,**344**
biotechnosciences	348,350
bioterrorism	338
blastocyst	279
blood	
movements	58
"pneumatized"	60
BOERHAAVE H.	*80*
BOYLE-MARIOTTE's law	88
BOYLE R.	**116**,*117*
air (vacuum) pump	77,95,*96*,99
three technologies	117
brain	25,56,86,127,168,274,287
	290,*291*,292,306,307,310,312,315,330
BUCHNER's reductionist approach	221

C

Caenorhabditis elegans	195,198,262
cancer (predisposition for)	293
cDNA (complementary DNA)	247,254
cell	68
amoeboid	160
biology	181,203,213,216,239,257,335
compartments	301
differentiation	194,*273*
division	*178*
enzymology	180
haploid	194
innate immunity	*157*
phagocytotic	196
programmed death	195
proteome	210
somatic	270,278
stem	196,*273*
cellular or innate immunity	157
centrifugation	
differential	*214*
isopycnic	*214*
centuries	
5th BC	93
13th	46,98
14th	43,127
15th	43,125,126
16th	61,63,75,124,125
17th	49,63-65,68,73,75,77,79
	88,99,113,126,128,130,135
18th	75,78,79,83,97,98,100
	109,112,117,119-124,131,136
19th	79,83,109,112
20th	84,109
chaperones	225
chemical	
library	296
technology	*101*
chemiosmotic	
mechanism	316
theory	321
chemistry	117
analytical	140,179
biological	142
combinatory	295
computer-aided	296
in silico	296
microbial	192
of gases	97,102,112
of life	323
of living beings	142
organic	139
pathological	323
physiological	142,166,323
pneumatic	102,109
quantitative	100

structural	140
synthetic	140,249
technical	44
chemosynthesis (bacterial)	179
China	15,40,124,132
scholars	93
chip	
DNA	253,254
protein	253,256
Christian	
cosmology	37
doctrine	37
dogma of Genesis	29
era	46
orthodoxy	37
tradition	119
West	132
chromatography	209
affinity	213
ascending	*212*
gel filtration	213
ion-exchange	213
partition	210,*212*
chromosomes	177,195
theory of heredity	187
Church	134
Catholic	33,126
Gallican	126
circulation	
major	*56,*61,62,135
minor (pulmonary)	30,53,*56,*60-62,135
of knowledge	**336**
of the blood	25,**50-53**,*56*-60,71,79
	88,99,118,127,130
civilization	325
medieval	125
classification	
binomial	67
natural system of	67
nosological	185
clinical trials	331
cloning	272, 277,324
animal reproductive	278,324
human	326
human reproductive	327
non-reproductive	324
of human being	327
reproductive	*279,*280
the bias of parthenogenesis	280
the specter of	277
therapeutic	277,*279,*280,282,328
CO_2	103,107,170,180
code	
genetic	192
histone	248

of bioethics	283
transgession of the genetic	264
combustion	112
compartmentation (conditioned)	321
complementary DNA (cDNA)	247,254
complex	
enzyme-substrate	304
macromolecular	222
composition elementary	179
computer technology	296
computing techniques	249
COMTE A.	*230*
concept	
Cartesian	121
cell signaling	299
evocative	321
generalizing	321
lab-in-a-cell	318
metaphorical	323
of "abhorrence of a vacuum"	94
of compartmentation	321
of covalence	143
of Darwinian selection	306
of DNA replication	321
of humanity	18
of ideas	17
of *impetus*	38
of organicism	233
of organism	120
of the animal-machine	118
of the chemistry of life	323
of the gene	322
of the man-machine	123
of the operon	321
of the program	323
of the reflex arc	320
of the reflex center	87
of gene expression regulation	321
of the Universe	38
of vaccination	183
premonitory	320
control	
experiment	149,155
of gene expression	262
controversies	
conceptual	82
philosophical-theological	34
COPERNICUS' heliocentrism	90
Council of Latran	33
creative imagination	336
creativity	338,342
cross-disciplinarity	338,339
culture (philosophical)	28
cybernetics	228,256
cytological map	322
cytology	72

D

Dalton's law	140
daphnia (infection of)	160
deduction	18
Descartes R.	*118*
cardinal principles	118
determinism	146
impact in the life sciences	139
developmental biology	184
diagnosis (pre-implantation genetic)	329
dialectics	11
diastase	165
Dictyostelium discoideum	198
differentiation of ES cells	274
diffraction (X-ray)	207
digestion	78
artificial	79
of DNA	246
of red blood cells	*159*
mechanism of	77
vessel	*40*
discovery	149
allostery	305
by chance	162
Harvey's	50
Metchnikoff's	217
of glycogenesis	156
of the blood circulation	50
of the experimental method	132
unexpected	155
disease	
Alzheimer's	292
cardiovascular	294
hereditary	286
infectious	285
lysosomal	335
Parkinson's	287,288,292
viral	286
dissection	25
of animals	30
of human cadavers	34,61,127
vivisection	25
DNA	
binding domain	301
chip	253,*254,255*
code	248
complementary	247
construction material	**259**
double helices	*245,260*
infectious agent	243
injection	*251*
manipulations	**266,270**
modifications	280
molecular tool	**249**
nanomachine	260,*261*
sequencing	349
self-replication	*245*
strands	244,*245*
transcription	*245*
transforming power	243
doctrine	
Bacon's	113
Cartesian	119
determinist of Claude Bernard	148
Galenic	46
mechanistic	229
of four humors	15
of "Nature Philosophy"	123
dogma	57,135
breakdown of	237
central of molecular biology	193,244
of phlogistics	136
one gene, one enzyme	341
revisions	237
drosophila (fruit fly)	184,187,195,322
drug library	**295**
dsRNA (double-strand RNA)	262
Du Bois-Reymond's induction sledge	*171*
Dutrochet's osmometers	*173*

E

electron microscope	200
transmission	*201*
electron microscopy	*204*
electrophoresis	209
two-dimensional	*211*
elements	
air	9,*12*,63
earth	*12*,63
fire	10,*12*,63,100
water	9,*12*,63
embryo development of the chick	71
embryogenesis	135
embryology	184
embryonic stem cells	→ ES cells
empiricism	5,39
Encyclopedias	136
endocrinology	167
energetics (cellular)	151
engineering	
genetic	249,267,343
instrumental	169
metabolic	306
sciences	344
enzyme	
activity	305
allosteric	305
reaction kinetics	174,304
restriction	*246*
epigenetic	
DNA modifications	280
events	301

Index of subjects

epigenetics 248
epistemologists 132
error
   in experimentation 134
   margin of 131
ES cells 196,273,274,282,288
   differentiation 274
   human (hES cells) 275
Escherichia coli (*E. coli*) 247,323
ESRF (European Synchrotron Radiation Facility) *208*,218
EST (Expressed Sequence Tags) 330
Europe 3,32,41, 61,124,126,133,349
experiment
   crucial 114,154,235
   decisive 150,153
   electrophysiology 172
   of artificial fertilization 73
   microspectroscopic *175*
   on man's oxygen requirement *108*
   on the release of oxygen *107*
   "provoked observation" 149
   swan-necked flask 153,*154*
   to see what happens 150,151,152
   tourniquet 53,54,62
experimental
   demonstration
      of protein interactions 298
   determinism 149
   embryology 184
   medicine 284
   models *190*
   physiology 146
   process 320
   proof 185
   protocol 117
   reductionism 180,219
   sciences 325
experimental method
   a new paradigm 242
   and society 324
   birth in the 17th and 18th centuries 49
   enigma of the discovery 132
   excesses 332
   expansion of 238
   explanatory logic for the birth of 124
   impact on the physical sciences 88
   instrumentation 131
   luck in 161
   metamorphosis 333
   taught and discussed 340
experimental science
   faltering steps 63
   roots of 7
   seen by the 17th and 18th centuries philosophers 112

experimentation
   animal 135
   physiological 169
   quantitative 100
   systematic 133
   the 21th century challenges 241
explanation (cause-and-effect) 99

F

F1 (ATPase sector) 204
Faculty
   of Arts 33,37
   of Theology 37
falsifiability 234
fermentation 109
   alcohol 110
   apparatus *109*
   of sugars 179
fertilization (artificial) 73,135
FLEMING's experiment *162*
fly
   eye *313*
   fruit 195
   vision 312
force
   electrical 98
   formative 121
   motive 121
functions (cognitive and behavioral) 306

G

GALEN
   doctrine 46
   theory 56,59
GALILEO 91
   experiments 92
   laws 88
gases
   composition of 109
   exchanges in living beings 102,103
gene
   definition 323
   development 195
   evolution of the concept 322
   mosaic 323
   regulatory 244
   somatostatin 249
   structural 244
   therapy 270,282
generation (spontaneous) 82,83
genetic
   cause of illness 285
   counseling 293
   engineering 243,249,*267*,317,343
   inheritance 271,351

map	294
recombination	250
genetic code	
expansion	*265*
transgression	**264**
Genetically Modified Organism	
(GMO)	266,324
plants (GMPs)	*267*,282
genome	*309*
annotation	252
bacterial	258
eukaryotic	198
explored	**243**
human	242,247,350
manipulated	**259**
manipulation of the expression	**262**
mitochondrial	329
sequence	198,199
genomics	7,**253**,258
comparative	199
glucose	156
acellular fermentation of	221
fermentation	180
glycogenesis	220
glycolysis	**221**,305
GMO (Genetically Modified	
Organism)	266,*267*,324
plants (GMPs)	*267*,282
Greece	
Ancient	7,29,37,65,133
pre-Socratic	14
Greek	
alchemists	*40*
cities	*9*
civilization	30
philosophers	4,29,30,45,133
philosophy	16
scholars	28
science	8,14,**23**,29
texts	30
thinkers	22,28,35,45
Grenoble Synchrotron (ESRF)	*208*

H

HALLER A. VON	*80*
HARVEY W.	*50*
discovery	**50**
experiment	**54**
theory	**51**,*56*
work	*52*
heart	**25**,**53**
vital spirits	*56*
movements	**58**
heat	
animal	**111**
principle of life	**15**

HELMHOLTZ H. VON	*151*
high-throughput	240
holism	219
homeogenes	
(or development genes)	195,199
in drosophila	184
homunculus	73
HOOKE'S compound microscope	*69*
human	
brain	*291*
drepanocytosis	270
embryonic stem cells (hES cells)	275
gene therapy	270
genome	198
pathologies	196,197,209
reproductive cloning	327
HUME D.	*120*
humoral or adaptive immunity	158
HUNTINGTON'S chorea	289,292
hybrid	
expertise	339
robots	312
hydra	84,272
regeneration of	84,*85*
hydrogen	103

I

iatrochemistry	43,78
iatromechanics	78
idea	
of a "quorum"	130
of irritability	79
of quintessence	42
of technoscience	345
of the pneuma	86
of the reflex	86
of the "thought collective"	237
imagination	336
creative	14,307,345
imaging	
brain	292
cell	301
medical	290
NMRI	*291*
ultrasound	290
immune	
disease	270
system of drosophila	195
immunity	
cellular or innate	157,158,184,196
humoral or adaptive	158,184,197
immunocompatible	281
impact (index)	336
in silico (virtual biology)	304
induction	18
in scientific reasoning	113

Influenzavirus AH1N1	286
instrumentation	131
manufacture of scientific	239
optical	173
instruments	134,136
balances	73,105
barometer	75
calorimeter	105,*106*
cytofluorimeters	220
digestion vessel	*40*
Fluorescence-Activated Cell Sorter (FACS)	220
gasometer	105,*106*
kymograph	*169*
macrorespirometer	*170*
of the 19th century physiologists	*171*
osmometers	172,*173*
polarimeter	*174*
spectroscope	*175*
thermometers	28,73,75
thermoscope	*27,75*
integrons	233
intelligence	336
artificial	228
interfering	
microRNAs	198,248
RNAs	262,*263*,289
isotopes	207
radioactive	215
stable	215
isotopic labeling	**215**

J

judgment	
analytical	120
synthetic	120

K

KANT's thought	121
KEPLER's laws	90
KOCH R.	*182*
bacillus	194
photomicrographs of bacteria	*182*
postulates	183,320

L

Lab In Cell (LIC)	221
language	
of science	111
vernacular	133
LAVOISIER A.L. DE	*105*
experiments	105,*107*
fermentation apparatus	*109*
gasometer and calorimeter	*106*
ideas	102
laboratory	*108*

laws	
BOYLE-MARIOTTE's	88
KEPLER's	90
MENDEL's	187
of electricity	141
of Nature	128
of physiology	168
of propagation of heat	141
lens	
apochromatic	176
condensing	177
life sciences	
idea of quantification	**186**
impact of determinism	**143**
impact of technology	**164**
new disciplines in the 19th c.	**180**
liposomes	202
LISTER J.	*181*
liver	25
natural spirits	*56*
glycogenic function	155
living beings	
mathematical modeling	**304**
patentability	**329**
understanding of	**324**
understanding of the functions	**298**
LOCKE J.	*119*
logic	18
luck	150,161
LUDWIG's kymograph	*171*
lungs	53,55,*56*,60,95
lysosomes	335

M

machine	
electrical	98
TURING	227
machinistics	230
macromolecular structures	**209**
macrophages	*157*,197
macroscopic structure	70
MAGENDIE F.	*145*
MALPIGHI M.	*68*
mathematics	29,134
mayfly larva	66
measurement	
instruments	188
of gas exchanges	169,*170*
of heat production	169
medicine	
clinical	285
curative	283
customized predictive	295
empirical	**284**
experimental	284,293
Hippocratic	45

predictive	283,293
Padua School of	127
Paris School of	127
preventive	283
regenerative	273
medieval	
alchemy	39,100,130
astronomy	89
civilization	32
magic	135
philosophy	113
scholasticism	46,58,88,99,113-117,131
surgery	43
teaching	28,33
memory (mechanisms)	196
MENDEL G.	*186*
laws	*187*
messenger RNA (mRNA)	255
silencing	*263*
metabolic	
balance	220
engineering	306
metabolism	
cell	179
reductionist approach	**221**
vectorial	321
metabolon	*309,310*
metagenomics	**258**
METCHNIKOFF E.	*158*
experiment	158
discovery	217
method	
birth of the scientific	39
electroencephalography	290
experimental	3,5,7,49,73
Fluorescent In Situ Hybridization (FISH)	256
inductive	113,233
magnetoencephalography	290
micromanipulation	*251*
scientific	7,16,17,36,348
transcriptomics	255
microorganisms (genetically modified)	253
microRNA (miRNA)	248,263
interfering	198
microscope	56,136
17th century	*69*
19th century optical	*178*
atomic force	205,*206*
electron	200,*201*
HOOKE's compound	*69*
optical	142, *201*
simple	72
VAN LEEUWENHOEK's simple	*69*

microscopy	
confocal	205,301
electron	*204*
fluorescence	205,317
optical	176,*204*
phase contrast	205
Middle Ages	22,29,39,60,133
alchemist's dispensary	*42*
philosophical and technological heritage	31
politico-economic context	31
technological revolution	39
miRNA (microRNA)	248,263
mitochondrial	
ATPase (ATP synthase)	204,318,319
mitochondrion	*204*
model organisms	**189**
aplysia	*190*,195
Arabidopsis thaliana	*190*,197
Ascaris megalocephala	**189**
Caenorhabditis elegans	*190*,195
Dictyostelium discoideum	*190*,194
drosophila (*D. melanogaster*)	*190*,195
Escherichia coli	192,193
mouse (*Mus musculus*)	*190*,197
Neurospora crassa	191
paramecium	198
planarium	195
rat (*Rattus norvegicus*)	197
Saccharomyces cerevisiae	194
Saccharomyces pombe	194
sickle cell mouse	270
Xenopus laevis	*190*
yeast	183,194
zebrafish (*Danio rerio*)	*190*,195
modeling (mathematical)	304,305
molecular	
dynamics	299
modeling	296
systematics	257
molecular biology	191,243
central dogma	*245*
morphogenesis	227
mechanistic explanation	225
movements of the arm	81
mRNA (messenger RNA)	255
silencing	*263*
multipotency	274
Mycobacterium tuberculosis	194,196
myoglobin	*207*
myopathies	287

N

nanomachine	
DNA	260,*261*
mitochondrial ATPase	*319*

INDEX OF SUBJECTS 389

Nanoscope II	206
nanotechnology	221
network	
arterial	59
of protein interactions	303
venous	59
neurons (snail brain)	315
Neurospora crassa	192,322,341
neutrophils	157,196
nitrogen	103,112
or foul air	108
NMR (Nuclear Magnetic Resonance)	207
NMRI (Nuclear Magnetic Resonance Imaging)	291
nomenclature (chemical)	110
nomenon	120
nosocomial (infections)	285
numerology	10

O

O_2	→ oxygen
obstacles	
epistemological	236
philosophical	237
pragmatic	236
religious	237
technical	236
verbal	236
optical	
microscope	201
of the 19th century	178
microscopy	204
tweezers	205
organicism	233
osmometer	173
osmotic (formations)	226,227
ovists	73
oxidation	
phenomenon of	105
phosphorylative	316
oxidative (stress)	194
oxygen	102-112,136,170

P

pancreatic fistula	163
paradigm	236,242
change	300
PARÉ A.	127
PARKINSON's disease	274,287,288,292
parthenogenesis	280
of aphids	73
PASCAL B.	95
PASTEUR L.	153
experiment	154
patentability (of living beings)	329
pathogenesis	239

pathology	209
penicillin	161
Penicillium notatum	161,162,163,297
peptidomics	257
period	
Alexandrine	23,27
Hellenistic	26
post-Hellenistic	28
Roman	26
phagocytosis	160,184
pharmacogenetics	295
pharmacogenomics	294
phenome	309
PHILO's thermoscope	27
philosophers	
Ancient Greek	45,88,136
Greek	4,10
Milesian	9,14
of science	135
of the 17th and 18th centuries	112
pre-Socratic	8,16
philosophy	
Aristotlian	114
atomist	45
experimental	113
logic in	16
mechanistic of living beings	122
of DESCARTES	118
of the sciences	120,325
positive	229
Pythagorian	45
phlogistics	108
phlogiston	101
photosynthesis	136
phylogenesis	253
physics	99,134,141
physiology	63,145,238
an experimental science	144
animal	97
birth of	73
experimental	168
operational	164
plant	80
the birth of	73
planarium	158,159,195,196
planets	8,17,19,89
elliptical orbits	89
plants	30
experiments on	104
genetically modified (GMP)	266,267
green	103
tissue	70
transgenesis	269
plasmid (bacterial)	249,267
PLATO's theories	16
plenists	13,94

pluripotent	274
pneuma	15,20,24,59,86
Polymerase Chain Reaction (PCR)	252
positive philosophy	229
Positron Emission Tomography (PET)	292
power	
refractive	99
theological and political	126
pragmatism	18
PRIESTLEY J.	*103*
principle	
BACON's	123
of causality	16,18
of determinism	146
of feedback	228
of finality	18
of inertia	93
of isolating proteins	*302*
of parsimony	38
of partition chromatography	*212*
of permission	329
of precaution	326
of production and detection of mutants	*192*
of serendipity	287
of the conservation of energy	141
teleonomic	19
the limit dilution	181
of two dimensional electrophoresis	*211*
vital	19,123
principles	
of Aristotlian physics	19
of DESCARTES	118
of Mendelian genetics	237
of scientific positivism	148
of statistics	186
of teaching	342
printing	124,133
invention of	32
prion	287,320
yeast Sup45	303
procedure	
"bottom-up" or synthetic	308
"top-down" or analytic	308
process	
deterministic	168
experimental	**235,320**
reductionist	240
protein	140
chips	**253**,*256*
Green Fluorescent Protein (GFP)	205,301
immune (complement)	282
protein-protein interactions	*300,302*
network of interactions	*303*
proteoliposomes	*202*

proteomics	7,253,*256*
analytical	*256,257*
structural	257
proto-oncogenes	293
PTOLEMY's geocentrism	88
publication (scientific)	336
pump (air or vacuum)	77,95,*96*

Q

quantification (in the life sciences)	186
quorum	129
scientific	102,129,133
sensing	130

R

radioactivity (artificial, natural)	143
rationality (scientific)	7,16
reasoning	
a posteriori	149
analogical	157
by analogy	157,158
deductive	16,35
inductive	35
of the Universals	35
scientific	113
reconstruction	
of ribosomes	224
of virus particles	223
REDI F.	*82*
reductionism (experimental)	**219**
reductionist	
approach	**221**
process	**240**
reflex centers	87
REGNAULT and REISET's macrorespirometer	*170*
Renaissance	31,39,47,63,82 124-129,132,133
replication	244,245,249,275 296,306,321,343
reprogramming	280
research	
fundamental	333,334,339,340
organization	334
respiration	95
a combustion	111
animal	108
in Man	107
nature of	109
restriction enzymes	246,252
revolution	
chemical	136
conceptual	143
scientific	49,129
technical	32,143

INDEX OF SUBJECTS

RNA	223
antisense	262
double-strand (dsRNA)	262,*263*
interference	**262**
messenger (mRNA)	244,255,*263*,330
microRNA (miRNA)	248,*263*
single-strand (siRNA)	262
RNAs	
interfering	*263*,289
non-coding	262
robot	
hybrid	**312**
neuromimetic	*313*
robofly	314
robotics	249

S

Saccharomyces cerevisiae	183,194,303
SARS (Severe Acute Respiratory Syndrome)	286
scanning (electron microscopy)	203
school	
Alexandrian	*40*,46
Episcopal	33
Greek pre-Socratic	8
Italian anatomy	43,61
SCID (Severe Combined ImmunoDeficiency)	197,270
science	
17th century	47,88
Arab	40
experimental	5,46,49,59,112
Greek	8
mechanical	135
of bioenergetics	136
of movement	92
political	116
so-called normal	236
start of modern	135
sciences	
chemical	136
experimental	325
life	88
natural	37
physical	88
scientific	
method	349
publication	338
quorum	**129**
realism	325
rationality	7,**16**
reasoning	**113**
researcher	332
revolution	49
semantics	324
screening high-throughput	296,297,344

SEMMELWEIS I.F.	*152*
sequencing	
human genome	247
serendipity	150,155,297
sickle-cell anemia	209,270
siRNA (single-strand RNA)	262
SPALLANZANI L.	*73*
societies (learned)	119,125,135
society	326
knowledgeable	**127**
problem of	327
sociopolitical crises	**124**
somatic	
adult cell	278
gene therapy	270
soul	
animalistic	20
rational	20
vegetative	20
"vital principle"	19
species (first definition of)	67
spectrometry (mass)	210
spectroscope	
first prism optical	175
microspectroscope	176
19th century prism	*175*
spectroscopy	174
Fluorescence Resonance Energy Transfer (FRET)	*261*,301
Phosphorus-31 NMR	*291*
spirit	
academic	242
critical and skeptical	39
entrepreneurial	242
liberal	39
spirits	
animal	25,56,59,63,79,80,86,87,88
natural	25,56,59,63
of wine	75,76,110
theory of	60
vital	25,56,59,60,63,77
STAHL G.E.	*100*
statistics	186
stem cells	84,196,**272**,*273*
adult	276
embryonic	*273*,274
stereochemistry	140,174
stochastic (phenomena)	311
stoichiometry (of the reaction)	77
structures	
macromolecular	**209**
three-dimensional	208,*207*
surgery (experimental)	167
syllogism	18
symbols for the four elements	*12*
synchrotron (X-radiation)	207

391

syndromes 186
system
   PTOLEMY's geocentric 89
   weighing 74
   double hybrid *300*
systematics 63,65,68
   molecular 257
systems biology 300,308,311

T

teaching
   medieval scholastic 22
   of biology in universities 341
   of philosophy 124
   of physics 124
   of the life sciences 333
   scholastic 31
   universities 342
technique
   brain imaging 292
   computing 249
   Fluorescence Resonance Energy
      Transfer (FRET) 261,301
   genetic recombination *250*
   Nuclear Magnetic Resonance
      (NMR) 207
   trapping 301
technological
   platforms 338
   revolution of the Middle Ages 39
technology
   18th century chemical *101,106*
   computer 345
   development of 237
   DNA chip *254,255*
   impact on the life sciences **164**
   industrial 23
   information 228,350
   of hybrid robots 314
   plant transgenesis 269
   recombinant DNA 247
   scientific 120
technosciences 308
teleological
   conception 21
telescope 136
   GALILEO's 90,99
theology 28,37,39,60
theory
   ARISTARCHUS' heliocentric 26
   ARISTOTLE's 38
   atomic 13,143
   cell 161
   chemiosmotic 321
   conceptualism 35
   EMPEDOCLES' of four elements 19

GALEN's 56,62,77
   geocentric of HIPPARCHUS 26
   HARVEY's *56,71*
   mechanistic 57
   of atomism 38
   of epigenesis 73
   of evolution 183
   of heredity 187
   of ideas 35
   of movement 91
   of phagocytosis 160
   of phlogistics 50,105
   of plenism 94
   of preformation 73
   of spirits 77
   of spontaneous generation 82
   of tetravalent carbon 141
   of the conservation of matter 45
   of the cosmos 89
   of the four elements 15,63
   of the "miasma" 320
   of the reflex movement *87*
   of the spontaneous generation 86
   of the three types of spirits 63
   of the "World of RNA" 248
   of transformism 67
   of Universals 38
   phlogistics 44
   Platonic 35
   plenist 20
   pluralist 13
   protoplasmic 232
   theogonic 29
thermodynamics 141
thermometer (spiral) 76
tomography (electronic) 203
totipotent 184,197,278
transcription
   activation 301
   of DNA 244,*245*
transcriptome 255,*309*
transfection 194,270,271
transgenic
   mice 251,289
   plant 197,266
translation (of messenger RNAs) 244
TREMBLEY's apparatus *85*
trial and error 234
TURING machine 227,307

U

ultracentrifugation 209,213
ultrastructure (of the living cell) 203
unipotent 274
Universals (quarrel of the) 35
Universities 33,35,128,135,237,340

V

vaccination	183
vacuists	13,94
vacuum	93,97
proof of the existence	93
VAN LEEUWENHOEK A.	72
simple microscope	69
vectors	
insects, animals	286
of bacteria	157
plasmids	249
viral	270
virus	286
venous	
blood	54,55
system	55,56
valves	54,55
VESALIUS	61
Vienna Circle	232,234
viral	
diseases	286
vectors	270
virtual biology	225,305
virus	
AIDS	296
Ebola	286
Marburg	286
particles	223
reconstitution process	224
Spanish flu	286
tobacco mosaic	223
vitalism	122,229
vitaminology	185
vivisection	164
abolition of	331
animal	64

W

WARBURG's manometric respirometer	*170*,188
World	
Ancient	15,20,26,28,46
Arab	31
Christian	31
inanimate	97
medieval	25
of living beings	348
of RNA	248
of the sky	19
sublunary	19
western	133

X

xenoantibodies	281
xenoantigens	282
xenograft	281
xenotransplantation	281,282
X-ray	142
crystallography	206,*207*
diffraction	207,223

Y

yeast	109,110,141,179,183,194 221,249,300,*303*,325,330
geneticists	198

Z

zebrafish	195,196,272
zymase	180,221

LIST OF ILLUSTRATIONS

CHAPTER I

Figure I.1 - The world of Greek cities,
with its trading posts in Asia Minor, Sicily and Italy 9

Figure I.2 - Symbols for the four elements, air, fire, earth and water,
represented on columns of the Benedictine cloister at Monreale, in Sicily 12

Figure I.3 - Diagram illustrating the principle of PHILO's thermoscope
(3rd century BC) 27

Figure I.4 - Diagram of a still and a digestion vessel
used by Greek alchemists of the Alexandrian school 40

Figure I.5 - An alchemist's dispensary in the Middle Ages 42

CHAPTER II

W. HARVEY 50

Figure II.1 - Frontispiece of HARVEY's work on the circulation of the blood 52

Figure II.2 - Swelling of the valves of the veins due to pressure of the arm 54

Figure II.3 - Circulation of the blood, according to HARVEY's theory (A)
and according to GALEN's older theory (B) 56

VESALIUS (from the portrait at the front of *La Fabrica*, 1543) 61

Figure II.4 - Mayfly larva (7 millimeters)
dissected by the Dutch anatomist Jan SWAMMERDAM 66

M. MALPIGHI 68

Figure II.5 - Microscopes used in the 17th century 69

Figure II.6 - The first analyses of the macroscopic structure
of biological objects 70

Figure II.7 - Development of the chick embryo 71

A. VAN LEEUWENHOEK (from a painting by Yohannes VERKODJE (1686),
Rijksmuseum, Amsterdam) 72

L. SPALLANZANI (from a portrait by G.B. BUSANI) 73

Figure II.8 - The balance in physiology 74

Figure II.9 - Spiral thermometer manufactured in Italy in the 17th century 76
A. VON HALLER (from an engraving by Sigmund FEUDENBERGER) 80
H. BOERHAAVE (from a painting by Jan WANDELAAR) 80
Figure II.10 - A mechanical vision of the movements of the arm 81
F. REDI (from a painting, Istituto e Museo di Storia della Scienza, Firenze) 82
Figure II.11 - The regeneration of hydras 85
Figure II.12 - Theory of the reflex movement according to DESCARTES 87
GALILEO (from a painting by Ottavio LEONI, 1624) 91
B. PASCAL 95
Figure II.13 - BOYLE's first air pump (1658) 96
G.E. STAHL 100
Figure II.14 - Chemical technology at the turn of the 18th century 101
J. PRIESTLEY (from a pastel by James SHARPLES (around 1797),
    National Portrait Gallery, Washington) 103
A.L. DE LAVOISIER (from an engraving by Miss BROSSARD-BEAULIEU,
    Musée des Arts et Métiers, Paris) 105
Figure II.15 - Gasometer and calorimeter used by LAVOISIER 106
Figure II.16 - Experiment on the release of oxygen by heating mercury oxide
    in a retort, carried out by LAVOISIER 107
Figure II.17 - Experiment concerning Man's oxygen requirements
    in LAVOISIER's laboratory 108
Figure II.18 - Apparatus designed by LAVOISIER for analysis of the gases
    given off during "winey" fermentation 109
F. BACON 113
Figure II.19 - Frontispiece of Francis BACON's philosophical work
    concerning a new concept of science, *Novum Organum Scientiarum* (1620) 115
R. BOYLE (from an engraving by William FAITHORNE) 117
R. DESCARTES (from a painting by Frans HALS (around 1640),
    Musée du Louvre, Paris) 118
J. LOCKE (from a painting by Gottfried KNELLER) 119
D. HUME (from a painting by Allan RAMSAY (1766),
    Scottish National Gallery, Edimburg) 120
I. KANT (from an anonymous portrait) 121
A. PARÉ (from an engraving by Etienne DELAULNE (1582),
    Bibliothèque Nationale, Paris) 127

Chapter III

F. Magendie (from a portrait by Paul Guérin, Collège de France, Paris) 145
Figure III.1 - Title Page of the first edition, in French, of the *Introduction to the Study of Experimental Medicine* by Claude Bernard 147
J. Müller 148
C. Bernard (from Claude Bernard Museum, Mérieux Foundation, Saint-Julien, France) 149
H. von Helmholtz (from a photograph) 151
I. F. Semmelweis (from en engraving by E. Dopy (1860), Institut für Geschichte der Medizin der Universität, Vienne) 152
L. Pasteur 153
Figure III.2 - Swan-neck flask experiment carried out by Pasteur 155
Figure III.3 - Internalization of choleric vibrions by the neutrophil or macrophage, two innate immunity cells 157
E. Metchnikoff (from a photograph by Ph. Nadar (around 1900 - 1910), Pasteur Institute, Paris) 158
Figure III.4 - Digestion of goose red blood cells by the Planarium 159
Figure III.5 - Photograph of the nutritive gel plate on which Alexander Fleming observed the antibiotic capabilities of a mold, *Penicillium notatum*, with respect to colonies of staphylococci 162
Figure III.6 - Pancreatic fistula produced in the dog 165
Figure III.7 - Measurement of gas exchanges. From macrorespirometry to microrespirometry 170
Figure III.8 - Types of instruments invented and used by 19th century physiologists 171
Figure III.9 - Two types of osmometer constructed by Dutrochet 173
Figure III.10 - The polarimeter in the 19th century, and its use in the analysis of optically active solutions 174
Figure III.11 - The spectroscope in the 19th century, and its use in the analysis of photosynthesis 175
Figure III.12 - The optical microscope in the 19th century, and its use in the study of cell division 178
J. Lister (from a photograph, Probert Encyclopedia Picture Gallery) 181
R. Koch 182
Figure III.13 - First photomicrographs of bacteria taken by Robert Koch in 1877 182

G. MENDEL (from a portrait signed E. L.,
   Mendelianum at the Moravian Museum, Brno) 186
Figure III.14 - Different experimental models used in biological research 190
Figure III.15 - Principle of production and detection of mutants
   in the microscopic mold *Neurospora crassa* 192
Figure III.16 - Comparison of the principle of the Transmission Electron
   Microscope and that of the Optical Microscope 201
Figure III.17 - Principle involved in cryofracturing biological membranes,
   used in electron microscopy and application to the highlighting
   of protein particles incorporated into liposomes 202
Figure III.18 - The change from optical microscopy to electron microscopy
   in the middle of the 20th century. The Mitochondrion 204
Figure III.19 - Atomic Force Microscope, Nanoscope II 206
Figure III.20 - Three-dimensional structure of sperm whale myoglobin
   resolved by X-ray crystallography 207
Figure III.21 - The era of enormous instruments used in biology.
   The Grenoble Synchrotron (ESRF) 208
Figure III.22 - Separation of the proteins of a biological
   extract by two-dimensional electrophoresis 211
Figure III.23 - Principle of partition chromatography on paper
   (ascending chromatography) 212
Figure III.24 - Separation of rat liver endocellular organelles
   by differential centrifugation and by isopycnic centrifugation 214
Figure III.25 - Deconstruction and reconstruction
   of tobacco mosaic virus particles 223
Figure III.26 - Osmotic formations imitating forms in Nature (plant material) ... 226
Figure III.27 - Osmotic formations imitating forms in Nature (cells) 227
A. COMTE 230

CHAPTER IV

Figure IV.1 - The central dogma of molecular biology 245
Figure IV.2 - Mode of action of restriction enzymes 246
Figure IV.3 - Genetic recombination technique 250
Figure IV.4 - Injection of DNA into individual cells
   by micromanipulation under the microscope 251
Figure IV.5 - Technology of DNA chips 254

LIST OF ILLUSTRATIONS

Figure IV.6 - Use of biochips in proteomics 256
Figure IV.7 - Building a cubic construction from DNA double helices
with sticky ends 260
Figure IV.8 - Construction of a DNA nanomachine 261
Figure IV.9 - Silencing of messenger RNA (mRNA) by interfering RNAs from
double strand RNA that is either synthetic (dsRNA) or natural (miRNA) 263
Figure IV.10 - Expansion of the genetic code 265
Figure IV.11 - Creation of a plant GMO by genetic engineering 267
Figure IV.12 - Diagram illustrating how to obtain differentiated cells
from stem cells 273
Figure IV.13 - Therapeutic cloning *versus* reproductive cloning 279
Figure IV.14 - Application of NMRI (Nuclear Magnetic Resonance Imaging)
to the study of the neuron activity in a normal human brain 291
Figure IV.15 - Evolution of ideas concerning the transfer of information
between endocellular proteins 299
Figure IV.16 - Principle of highlighting protein-protein interactions
using the double hybrid system 300
Figure IV.17 - Principle of isolating proteins that are interactive
with respect to a protein of interest immobilized on a solid medium 302
Figure IV.18 - Network of protein interactions in yeast 303
Figure IV.19 - The post-genomic era: from the genome to the phenome 309
Figure IV.20 - Neuromimetic robot based on the operation of fly's eye 313
Figure IV.21 - Production of an electrical interaction
between snail brain neurons and a transistor 315
Figure IV.22 - Mitochondrial ATPase, rotational nanomachine 319

The portraits (pencil drawings) were done by Angel FELICES.

GLOSSARY

A

Active site: region of an enzyme that binds the substrate and catalyses its transformation.

Adaptation: modification by which a cell or an organism adapts to outside constraints.

Adenine: purine base that is a component of nucleic acids (DNA and RNA) and adenine nucleotides (AMP, ADP and ATP).

Adenosine monophosphate (AMP): nucleotide formed by the association of adenine, ribose and phosphate.

Adenosine triphosphate (ATP): molecule formed by the association of adenine, ribose and three phosphate residues, involved in energy storage and transfer in cell metabolism.

Adenovirus: DNA virus with an affinity for the lymphoid tissue.

Agar (or **agar-agar**): polysaccharide extracted from certain algae and used as a solid culture medium, for example in PETRI dishes (also called gelose).

Agarose: chromatography medium derived from agar-agar.

Agrobacterium tumefaciens: bacterium often used to obtain a plant GMO.

Allele: one of the forms of a gene occupying a locus on a chromosome.

Allosteric site: region of an enzyme that binds a ligand that regulates the activity of the enzyme.

Allosteric: word used to describe an enzyme that, in addition to its active site (which fixes the substrate of the enzyme), has another site (called the allosteric site) that is able to bind a regulating ligand. The binding of this ligand on the allosteric site modifies the conformation of the enzyme and its activity.

Amino acid: organic molecule represented by the formula $R-C_\alpha H(NH_2)COOH$ and carrying an amine group, $-NH_2$ and a carboxyl group, $-COOH$ on the same terminal carbon atom, C_α. Amino acids are the elementary structural units of proteins.

Anabolism: synthesis of metabolites inside a cell from small molecules.

Anaerobiosis: life process in which the electron acceptor is a molecule other than oxygen.

Anastomosis: communication between vessels or nerves.

Angiography: radiography of vessels after injection of a contrast product.

Angström (Å): unit of length equal to 10^{-10} meters or 0.1 nanometers.

Anthrax: contagious infectious disease that is common to certain animals and Man, caused by the anthrax bacillus *Bacillus anthracis*.

Antibiotics: organic compounds produced by microorganisms and able to inhibit the growth of other microorganisms.

Antibody: defense protein synthesized in an organism in response to the injection of a foreign substance called an antigen. Antibodies are synthesized in B lymphocytes that are differentiated into plasmocytes.

Antigen: foreign substance that is able to trigger the synthesis of a specific antibody, in vertebrates.

Antisense RNA: RNA that is complementary to a messenger RNA transcribed from a gene: by hybridization with this messenger RNA it blocks the function.

Aplysia or sea hare or sea slug: gastropod mollusk.

Apoptosis or programmed cell death: physiological self-destruction of living cells.

Arabidopsis thaliana **(or mouse ear cress)**: small flowering plant used as a model eukaryotic organism.

Archeobacterium *(Archaea)*: prokaryote belonging to the *Archaea* phylogenetic group, distinct from that of the eubacteria (*Bacteria*) and comprising, in particular, halophiles, thermophiles and methanogenic bacteria.

Autoradiography: technique for detecting radioactivity, in which a sheet of paper containing the radioactive material is applied to a photographic film.

Autosomes: chromosomes other than sex chromosomes.

B

β-Galactosidase: enzyme that catalyzes the cleavage of lactose into two, six-carbon-atom sugars, galactose and glucose.

Bacillus: rod-shaped bacterium

Bacterial colony or clone: group of identical bacteria arising from a single bacterium.

Bacterial transformation: Basic technique to introduce a foreign plasmid into a bacterium. Term also used to indicate modification of the genome of a bacterium by incorporation of added DNA.

Bacteriophage: virus that infects bacteria.

Bacterium: unicellular prokaryotic microorganism.

Bioenergetics: see *Cell Energetics*.

Bioinformatics (or **Biocomputing**): analysis of DNA and protein sequences using computer technology. By extension, these terms also refer to the science of information processing as applied to the understanding of how living beings function, using programs implemented on computers.

Biology: science of living beings, or biosciences.

Biometrics: application of mathematics to the statistical study of the resemblances and differences between living beings within a given species.

Biotechnology: application of biological processes to industrial procedures in the areas of food technology, agribusiness, pharmaceutics, etc.

Blastocysts: multicellular structures resulting from the division of a fertilized ovum at the first stage of embryogenesis

Blastomere: cell that appears during the first divisions of a fertilized egg.

Blastopore: transient opening on the surface of an embryo at the gastrula stage, through which the internal cavity (archanteron) communicates with the outside.

Blastula: hollow sphere made up of cells at the beginning of the division of the fertilized egg.

Breathable air (vital air): oxygen.

C

C terminal end: end of a polypeptide chain that carries the last amino acid with its carboxyl group as a free α (carboxyl fixed on the C_α).

Cadherin: protein involved in the interactions between cells in eukaryotes.

Caenorhabditis elegans: small nematode used as a model in cell biology.

Carbon assimilation: incorporation of carbon dioxide, CO_2 into organic molecules, for example, by photosynthesis.

Cardiomyocyte: cardiac muscle cell.

Cell biology: branch of Biology that brings together cytology and cell physiology.

Cell cycle: reproductive cycle of the cell.

Cell energetics (or Bioenergetics): discipline of the life sciences that studies the biosynthetis processes by which ATP is generated by cell respiration or photosynthesis and their impact on cell metabolism.

Cell junction: connective structure between two cells.

Cell line: population of cells of animal or plant origin able to divide indefinitely in culture.

Cellular (or innate) immunity: immunity involving the phagocytosis of bacteria by specialized cells: blood neutrophils, tissue macrophages.

Cerebral tomography: imaging technique used to explore zones of the brain, which takes pictures in successive planes.

Chaperone: protein molecule involved in controlling the folding of the primary structure of other proteins, in order to allow them to acquire their specific three-dimensional structure.

Chemical (or drug) library: collection of molecular structures.

Chemosynthesis: synthesis of organic compounds using the energy liberated by chemical reactions between inorganic molecules (minerals).

Chimera: organism carrying genetic characteristics arising from two different genotypes.

Chiral: word to describe a molecule that contains an asymmetrical center and can form two, non-superimposable, mirror images.

Chloroplast: photosynthetic eukaryotic organelle that contains chlorophyll, where the reactions of photosynthesis are catalyzed.

Chromatin: filamentary nucleoprotein complex, comprising mainly DNA and histone, and present in eukaryotic cells.

Chromatography: technique for separating different molecular species based on their differential adsorption by a solid medium or their distribution between liquid and gas phases.

Chromosome: structure comprising a molecule of DNA and associated proteins.

Clone: population of cells arising from division of a single cell.

Cloning: technique involving the transfer of one or more genes into the genome of a cell, the transferred gene(s) then being transmitted in the cell line arising from the iterative divisions of this transformed cell.

Codon: sequence of three nucleotides in a molecule of DNA or messenger RNA, corresponding to a specific amino acid.

Complement: protein complex present in the blood serum that can be activated by antigen-antibody complexes, and, in its activated form, favors the phagocytosis of bacteria.

Confocal microscopy: microscopy in which the object is lit by a fine laser beam that scans the object on a single level, thus producing an optical cross-section.

Cryofracturing: technique that involves freezing and fracturing cell membranes in order to separate the two sheets and reveal the intrinsic proteins by electron microscopy.

Cytochromes: proteins that are part of electron transport chains and contain a prosthetic group comprising four pyrrole rings and an iron atom (heme). The iron in the cytochromes changes alternately from the ferrous to the ferric state during the transfer of electrons.

Cytokines: protein factors secreted by cells, and involved in the control of growth and the functioning of other cells.

Cytosine: pyrimidic nitrogenous base that is a component of nucleic acids (DNA, RNA) and nucleotides (CMP, CDP, CTP).

Cytosol: contents of the cytoplasmic compartment, not including the endocellular organelles. "Cytosol" is also used as an operational term used to designate the fraction of a cell homogenate that is not sedimented out by centrifugation at 100 000 g for one hour.

D

Dalton: unit of molecular mass equal to the mass of a gram-atom of hydrogen.

Darwinism: theory of evolution by natural selection.

Denaturation: unfolding of a protein polypeptide chain, associated with the loss of activity of this protein. Denaturation is a result of the loss of secondary and tertiary structures.

Deoxyribose Nucleic Acid (DNA): nucleic acid in which pentose is the deoxyribose and the four cyclic bases are adenine, guanine, cytosine and thymine.

Dephlogisticated air: oxygen.

Detergent: amphipathic molecule used in Biology to solubilize membrane proteins.

Determinism: philosophical principle stipulating that any phenomenon depends on conditions (causes) that exist beforehand (or simultaneously).

Developmental biology: branch of Biology based on embryology, bringing together the techniques of cell biology and molecular biology.

Dialysis: process by which molecules of different sizes are separated by diffusion of the small size molecules through a semi-permeable membrane.

Diastase: enzyme.

Dictyostelium discoïdeum: ameba capable of differentiation, used as a model system in cell biology.

Diploid: term to describe the status of a somatic (non-germ) cell that contains a pair of each of the chromosomes typical of the species, i.e., an even total number, written $2n$.

Disulfide bridge (or **disulfide bond**): covalent bond linking two sulfur atoms.

DNA chip: small piece of silicon or plastic, to which DNA fragments (oligonucleotides) are fixed.

DNA polymerase: enzyme that catalyzes the synthesis of DNA on a DNA template using deoxyribonucleoside-5'triphosphate precursors.

Domain (in a protein): region of a protein characterized by a specific tertiary structure and endowed with a specific function.

Dominant: opposite of recessive. When paired with a recessive allele, the dominant gene is expressed in the phenotype of an organism whereas the recessive one is not.

Double helix: double-stranded helicoidal structure of DNA demonstrated by CRICK and WATSON in 1953.

Drosophila melanogaster: scientific name for the fruit fly.

E

Ectoderm: upper layer of cells of the animal embryo which develop into the epidermis and nerve tissue in the adult.

Electrophoresis: technique for separating positively or negatively charged molecular species by application of an electrical field.

Electroporation: creation of pores in a membrane under the influence of a high voltage electrical current.

Embryogenesis: development of the animal or plant embryo.

Enantiomer: one of the two forms of a chiral molecule.

Endocellular organelles: intracellular microstructures characterized by a specific morphology and specific metabolic functions.

Endoderm: layer of cells that is internalized at the time of gastrulation in the animal embryo. These cells subsequently develop into the cells of the digestive tract and associated organs.

Endoplasmic reticulum: intracytoplasmic membrane network in eukaryotic cells, having many metabolic functions. A distinction is made between the smooth reticulum and the rough reticulum, the latter being edged with ribosomes.

Endosome: organelle formed from the plasma membrane by endocytosis, which transfers newly ingested materials to the lysosomes.

Enzyme: protein that catalyzes a defined chemical reaction. It should be noted that enzyme-like activity has also been found in certain RNAs (ribozymes).

Epigenesis: formation of new structures during embryogenesis.

Epigenetic: describes phenotypical modifications that depend on the environment and occur after gene expression.

Epistemology: philosophical discipline that asks questions about the origin of the sciences and the conditions under which they have developed.

Equatorial plate: virtual plane inside a eukaryotic cell where the chromosomes are located during the stage of mitosis known as the metaphase.

ES cells: Embryonic Stem cells.

Escherichia coli: Enteric (intestinal) bacterium used as a model prokaryotic organism.

Etiology: study of the causes of a disease.

Eugenics: theory that recommends the application of methods that aim to improve the genetic inheritance, in Humankind.

Eukaryote: Organism whose cells contain a nucleus delimited by an envelope or membrane.

Exon: segment of a gene coding for a peptide chain.

F

Femto-: prefix which, put in front of the name of a unit, indicates that the latter is divided by 10^{15}.

Fermentation: breakdown of organic molecules in the absence of oxygen (under anaerobiosis). For example, alcohol fermentation corresponds to the transformation of glucose into ethanol in an oxygen-deprived medium.

Finalism: doctrine according to which evolution tends towards a certain end.

Fixed air: carbon dioxide.

Foul air (mephitical air): nitrogen.

G

Gametes: sex cells of the male (spermatozoa) and the female (ova). The union of a male and a female gamete gives rise to a fertilized egg (zygote).

Gastrula: animal embryo at the development stage where there is an invagination of the upper layer of cells to form an internal cavity.

Gel filtration chromatography: separation of different molecular species according to their size by passage through a gel column acting as a molecular sieve.

Gene: unit of heredity carried on chromosomal DNA and transmitted from generation to generation by gametes.

Genetic code: all the nucleotide triplets (codons) in a DNA that contain genetic information translated into amino acids in the proteins.

Genetic engineering: all of the procedures whose purpose is the manipulation of the genome and its modification. Methods for experimentation on genes.

Genetic map: diagram showing the position of the genes on a chromosome.

Genome: all of the genes of a cell.

Genomics: science that studies the genome.

Genotype: genetic makeup of an organism.

Germ (sexual) cells: cells that give rise to gametes.

GFP (or "Green Fluorescent Protein"): protein that emits a green fluorescence, originally isolated from the jellyfish *Aequoria victoria*.

Giga-: prefix which, put in front of the name of a unit, multiplies the latter by 10^9.

Glycogen: polysaccharide formed by the association of molecules of glucose.

Glycogenesis: production of glucose in the liver, from a precursor.

Glycogen-forming amino acid: amino acid whose metabolic modifications lead to the formation of glucose in the liver.

Glycolysis: metabolic pathway for the breakdown of glucose into pyruvate and lactate.

GMO (Genetically Modified (or transgenic) Organism): a living organism whose genetic inheritance has been modified by transgenesis, i.e., by the introduction into its genetic material of a gene coming from another organism.

GOLGI apparatus: set of intracytoplasmic vesicles involved in the secretion of modified proteins, particularly by glycosylation.

Guanine: purine base of nucleic acids (DNA, RNA) and nucleotides (GMP, GDP, GTP).

H

Half life: time necessary for the disappearance of half of the molecules of a given compound.

Haploid: term to describe the status of a cell having a single set of chromosomes, i.e., half of the number of chromosomes for the species. This is the status of the germ cells: spermatozoa and ova. This term is opposed to the term **diploid** (see above).

HeLa cells: strain of epithelial cells continuously cultured since its isolation in 1951 from a patient suffering from uterine cervical carcinoma. (The designation HeLa is derived from the name of the patient, Henrietta LACKS). Cells used in biology because of their rapid division in culture.

Hematopoiesis: genesis of blood cells in the bone marrow.

Heme: molecule comprising four pyrrole rings and an iron atom.

Hemoglobin: heminic protein of the red blood cells that carries out the transport of oxygen.

Hepatocyte: cell of the hepatic parenchyma.

Heterocaryon: result of the fusion of two nuclei belonging to two different living species.

Heterozygote: diploid cell carrying two different alleles of a given gene.

Homeobox: DNA sequence that codes for a peptide sequence of around sixty amino acids, called the homeodomain, which is able to bind to specific regions of the nuclear DNA.

Homeogene: gene that controls the positioning of the organs in multicellular organisms.

Homeostasis: ability of living beings to maintain certain biological constants (for example, the concentration of glucose in the blood).

Homeotic gene: gene involved in the body's organizational plan.

Homology: similarity in the structure of two or more proteins having a common origin in evolution.

Homozygote: diploid cell having two identical alleles of a given gene.

Hormone: chemical messenger produced by one cell type and affecting the activity of another cell type.

Humoral (or **adaptive**) **immunity**: immunity involving serum factors secreted by specialized cells in vertebrates, for example antibodies synthesized by B lymphocytes differentiated into plasmocytes.

Hybridization: cross-breeding of individuals belonging to two natural populations. Also designates the process by which two complementary strands of a nucleic acid form a double helix during their association.

I

Immunoglobulin: antibody.

Infusoria: ciliated protozoans. Their name comes from the fact that they were first found in hay infusions.

Insulin: polypeptide hormone secreted by specialized pancreatic cells. Contributes to the control of glucose metabolism.

Interfering RNA: RNA molecule made up of around twenty nucleotides that is able to associate with a messenger RNA and blocks its translation into protein.

Intron: intervening DNA sequence that interrupts the coding sequence of a gene.

Isoelectric point: pH value at which the positive and negative charges carried by the amino acid residues of a protein cancel each other out so that the protein has no net charge.

Isofocalization: migration of a charged macromolecule in an electrical field and in a pH gradient to the pH corresponding to its isoelectric point.

K

Karyokinesis: division of the cell nucleus during mitosis.

Karyotype: all of the chromosomes of a cell arranged according to their size, shape and number.

Kilo-: prefix which, put in front of a unit of measurement, indicates that the latter is multiplied by a thousand.

Kilobase: unit equal to one thousand bases, used to measure the length of DNA fragments.

Kinase: enzyme that transfers the terminal phosphate group of ATP onto other molecules. For example, creatine kinase catalyzes the synthesis of creatine phosphate from ATP and creatine by transferring the terminal phosphate of the ATP to the creatine.

L

Laser light: beam of coherent light produced by a laser generator.

Lipid: biomolecule that is insoluble in water, having fatty characteristics.

Locus: location on a chromosome of a gene that determines a particular trait in a phenotype.

Lymphocyte: white blood cell that is part of the immune system.

Lysogenic: term describing the status of a bacterium that has the DNA of a virus (bacteriophage) integrated into its genome, inactive because it is repressed.

Lysosome: organelle of eukaryotic cells surrounded by a single membrane, containing hydrolases that are active at acid pH (approximately pH 5).

M

Macrophage: phagocytotic cell located in the tissues.

Magnetoencephalography: examination of cerebral activity by a procedure involving a magnetic field.

Materialism: philosophical theory, which postulates that material processes are at the origin of life and consciousness. The materialism of DEMOCRITUS, LEUCIPPUS and EPICURUS was taken up again in the 18th century by DIDEROT and DE LA METTRIE.

Medical imaging: all diagnostic procedures that use radiography, ultrasound, MRI, etc.

Meiosis: type of cell division involved in the formation of ova and spermatozoa, which produces four haploid daughter cells from an initial diploid cell.

Mendelian heredity: heredity that depends on genes carried by the nuclear DNA.

Mesoderm: cell layer in the gastrula between the ectoderm and the endoderm.

Metabolism: all of the reactions involving the synthesis and breakdown of organic molecules in a cell.

Metabolite: in metabolism, an organic molecule that appears as an intermediary in reactions catalyzed by enzymes.

Metabolome: all of the enzyme reactions of the metabolism carried out in a cell.

Metabolon: group of structurally associated enzymes that participate in a sequence of reactions in a metabolic chain.

Metagenomics (also called **ecogenomics**): analysis of the nucleotide sequences of all the genomes of different species of microorganisms present in a given environment, independently from the isolation and culture of each species.

Metaphysics: The branch of philosophy that examines the nature of reality, including the relationship between mind and matter, substance and attribute, fact and value. A priori speculation upon questions that are unanswerable to scientific observation, analysis, or experiment

Metazoan: multicellular animal organism (as opposed to a protozoan, which is unicellular).

Methylation: addition of a methyl group ($-CH_3$) to an organic molecule.

Micro-: prefix which, put in front of the name of a unit of measurement, indicates that the latter is divided by a million (10^6).

Microplate: plastic plate having rows of small cavities known as wells, and used in an automated manner for chemical reactions. Plates with 96 wells or more, with a capacity of between 0.1 and 0.5 mL, are commonly used in immunochemistry, genomics, proteomics, etc.

MicroRNA: interfering microRNA, a natural product of eukaryotic cells involved in regulation of the translation of messenger RNA into protein.

Microsomes: heterogeneous particles collected by differential centrifugation of a cellular homogenate, made up of fragments of endoplasmic reticulum, GOLGI apparatus and plasma membrane.

Microtubule: long, cylindrical structure made up of tubulin.

Mitochondrial heredity: heredity that depends on genes carried by the mitochondrial DNA.

Glossary

Mitochondrion (plural, **mitochondria**): cytoplasmic organelle of eukaryotic cells, which has an enzyme system that is able to couple oxidation reactions and ATP synthesis.

Mitosis: division of a eukaryotic cell that leads to the formation of two daughter cells from a parent cell, with the daughter cells having the same number of chromosomes as the parent cell.

Mitotic spindle: structure made up of microtubules and associated proteins, which directs the migration of the chromosomes in the daughter cells during mitosis.

Molecular biology: is the study of biology at a molecular level, particularly the interactions between DNA, RNA and protein synthesis. It overlaps the fields of biochemistry and genetics.

Molecular sieve: resin that makes it possible to separate molecules according to their size.

MRI (Magnetic Resonance Imaging): set of radiographic techniques based on nuclear magnetic resonance.

mRNA (or messenger Ribonucleic Acid): ribonucleic acid transcribed by complementarity of bases from a deoxyribonucleic acid sequence corresponding to a gene. (The adenine, guanine, cytosine and uracil of the RNA are complementary to the thymine, cytosine, guanine and adenine in the DNA, respectively).

Multipotency: characteristic of cells present in already differentiated tissues that allows them the possibility of giving rise to a restricted number of cell lines.

Mutation: change in genetic material that is transmitted by inheritance, resulting from the replacement, deletion or duplication of one or more DNA bases.

Myopathy: disease of the muscular tissue leading to a deficiency in movement.

N

N terminal end: end of a polypeptide chain that carries the first amino acid with its amine group as a free α (amine fixed on the C_α).

Nano-: prefix which, put in front of a unit of measurement, indicates that the latter is divided by 10^9.

Native protein: protein in a conformation corresponding to its active state, as opposed to a denatured protein that is no longer active because its three-dimensional structure has been modified.

Natural selection: basis of the theory of evolution formulated by DARWIN.

Neuron: cell that ensures that nerve impulses are conducted.

Neurosciences: sciences relating to the study of the nervous system.

Neutrophils: white blood cells involved in the phagocytosis of bacteria after migration to the site of infection.

Nitrogen assimilation: incorporation of atoms of nitrogen into proteins synthesized by microorganisms able to reduce atmospheric nitrogen to ammonia.

Nominalism: doctrine according to which the only existing reality is that of individuals or things designated by a name. Nominalism opposed the doctrine of the "Universals", based on the reality of general concepts ("universals").

Nuclear Magnetic Resonance (NMR): resonance of the nucleus of an atom belonging to a molecule, due to transitions between energy levels, when a certain electromagnetic radiation frequency and a certain magnetic field intensity is applied to this molecule.

Nucleic acid: linear macromolecule formed by association of mononucleotides that are made up by the association of a cyclic base (heterocycle), a pentose and a phosphate.

Nucleoside: compound made up of a purine or pyrimidic base linked by a covalent bond to a pentose (ribose or deoxyribose).

Nucleotide: phosphate nucleoside.

Nucleus: in eukaryotes, organelle surrounded by an envelope or membrane, which contains the chromosomes.

O

Oligonucleotide: small fragment of DNA or RNA containing a sequence of several nucleotides.

Oncogene: gene present in a tumorigenic virus (v oncogene) or arising by mutation of a protooncogene in a eukaryotic cell (c oncogene).

Ontogenesis: development of an individual from the fertilized egg.

Oocyte (or **ovocyte**): female gamete that has not reached maturity.

Operon: genetic unit comprising coding DNA sequences and regulatory DNA sequences (operator and promoter).

Osmometer: apparatus used to measure the osmotic pressure.

Osmosis: passage of water through a semi-permeable membrane separating two solutions of different concentrations.

Ovum (plural **ova**): mature female gamete.

Oxidoreduction (Redox): term to describe a reaction in which electrons are transferred from a donor to an acceptor.

P

Parthenogenesis: reproduction from a non-fertilized ovum.

Partition chromatography (on filter paper): in vertical chromatography on paper, the lower end of the paper on which a spot of the mixture of molecules has been deposited is bathed in a solvent phase. The solvent rises up the paper by capillarity. The molecules of the mixture are carried with the solvent and migrate more or less quickly, depending on their affinity for the solvent or the water that impregnates the paper.

Pathogenic: term meaning "disease-causing".

PCR (Polymerase Chain Reaction): technique that makes it possible to amplify specific regions of DNA.

Peptide bond: covalent bond between the carboxyl group of an amino acid and the amine group of a second amino acid.

Peroxisome: cytoplasmic organelle in eukaryotic cells carrying the enzymes that catalyze oxidation reactions producing peroxide and other reactions that destroy this peroxide.

Phagocytosis: process by which certain cells internalize and digest foreign particles.

Phase contrast microscopy: microscopy that transforms differences in refraction index into differences in intensity that can be perceived by the eye.

Phenome: all of the phenotypical characteristics that characterize a living being.

Phenotype: set of apparent characteristics of an organism that depend on its genetic make up (genotype).

Phenotypic(al): relating to the phenotype.

Phlogistics: outdated theory according to which any inflammable body is supposed to contain a substance, phlogiston, which is released when the body burns.

Phosphorylative oxidation: process by which the energy arising from the oxidation of metabolites is used for the synthesis of ATP from ADP and mineral phosphate.

Photons: particles associated with light radiation.

Photophosphorylation: process of phosphorylation of ADP into ATP in green plants and phototrophic microorganisms that use light as an energy source.

Photosynthesis: process by which green plants and phototrophic prokaryotes use light energy to synthesize organic molecules after incorporation of CO_2 from the air.

Phylogenesis: study of the formation and evolution of living species (term created by HAECKEL in 1868).

Physiology: branch of Biology that studies the functions of organs.

Pico: prefix which, put in front of the name of a unit of measurement, indicates that the latter is divided by 10^{12}.

Planarium: small freshwater flatworm.

Plasma membrane: membrane that contains the cell cytoplasm.

Plasmid: small DNA molecule that replicates independently of the chromosome in a bacterium.

Pluripotency: characteristic of a cell arising from a fertilized ovum, which has lost the possibility of giving rise to a complete individual but has kept the ability to differentiate into different cell types.

Polarimeter: instrument used to determine the degree of rotation of polarized light caused by an organic substance carrying asymmetric carbon in solution.

Polyacrylamide: resin resulting from the polymerization of acrylamide, used in gel form for the separation of macromolecules by electrophoresis.

Polyploid: term to describe the status of a cell that contains more than two sets of homologous chromosomes.

Positivism: philosophical doctrine that bases understanding on facts, on what is perceived by the senses and on experience, and refutes all metaphysical arguments.

Positon (or **positron**): positive anti-particle of the electron.

Pragmatism: doctrine according to which the truth is that which can be verified and practiced by the scientific community.

Primary structure: sequence of amino acids in a protein.

Prion: protein-like infectious agent, apparently without any nucleic acid.

Prokaryote: microorganism whose DNA is not incorporated into a well-defined nucleus.

Prophage: form of non-infectious bacteriophage.

Prophylaxis: set of measures designed to prevent the appearance or spread of an infectious disease.

Protein: macromolecule made up of one or more polypeptide chains possessing a characteristic amino acid sequence and molecular weight.

Proteome: all of the proteins contained in a cell.

Protist: unicellular eukaryote.

Protooncogene: cellular gene that can be converted into an oncogene by mutation.

Pure culture: culture containing only one type of microorganism descending from a single ancestor.

Puric bases (adenine and guanine): cyclic nitrogenous bases containing a hetrocyclic purine ring found in nucleic acids and nucleosides.

Pyrimidic bases (cytosine, uracil and thymine): cyclic nitrogenous bases containing a heterocyclic pyrimidine ring found in nucleic acids and nucleosides.

Q

Quaternary structure: protein structure resulting from the association of several polypeptide chains that are designated as protomers, monomers or sub-units.

R

Racemic: term to describe an equimolar mixture of two enantiomers, a mixture which is thus inactive with respect to polarized light.

Rationalism: theory according to which everything that exists has a reason for being and, in principle, must be discovered and understood by reason.

Recessive: term to describe a hereditary characteristic that only manifests itself when the dominant allele is absent.

Recombinant DNA: DNA molecule produced by genetic engineering from messenger RNA in the presence of a reverse transcriptase.

Replication: operation by which a DNA sequence is replicated by complementarity, from a DNA template

Repressor: a protein which binds to the operator of an operon to prevent the transcription of the latter.

Reproductive cloning: identical reproduction of a living species.

Restriction enzyme: type of enzyme commonly found in bacteria, which is able to cleave DNA that is foreign to these bacteria, thus protecting them against viral attacks.

Retrovirus: RNA virus whose RNA is transcribed reversibly into DNA by a specific enzyme, reverse transcriptase.

Reverse transcriptase: enzyme that catalyzes the formation of complementary DNA by copying from an RNA template.

Reverse transcription: process by which the information contained in the RNA is transferred to the DNA.

RFLP (Restriction Fragment Length Polymorphism): a technique that makes it possible to differentiate DNA by the size of the fragments obtained after cutting with restriction enzymes.

Ribonuclease: enzyme that is able to hydrolyze the internucleotide bonds in RNA.

Ribonucleic acid (RNA): nucleic acid in which pentose is the ribose and the four cyclic bases are adenine, guanine, cytosine and uracil.

Ribose: a pentose (an ose with 5 carbons) present in RNA.

Ribosomes: intracellular particles made up of RNA and proteins on which the messenger RNA is positioned during its translation into a protein.

RNA polymerase: enzyme that catalyzes the association of ribonucleotides into an RNA molecule by transcription from a DNA matrix. In eukaryotes, there are three RNA polymerases that are assigned to the synthesis of ribosomal RNAs, messenger RNAs and transfer RNAs, respectively.

RNA: ribonucleic acid.

RNAome: by convention, all non-coding RNAs, as opposed to the transcriptome, which represents all coding RNAs.

S

Saccharomyces cerevisiae: species of yeast used as a model eukaryotic organism.

Scanning Electron Microscopy (SEM): microscopy that uses the electrons refracted by the surface of an object to produce an image.

Scholasticism: school of philosophy in the Middle Ages (from the 9th to the 13th centuries), which held that theology is the foundation of philosophy.

Scientism: type of philosophy that seeks to prove that the science created by Man can study Man as a whole, and that it will resolve all questions that have a meaning, including sociological ones.

SDS: sodium dodecyl sulfate.

Secondary messenger: small molecule formed in the cytoplasm of a cell in response to an extracellular signal.

Secondary structure: first level of organization of a polypeptide chain by winding into a helix (α helix) or folding into sheets (β sheets).

Serendipity: "happy chance". Term that refers to the discovery of something important that has not been looked for.

Sickle Cell anemia (or **Drepanocytosis** or **Falciform anemia**): hereditary illness caused by a modification of the hemoglobin molecule in the red blood cells.

Somatic cells: eukaryotic cells of metazoans (animals), other than germ or sex cells.

Stem cell: cell able to divide and differentiate into all the different cell types of an organism.

Stereochemistry: a sub-discipline of Chemistry that makes connections between the spatial arrangement of atoms within molecules and the properties of these molecules.

Striatum: double mass of cerebral gray matter located between the two cerebral hemispheres.

Structural biology: branch of Biology that specializes in the study of the three-dimensional structures of macromolecules.

Subcellular fractioning: procedure for separating endocellular organelles in a tissue homogenate by successive centrifugations at increasing speeds.

Synapse: region located between two neurons or between a neuron and a muscle cell, into which neurotransmitter molecules are released.

Systematics: science of the classification of living beings.

T

Taxonomy (or **taxinomy**): (term created by CANDOLLE in 1813). Science of the classification of living beings and extinct organisms.

Telengiectasia: dilatation of a blood vessel in Man.

Teleology: the philosophical study of the design, purpose, directive principle or finality in nature or human creation, a doctrine which postulates that evolution has occurred in a goal-oriented direction.

Tertiary structure: three-dimensional structure of a peptide chain including α helices and β sheets, involving bonds of different kinds; disulfide bridges, ionic bonds, hydrophobic interactions.

Therapeutic (non-reproductive) cloning: use of the cloning method with the aim of producing differentiated tissue that can be used in reparative medicine.

Thymine: a cyclic nitrogen base involved in the structure of DNA.

Thyroid: an endocrine gland, located in front of the trachea, which secretes several hormones, including thyroxin.

Totipotency: characteristic of a cell arising from the very first stage of the division of a fertilized ovum, which is able to form a complete individual after implantation *in utero*.

Transcription factor: regulation protein that binds to DNA sequences, generally upstream of the transcribed region, and regulates the transcription of genes.

Transcription: process by which genetic information is transferred from DNA to RNA.

Transcriptome: all of the messenger RNAs synthesized from the DNA of the genome.

Transfection: introduction of a molecule of foreign DNA into a eukaryotic cell.

Transfer ribonucleic acid (tRNA): small size RNA that plays the role of adaptor in the translation of messenger RNA into a sequence of amino acids in a protein. Part of the tRNA fixes a specific amino acid and another part of the tRNA recognizes a trinucleotide sequence (codon) that codes for this amino acid in the messenger RNA.

Transgenesis: operation in which a transgenic organism is created.

Transgenic Mouse: mouse whose genetic inheritance has been modified by the introduction of one or more foreign genes.

Transgenic: term used to describe an organism whose genome has incorporated and expressed genes from another species. Transgenic organisms are created by genetic engineering techniques, using appropriate DNA vectors to insert the foreign gene into the fertilized egg.

Translation: in molecular biology, process by which the information contained in a messenger RNA is translated in terms of amino acid sequences into a protein.

Transmission Electron Microscopy (TEM): microscopy that produces images from electrons transmitted through the object.

Transposon: mobile DNA sequence that can become inserted in different regions of a chromosome.

U

Ultrasound: medical exploratory method that makes use of the reflection of ultrasonic waves by the organs.

Unipotency: characteristic of cells present in differentiated tissue, which are limited to the ability to produce a single type of cell.

Universals: general concepts that supposedly existed before things, in the form of eternal ideas or innate ideas. The concept of the existence of such universals, which was referred to as "realism", was part of medieval scholastic teaching, and followed on from the Platonic theory of Ideas.

Ureogenesis: synthesis of urea.

V

Virus: infectious nucleoprotein complex made up of a single type of nucleic acid, DNA or RNA, which has to make use of either a prokaryotic or a eukaryotic cell for its reproduction.

Vitalism: doctrine that attributes specific laws to the vital processes, these laws being independent of the fundamental principles of physics and chemistry.

Vitamin: organic substance, small quantities of which are indispensable to the growth and survival of higher animals. Most vitamins are constituents of coenzymes.

Vitriol (oil of): sulfuric acid.

X

Xenograft: transplantation of tissue from one animal species to another.

Xenopus laevis (or **African clawed toad**): toad from South Africa.

X-ray crystallography: study of crystal structures by diffraction of X-rays.

X-ray diffraction: modification of the propagation direction of an X-ray when it meets the atoms of a molecule.

Z

Zygote: diploid cell formed by the merging of a male gamete and a female gamete.

Zymase: old term used to describe the active principle in yeast responsible for alcohol fermentation.